CORRECTIONS

D1298995

Dedication

Mary K. Stohr: To my dad and mom, Robert (Stanley) Stohr and the late Elizabeth (Betty) Stohr. They were so skilled and loving when "correcting" and guiding their own eight children (I was third) that much of what I know about love, forgiveness, and life comes from them. I will be forever grateful for the gift they were, and are, as parents. I would also like to acknowledge my husband, Craig Hemmens, and our daughter, Emily Rose Stohr-Gillmore, for their love and support; I could do nothing well without them.

Anthony Walsh: To all my family near and far, and especially to my drop-dead gorgeous wife, Grace, a.k.a. "the face." I also want to dedicate it to Mary and her husband, Craig. We have been colleagues, friends, and coauthors for so many years that I don't know what I'll do without them now that they have moved to Missouri and greener pastures.

CORRECTIONS

THE ESSENTIALS

MARY K. STOHR
Missouri State University

ANTHONY WALSH
Boise State University

Los Angeles | London | New Delhi
Singapore | Washington DC

Los Angeles | London | New Delhi
Singapore | Washington DC

FOR INFORMATION:

SAGE Publications, Inc.
2455 Teller Road
Thousand Oaks, California 91320
E-mail: order@sagepub.com

SAGE Publications Ltd.
1 Oliver's Yard
55 City Road
London EC1Y 1SP
United Kingdom

SAGE Publications India Pvt. Ltd.
B 1/I 1 Mohan Cooperative Industrial Area
Mathura Road, New Delhi 110 044
India

SAGE Publications Asia-Pacific Pte. Ltd.
33 Pekin Street #02-01
Far East Square
Singapore 048763

Acquisitions Editor: Jerry Westby
Editorial Assistant: Erim Sarbuland
Production Editor: Karen Wiley
Copy Editor: Teresa Herlinger
Permissions Editor: Karen Ehrmann
Typesetter: C&M Digitals (P) Ltd.
Proofreader: Kristin Bergstad
Indexer: Wendy Allex
Cover Designer: Gail Buschman
Marketing Manager: Erica DeLuca

Copyright ©2012 by SAGE Publications, Inc.

All rights reserved. No part of this book may be reproduced or utilized in any form or by any means, electronic or mechanical, including photocopying, recording, or by any information storage and retrieval system, without permission in writing from the publisher.

Printed in Canada

Library of Congress Cataloging-in-Publication Data

Stohr, Mary K.

Corrections : the essentials/Mary K. Stohr, Anthony Walsh.

p. cm.
Rev. ed. of: Corrections : a text/reader/[edited by] Mary Stohr, Anthony Walsh, Craig Hemmens. c2009.

Includes bibliographical references and index.

ISBN 978-1-4129-8699-1 (pbk.)

1. Corrections—United States. 2. Criminal justice, Administration of—United States. 3. Punishment—United States. I. Walsh, Anthony, 1941- II. Corrections. III. Title.

HV9275.C633 2012
365'.973—dc23   2011028304

This book is printed on acid-free paper.

11 12 13 14 15 10 9 8 7 6 5 4 3 2 1

Brief Contents

Detailed Contents

Preface

There are plenty of excellent corrections books available for use, but we think this particular book fills a niche for professors and students in that it is comprehensive and relatively inexpensive. These twin ideas became our goals and guided our development and writing of this textbook. We wanted to cover the most interesting and compelling information currently available on all aspects of corrections, while also keeping the page limit within reason and the book published as a paperback. We hope that readers will find this work both informative and accessible.

The information in this textbook is what you might expect from major texts. However, beyond the facts, figures, and concepts commonly contained in textbooks, this book also showcases the history and research on a number of aspects of corrections. We chose, despite the relative brevity of the book, to include two chapters on history, rather than one, as a historical perspective provides the framework for all that follows, in corrections as in so many other social, political, and cultural initiatives and enterprises. We also believe the presentation of research findings from academic, government, and journalistic sources will provide the context for understanding policy decisions and their consequences, both past and present.

Other special features of the book, which are designed to develop "perspective," include brief "comparative" corrections sections that highlight what other countries are doing in terms of correctional operation. This glimpse of corrections internationally is meant to provide readers with another way of viewing correctional practice in the United States, while also giving them some insight into how alternative practices might work.

This book can serve as a primary text for an undergraduate course in corrections, or as a supplemental text for a graduate course. The topical areas covered are comparable to other major texts, with the exceptions noted in the above regarding the inclusion of enhanced history, research, and the comparative perspectives. Undergraduates, we hope, will find it informative and enlivening. Graduate students might use it as an introduction, overview, and backdrop for other more topically specialized books or articles. Discussion questions appear at the end of each chapter and might be used by both types of students to spur thought about, and critique of, corrections.

❖ Structure of the Book

The structure of the book is much like that found in other textbooks on corrections. We begin with an overview of corrections and some key concepts. We do include two chapters on history, though many textbooks have only one. We then follow the flow of the corrections system, from sentencing, to jails, to probation, to prisons. We then stop and examine the correctional experience for staff as we did the experience for inmates and probationers in the preceding chapters. We finish the system description with a discussion of parole and reentry. In the three chapters that follow, we address the reality for women, minorities, and juveniles in corrections. We then focus attention on legal issues and correctional programming in corrections. We end with a look to the future of corrections and what developments we might expect in the coming years.

❖ Ancillaries

To enhance the use of this book, we have developed high-quality ancillaries for instructors and students.

Instructor Teaching Site. A password-protected site, available at www.sagepub.com/stohressentials, features resources that have been designed to help instructors plan and teach their course. These resources include test banks, PowerPoint slides, lecture outlines, class activities, links to SAGE journal articles, Web resources, video links, and figures and tables from the text.

Student Study Site. A Web-based study site is available at www.sagepub.com/stohressentials. This comprehensive site features Web quizzes, flashcards, video clips, links to SAGE journal articles, Web resources, audio resources, section summaries, corrections state rankings, and Department of Corrections websites for all 50 states.

Acknowledgments

We would like to thank executive editor Jerry Westby. Jerry's faith, hard work, and incredible patience made this book both possible and fun to write. We also would like to thank Jerry's developmental editors, Denise Simon and Erim Sarbuland, who helped shepherd the book through the review process and whose gentle prodding ensured the deadlines would be met. Our copy editor, Teresa Herlinger, ensured that the sentences were clean and the spelling correct.

We would also like to acknowledge each other. We were colleagues at Boise State University for many years, the last several (before Mary moved to Missouri State in the summer of 2011) with offices right next door to each other. We have come to appreciate the work and perspectives of the other. This work was a true collaboration between us and reflects our shared belief in the possibilities for decency and justice as that is elaborated upon by social institutions and their workers, and by individuals willing to change.

We are also grateful to the reviewers who took the time to review early drafts of our work and who provided us with helpful suggestions for improving the chapters and the book as a whole. There is no doubt that their comments made the book much better than it would have otherwise been. Heartfelt thanks to the following experts: Deborah Baskin, California State University, Los Angeles; Stephen Costanza, Central Connecticut State; Marie Griffin, Arizona State University; Kate Hanrahan, Indiana University of Pennsylvania; Laura Hansen, University of Massachusetts, Boston; Robert Hormant, University of Detroit; Jessie Krienert, Illinois State University; Cathy Levey, Goodwin College; Michael Montgomery, Tennessee State University; Sheree Morgan, St. Cloud State University; Mario Paparozzi, University of North Carolina, Pembroke; Danielle Rudes, George Mason University; Bill Sexson, Northwestern State University, Louisiana; Brenda Vose, University of North Florida; Charles E. Wilson, University of Detroit Mercy; and Vanessa Woodward, University of Southern Mississippi.

Photo Credits

Photo 1.1. Ablestock.com/Ablestock.com/ Thinkstock.

Photo 1.3. Source via Wikimedia Commons : National Portrait Gallery, London.

Photo 2.1. Courtesy of the Beinecke Rare Book & Manuscript Library, Yale University.

Photo 2.2. Library of Congress Prints and Photographs Division.

Photo 2.3. Photos.com/Thinkstock.

Photo 2.4. Photos.com/Thinkstock.

Photo 2.6. Photos.com/Thinkstock.

Photo 3.1. I. N. Phelps Stokes. Collection of American Historical Prints.

Photo 3.2. Mike Graham/Flicker.

Photo 3.3. Photos.com/Thinkstock.

Photo 3.4. New York State Archives. Education Dept. Division of Visual Instruction. Instructional lantern slides.

Photo 3.5. Brett Weinstein.

Photo 3.6. Library of Congress.

Photo 4.1. Comstock/Thinkstock.

Photo 4.2. Jupiter Images/Liquid Library/Thinkstock.

Photo 5.1. Wikimedia/WhisperToMe.

Photo 5.2. Thinkstock Images/Comstock.

Photo 5.3. California Department of Corrections and Rehabilitation.

Photo 5.4. Thinkstock Images/ Comstock.

Photo 5.5. © James Leynse/Corbis.

Photo 5.6. Courtesy of Burgen County Sheriff's Office/ Burgen County Jail.

Photo 6.1. Courtesy of U.S. Probation Office- Western District of Kentucky.

Photo 6.2. © Joan Barnett Lee/ZUMA Press/Corbis.

Photo 7.1. John Foxx/Stockbyte/Thinkstock.

Photo 7.2. © George Steinmetz/Corbis.

Photo 7.3. California Department of Corrections and Rehabilitation.

Photo 7.4. U.S. Immigration and Customs Enforcement.

Photo 7.5. Photodisc/Thinkstock.

Photo 7.6. © Axel Koester/Corbis.

Photo 8.1. Steve Daggar.

Photo 8.2. © Bettmann/Corbis.

Photo 8.3. © istockphoto.com/TomGordonsLife.

Photo 9.1. © Philip Zimbardo.

Photo 9.2. © Shawn Baldwin/Corbis.

Photo 9.3, Thinkstock Images/Comstock.

Photo 10.1. Library of Congress.

Photo 10.2. Thinkstock Images/Comstock.

Photo 10.3. California Department of Corrections and Rehabilitation.

Photo 11.1. Thinkstock Images/ Comstock.

Photo 11.4. Thinkstock Images/Comstock.

Photo 12.1. Ulrick Tofte/Digital Vision/Thinkstock.

Photo 12.2. © Reuters/Corbis.

Photo 12.3. © Bill Gentile/Corbis.

Photo 13.1. © Bill Gentile/Corbis.

Photo 13.2. NYS Department of Correctional Services.

Photo 13.3. California Department of Corrections and Rehabilitation.

Photo 14.1. Comstock/Comstock/Thinkstock.

Photo 15.1. Thinkstock Images/Comstock/Thinkstock .

CHAPTER 1

The Philosophical and Ideological Underpinnings of Corrections

❖ Introduction: What Is Corrections?

The primary responsibility of the government of any country or state is to protect its citizens from those who would harm them. The military protects us from evildoers from beyond our shores, and the criminal justice system protects us from them within our shores. The criminal

Photo 1.1

A multilevel cellblock of a large American prison

justice system is composed of a number of subsystems, broadly categorized into law enforcement, the courts, and corrections—the so-called Catch 'em, convict 'em, and correct 'em trinity. Corrections is thus a system embedded in a broader collection of public protection agencies and programs, one that comes into play after the accused has been caught by law enforcement, prosecuted, and convicted by the court system.

Corrections is a generic term covering a wide variety of functions carried out by government (and increasingly private) agencies having to do with the punishment, treatment, supervision, and management of individuals who have been accused (in the case of some jail inmates) or convicted of criminal offenses. These functions are implemented in prisons, jails, and other secure institutions, as well as in community-based correctional agencies such as probation and parole departments. Corrections is also the name we give to the field of academic study of the theories, missions, policies, systems, programs, and personnel that implement those functions, as well as the behaviors and experiences of its unwilling customers. As the term implies, the whole correctional enterprise exists to "correct," "amend," or "put right" the clientele. This is a difficult task because the behavior, values, and attitudes to be corrected have typically festered for many years in atrocious environments. The experiences of many offenders (not all, of course) devoured their childhood and youth and marred their characters to the point where many of them have a psychological, emotional, or financial investment in their current lifestyles and have no intention of being "corrected" (Andrews & Bonta, 2007; Walsh & Stohr, 2010).

Cynics may think that the correctional process should be called the "punishment process" (Logan & Gaes, 1993). The correctional enterprise is primarily about punishment, which almost everyone agrees is an unfortunate but necessary part of life. Earlier scholars were more accurate in calling what we now call corrections **penology,** which means the study of the processes adopted for the punishment and prevention of crime. No matter what we call our prisons, jails, and other systems of formal social control, we are compelling people to do what they do not want to do, and such arm-twisting is experienced by them as punitive regardless of what name we call it.

❖ The Theoretical Underpinnings of Corrections

Just as all theories of crime contain a view of human nature, so do all models of what to do with individuals who commit it. Some thinkers (mostly influenced by sociology) formulate their theories and arguments as consistent with the assumption that human nature is socially constructed; that is, the human mind is basically a "blank slate" at birth and subsequently formed by cultural experiences. Those holding this position tend to see human nature as essentially good and believe that people must learn to be antisocial. Others (mostly influenced by evolutionary biology and the brain sciences) argue that there is an innate human nature that evolved, driven by the overwhelming concerns of all living things: to survive and reproduce. This perspective does not deny that specific behaviors are learned, but maintains that certain traits evolved in response to survival and reproductive challenges. They further maintain that some of these traits, such as high aggressiveness and low empathy, are also useful in pursuing criminal goals, and that human nature is essentially selfish (not "bad," just self-centered) and people must learn to be prosocial rather than antisocial.

For those who assume that human nature is basically good, the task is to discover why social animals commit antisocial acts. If human nature is socially constructed, the presence of antisocial characters among us reflects defective social construction, not defective human materials. We must therefore search for flaws and defects in society and not in the individual products of society. If we wish to reduce crime, then we must change society, not the individual. For some people in this tradition, punishment is seen as the vindictive infliction of pain on society's own creations.

The opposite position maintains that ever since humans first devised rules of conduct, they have wanted to break them. Of course, most of us conform to the rules of our social groups most of the time and feel shamed, embarrassed, and guilty when we violate them. But the straight and narrow road does not always come naturally, so the task is not to understand why some people commit crimes, but rather why most of us do not. After all, crime affords immediate gratification of desires with very little effort, or, as Gottfredson and Hirschi (2002) put it, "money without work, sex without courtship, revenge without court delays" (p. 210). The point we are making is that the assumptions about human nature we hold strongly influence our ideas about how we should treat the accused or convicted once they enter the correctional system.

❖ A Short History of Correctional Punishment

Legal **punishment** may be defined as the state-authorized imposition of some form of deprivation—of liberty, resources, or even life—upon a person justly convicted of a violation of the criminal law. The earliest known written code of punishment is the Code of Hammurabi, created about 1780 B.C. This code expressed the well-known concept of *lex talionis* (the law of equal retaliation), which is further enunciated in the Mosaic Code, the ancient law of the Hebrews, as "an eye for an eye, a tooth for a tooth." These laws codified the natural inclination of individuals harmed by another to seek revenge, but they also recognized that personal revenge must be restrained if society is not to be fractured by a cycle of tit-for-tat blood feuds. Blood feuds (revenge killings) perpetuate the injustice that "righteous" revenge was supposed to diminish. As Susan Jacoby (1983) put it,

> The struggle to contain revenge has been conducted at the highest level of moral and civic awareness at each stage in the development of civilization. The self-conscious nature of the effort is expectable in view of the persistent state of tension between uncontrolled vengeance as destroyer and controlled vengeance as an unavoidable component of justice. (p. 13)

"Controlled vengeance" is about the state taking responsibility for punishing wrongdoers from the individuals who were wronged. Nevertheless, early state-controlled punishment was typically as uncontrolled and vengeful as what any grieving parent might inflict on the murderer of his or her child. Prior to the 18th century, all human beings were considered born sinners because of the Christian legacy of Original Sin. Cruel tortures used on criminals to literally "beat the devil out of them" were justified by the need to save sinners' souls. Earthly pain was temporary and certainly preferable to an eternity of torment if sinners died unrepentant. Punishment was often barbaric, regardless of whether those ordering it bothered to justify it with such arguments or even believed them themselves.

It was not only the poor who suffered almost arbitrary arrest and punishment. Under the notorious French system of *lettres de cachet* ("letters with a seal"), stamped with the official seal of the king, anyone could be arrested and imprisoned without formal charges or trial. Rich individuals could buy such a letter from the king's ministers to get rid of some bothersome person. The idea of due process was a totally foreign concept to all until fairly recently. Persons arrested under the authority of a *lettre de cachet* had no right to know why they were imprisoned, no right to confront their accuser, no right to legal counsel, no right to a trial, and no right to appeal. All this was perfectly legal under the *Code Louis* of 1670, which was the legal code of France until the *Code Napoleon* of 1804 (Walsh & Hemmens, 2011).

The practice of brutal punishment and arbitrary legal codes began to wane in the late 18th century with the beginning of a period historians call the **Enlightenment,** which was essentially a major shift in the way people began to view the world and their place in it. It was also marked by the narrowing of what Thompson (1975) called the "mental distance" between people such that lawmakers began to expand their circles of individuals they considered to be "just like us." Perhaps the first person to apply Enlightenment thinking to crime and punishment was the English playwright, author, and judge, Henry Fielding (1707–1754). Fielding believed that the cause of robbery (robbery was being fueled by London's gin epidemic much the way it was fueled by crack in American cities in the 1980s) was poverty and called for a "safety net" for the poor (free housing and food) as a crime prevention strategy. Many of his suggestions were implemented and were remarkably successful by most accounts (Sherman, 2005).

The Emergence of the Classical School

Enlightenment ideas eventually led to a school of penology that has come to be known as the **Classical School.** More than a decade after Fielding's (1751/1967) book, Italian nobleman and professor of law *Cesare Bonesana, marchese di Beccaria* (1738–1794) published what was to become the manifesto for the reform of judicial and penal systems throughout Europe, *Dei Delitti e della Pene (On Crimes and Punishment)* (1764/1963). The book was a passionate plea to humanize and rationalize the law and to make punishment just and reasonable. Beccaria did not question the need for punishment, but he believed that laws should be designed to preserve public safety and order, not to avenge crime. He also took issue with the common practice of secret accusations, arguing that such practices led to general deceit and alienation in society. He argued that accused persons should be able to confront their accusers, to know the charges brought against them, and to be granted a public trial before an impartial judge as soon as possible after arrest and indictment.

Punishments should be proportionate to the harm done to society, should be identical for identical crimes, and should be applied without reference to the social status of either offender or victim. Beccaria championed the abolition of the death penalty and believed that punishments should only minimally exceed the level of damage done to society. Punishment, however, must be certain and swift to make a lasting impression on the criminal and to deter others. To ensure a rational and fair penal structure, punishments for specific crimes must be decreed by written criminal codes, and the discretionary powers of judges must be severely limited. The judge's task was to determine guilt or innocence, and then to impose the legislatively prescribed punishment if the accused is found guilty.

Beccaria's work was so influential that many of his recommended reforms were implemented in a number of European countries within his lifetime (Durant & Durant, 1967). Such radical change over such a short period of time, across many different cultures, suggests that Beccaria's rational reform ideas tapped into and broadened the scope of emotions such as sympathy and empathy among the political and intellectual elite of Enlightenment Europe. An influential early commentator, *Alexis de Tocqueville* (1805–1859), noticed the diffusion of these emotions

Photo 1.2

Italian nobleman and professor of law Cesare Beccaria (1738–1794) published what was to become the manifesto for the reform of judicial and penal systems throughout Europe, *Dei Delitti e della Pene (On Crimes and Punishment)* (1764/1963).

across social classes beginning in the Enlightenment and attributed the "mildness" of the American criminal justice system to the country's democratic spirit (1838/1956). We tend to feel empathy for those whom we view as being "like us," and this leads to sympathy, which may translate the vicarious experiencing of the pains of others into an active concern for their welfare. With cognition and emotion jelled into the Enlightenment ideal of the basic unity and worth of humanity, justice became both more refined and more diffuse (Walsh & Hemmens, 2011).

Another prominent figure was British lawyer and philosopher *Jeremy Bentham* (1748–1832). His major work, *Principles of Morals and Legislation* (1789/1948), is essentially a philosophy of social control based on the **principle of utility,** which posits that human actions should be judged moral or immoral by their effect on the happiness of the community. The proper function of the legislature is thus to make laws aimed at maximizing the pleasure and minimizing the pain of the largest number in society—"the greatest good for the greatest number" (p. 151).

If legislators are to legislate according to the principle of utility, they must understand human motivation, which for Bentham (1789/1948) was easily summed up: "Nature has placed mankind under the governance of two sovereign masters, pain and pleasure. It is for them alone to

Photo 1.3

Jeremy Bentham's (1748–1832) major work, *Principles of Morals and Legislation* (1789/1948), is essentially a philosophy of social control based on the principle of utility, which posits that human actions should be judged moral or immoral by their effect on the happiness of the community.

point out what we ought to do, as well as to determine what we shall do" (p. 125). This was essentially the Enlightenment concept of human nature, which was seen as hedonistic, rational, and endowed with free will. The classical explanation of criminal behavior, and how to prevent it, can be derived from these three assumptions about human nature. Bentham devoted a great deal of energy (and his own money) to arguing for the development of prisons as punitive substitutes for torture, execution, or transportation. He even designed a prison in the 1790s called the *panopticon* (panoptic means "all seeing"), which will be discussed more fully in the next chapter.

The Emergence of Positivism: Should Punishment Fit the Offender or the Offense?

Just as classicism arose from the 18th century humanism of the Enlightenment, positivism arose from the 19th century spirit of science. Classical thinkers were "armchair" philosophers in the manner of the thinkers of classical Greece (hence the term *classical*), while **positivists** took upon themselves the methods of empirical science, from which more "positive" conclusions could be drawn (hence the term *positivism*). An essential assumption of positivism is that human actions have causes, and that these causes are to be found in the uniformities that typically precede those actions. The search for causes of human behavior led positivists to dismiss the classical notion that humans are free agents who are alone responsible for their actions.

Early positivism went to extremes to espouse a hard form of determinism such as that of Lombroso's "born criminal." Nevertheless, positivism slowly moved the criminal justice system away from a singular concentration on the criminal act as the sole determinant of the type of punishment to be meted out, and toward an appraisal of the characteristics and circumstances of the offender as an additional determinant. Because human actions have causes, many of which are involuntary, i.e., out of their control, the concept of legal responsibility was called into question. For instance, Italian lawyer *Raffaele Garofalo* (1852–1934) believed that because human action is often evoked by circumstances beyond human control, the only thing to be considered at sentencing was the offenders' "peculiarities," or risk factors for crime.

Garofalo's (1885/1968) only concern for individualizing sentencing, however, was the danger offenders' posed to society, and his proposed sentences ranged from execution for what he called the *extreme criminal* (whom we might call psychopaths today), to transportation to penal colonies for *impulsive criminals,* to simply changing the law to deal with what he called *endemic criminals* (those who commit what we today might call victimless crimes). German criminal lawyer Franz von Liszt, on the other hand, campaigned for customized sentencing according to the rehabilitative potential of offenders, which was to be based on what scientists find out about the causes of crime (Sherman, 2005). This ideal of individualized sentences tailored to the characteristics of individuals meant that judges were to enjoy wide sentencing **discretion,** which argued against the classical ideal of predetermined statutory sentences imposed on all who commit the same crime without any consideration at all for individual differences.

❖ The Function of Punishment

Although most corrections scholars agree that punishment functions as a form of social control, some view it as a barbaric throwback to pre-civilized times (Menninger, 1968). But can you imagine a society where punishment did not exist? What would such a society be like? Could it survive? If you cannot realistically imagine such a society, you are not alone, for the desire to punish those who have harmed us or otherwise cheated on the social contract is as old as the species itself. Punishment aimed at discouraging cheats is observed in every social species of animals, leading evolutionary biologists to conclude that punishment of cheats is a strategy designed by natural selection for the emergence and maintenance of cooperative behavior (Alcock, 1998; Walsh, 2000). Cooperative behavior is important for all social species and is built on mutual trust, which is why violating that trust evokes moral outrage and results in punitive sanctions.

Brain imaging studies show that when subjects punish cheats, they have significantly increased blood flow to areas of the brain that respond to reward, suggesting that punishing those who have wronged us provides both emotional relief and reward (de Quervain et al., 2004; Fehr & Gachter, 2002). These studies imply that we are hardwired to "get even," as suggested by the popular saying, "Vengeance is sweet."

Sociologist Émile Durkheim (1858–1917) contended that punishment is functional for society in some ways. Durkheim (1893/1964) considered crime normal in the sense that it exists in every society and that criminal behavior is in everyone's behavioral tool kit. Punishing criminals maintains solidarity because the rituals of punishment reaffirm the justness of the social norms and allow citizens to express their moral outrage when others transgress those moral norms. Durkheim also recognized that we can temper punishment with sympathy. He observed that over the course of social evolution, humankind had moved from *retributive* justice (characterized by cruel and vengeful punishments) to *restitutive* justice (characterized by reparation—"making amends").

Retributive justice is driven by the natural passion for punitive revenge that "ceases only when exhausted . . . only after it has destroyed" (Durkheim, 1893/1964, p. 86). Durkheim goes on to claim that **restitutive justice** is driven by simple deterrence and is more humanistic and tolerant, although it is still "at least in part, a work of vengeance" (pp. 88–89). Both forms of justice satisfy the human urge for social regularity by punishing those who violate the social contract, but retributive justice oversteps its adaptive usefulness and becomes socially destructive. For Durkheim, restitutive responses to wrongdoers offer a balance between calming moral outrage on the one hand, and exciting the emotions of empathy and sympathy on the other.

Philosophies of and Justifications for Punishment

A philosophy of punishment involves defining the concept of punishment and the values, attitudes, and beliefs contained in that definition. Most clearly it involves justifying the imposition of a painful burden on unwilling subjects. When we speak of justifying something we are doing, we typically mean that we provide reasons for doing it both in terms of morality ("It's the right thing to do") and in terms of the goals we wish to achieve ("Do this, and we'll get that"). Philosophers, legal scholars, and criminologists have traditionally identified four major objectives or justifications for the practice of punishing criminals: retribution, deterrence, rehabilitation, and incapacitation. Criminal justice scholars have recently added a fifth purpose to the list: reintegration. Before we discuss these objectives, we must emphasize that all theories and systems of punishment are based on conceptions of basic human nature, and thus to a great extent on ideology. The view of human nature on which the law in every country relies today is the same view enunciated by classical thinkers Beccaria and Bentham, namely, that human beings are hedonistic, rational, and possessors of free will.

Hedonism is a doctrine that maintains that all life goals are desirable only as means to the end of achieving pleasure or avoiding pain. It goes without saying that pleasure is intrinsically desirable and pain is intrinsically undesirable, and that we all seek to maximize the former and minimize the latter. We are assumed to pursue these goals in rational ways, that is, in ways that are consistent with logic. That is, rationality involves a logical "fit" between the goals people strive for and the means they use to achieve them. For the classical scholar, the ultimate goal of any human activity is self-interest, and self-interest governs our behavior whether it takes us in prosocial or antisocial directions.

Hedonism and rationality are combined in the concept of the **hedonistic calculus**, a method by which individuals are assumed to logically weigh the anticipated benefits of a given course of action against its possible costs. If the balance of consequences of a contemplated action is thought to enhance pleasure and/or minimize pain, then individuals will pursue it; if not, they will not. If people miscalculate, as they frequently do, it is because they are ignorant of the full range of consequences of a given course of action, not because they are irrational or stupid.

The final assumption about human nature is that humans enjoy a free will that enables them to purposely and deliberately choose to follow a calculated course of action. If people seek to increase their pleasures illegally, they do so freely and with full knowledge of the wrongness of their acts. It is only with the concept of free will that we can justifiably assign praise and blame to individual actions. Because criminals know what is right and what is wrong and choose the latter, society has a perfectly legitimate right to punish those who harm it.

Retribution: Retribution is the justification for punishment underlined by the concept of *lex talionis*. It is a "just deserts" model that demands that criminals' punishments match the degree of harm they have inflicted on their victims, that is, what they justly deserve. Those who commit minor crimes deserve minor punishments, and those who commit more serious crimes deserve more severe punishments. This is perhaps the most honestly stated justification for punishment because it both taps into our most primitive punitive urges and posits no secondary purpose for it, such as rehabilitation or deterrence. California is among the states that have explicitly embraced this justification in their criminal codes (California Penal Code Sec. 1170a): "The Legislature finds and declares that the purpose of imprisonment for a crime is punishment" (cited in Barker, 2006, p. 12). This model of punishment avers that it is right to punish criminals regardless of any secondary purpose that punishment may serve, simply because justice demands it.

Some scholars consider retribution to be nothing more than primitive revenge, and therefore morally wrong (Tutu, 1999). However, retribution as presently conceived is not Durkheimian revenge that "ceases only when exhausted." Rather, it is constrained revenge supposedly curbed by proportionality and carried out by allegedly neutral parties bound by laws mandating respect for the rights of individuals against whom it is imposed. Logan and Gaes (1993) go so far as to claim that only retributive punishment "is an affirmation of the autonomy, responsibility, and dignity of the individual" (p. 252). By holding offenders responsible and blameworthy for their actions, we are treating them as free moral agents, not as mindless rag dolls being blown around by the winds of negative forces in the environment.

Deterrence: A more complex justification for punishment is deterrence, that is, the prevention of crime by the threat of punishment. The principle that people respond to incentives and are deterred by the threat of punishment is the philosophical foundation behind all systems of criminal law. Deterrence may be either specific or general.

Specific deterrence refers to the effect of punishment on the future behavior of persons who experience the punishment. For specific deterrence to work, it is necessary that a previously punished person make a conscious connection between an intended criminal act and the punishment suffered as a result of similar acts committed in the past. Unfortunately, it is not always clear that such connections, if made, have the desired effect, either because memories of the previous consequences were insufficiently potent or because they were discounted.

Committing further crimes after being previously convicted and punished is called **recidivism** ("falling back" into criminal behavior), which is a lot more common among ex-inmates than rehabilitation. Recidivism refers only to crimes committed after release from prison and does not apply to crimes committed while incarcerated. Nationwide in the United States, about 33% of released prisoners recidivate within the first 6 months after release, 44% within the first year, 54% by the second year, and 67.5% by the third year (Robinson, 2005, p. 222), and these are just the ones who are caught. Among those who do desist, a number of them cite the fear of additional punishment as a major factor (Wright, 1999).

As the classical scholars remind us, the effect of punishment on future behavior depends on its certainty, celerity (swiftness), and severity. In other words, there must be a relatively high degree of certainty that punishment will follow a criminal act, the punishment must be administered very soon after the act, and it must be quite harsh. The most important of these is certainty, but as we see from Figure 1.1 showing clearance rates for major crimes, the probability of being arrested is very low, especially for property crimes—so much for certainty. Factoring out the immorality of the enterprise, burglary, for instance, appears to be a very rational career option for the capable criminal.

If a person is caught, the wheels of justice grind very slowly. Evidence has to be collected and evaluated, juries must be selected, and the court dockets are consistently overloaded. Typically, many months pass between the act and the imposition of punishment—so much for celerity. This leaves the law with severity as the only element it can realistically manipulate (it can increase or decrease statutory penalties almost at will), but it is unfortunately the least effective element (National Center for Policy Analysis, 1998). Studies from the United States and the United Kingdom find substantial negative correlations (as one factor goes up, the other goes down) between the likelihood of conviction (a measure of certainty) and crime rates, but much weaker correlations in the same direction for the severity of punishment; that is, increased severity leads to lower offending rates (Langan & Farrington, 1998).

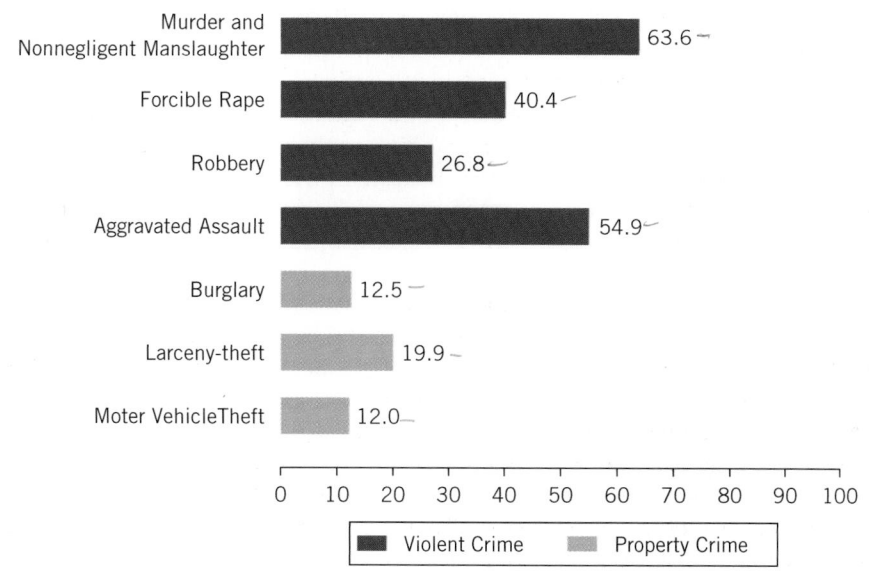

Figure 1.1

Percentage of Crimes Cleared by Arrest or Exceptional Means* in 2008

*A crime cleared by "exceptional means" occurs when the police have a strong suspect but something beyond their control precludes a physical arrest (e.g., death of suspect).

Source: Federal Bureau of Investigation [FBI] (2009).

The effect of punishment on future behavior also depends on the **contrast effect,** defined as the contrast or comparison between the possible punishment for a given crime and the usual life experience of the person who may be punished. For people with little or nothing to lose, arrest and punishment may be perceived as merely an inconvenient occupational hazard, an opportunity for a little rest and recreation, and a chance to renew old friendships. But for those who enjoy a loving family and the security of a valued career, the prospect of incarceration is a nightmarish contrast. Like so many other things in life, deterrence works least for those who need it the most (Austin & Irwin, 2001).

General deterrence refers to the preventive effect of the threat of punishment on the general population; it is thus aimed at *potential* offenders. Punishing offenders serves as examples to the rest of us of what may happen if we violate the law. As Radzinowicz and King (1979) put it more than 30 years ago, "People are not sent to prison primarily for their own good, or even in the hope that they will be cured of crime. . . . It is used as a warning and deterrent to others" (p. 296). The existence of a system of punishment for law violators deters a large but unknown number of individuals who might commit crimes if no such system existed.

What is the bottom line on the effectiveness of deterrence? Are we putting too much faith in the ability of criminals and would-be criminals to calculate the costs and benefit of engaging in crime? Although many violent crimes are committed in the heat of passion, or under the influence of mind-altering substances, there is quite a bit of evidence underscoring the classical notions that individuals do (subconsciously at least) calculate the ratio of expected pleasures to possible pains when contemplating a course of action. Nobel Prize–winning economist Gary Becker (1997) is a major adherent of the position. He dismisses the idea that criminals lack the knowledge and the foresight to take punitive probabilities into consideration when deciding whether or not to continue committing crimes. He says, "Interviews of young people in high crime areas who do engage in crime show an amazing understanding of what punishments are, what young people can get away with, how to behave when going before a judge" (p. 20). Becker also compared crime rates in Great Britain and the United States and demonstrated that crime rates rose in the UK as its penal philosophy became more and more lenient, and that they fell in the United States as its penal philosophy became more and more punitive.

Deterrence theorists do not view people as calculating machines doing their mental math before engaging in any activity. They are simply saying that behavior is governed by its consequences. Our rational calculations are both subjective and bounded; we do not all make the same calculations or arrive at the same game plan when pursuing the same goals. Think how the contrast effect would influence the calculations of a zero-income, 19-year-old high school dropout with a drug problem as opposed to a 45-year-old married man with two children and a $90,000 annual income. We all make calculations with less than perfect knowledge, with different mind-sets, different temperaments, and different cognitive abilities, but to say that criminals do not make such calculations is to strip them of their humanity and to make them pawns of fate.

More general reviews of deterrence research indicate that legal sanctions do have "substantial deterrent effect" (Nagin, 1998, p. 16; see also Wright, 1999), and some researchers have claimed that increased incarceration rates account for about 25% of the variance in the decline in violent crime over the last decade or so (Spelman, 2000; Rosenfeld, 2000). Of course, this leaves 75% of the variance to be explained by other factors, such as an improved economy. Unfortunately, even for the 25% figure, we cannot determine if we are witnessing a *deterrent* effect (i.e., has violent crime declined because more would-be violent people have perceived a greater punitive threat?) or an *incapacitation* effect (i.e., has violent crime declined because more violent people are behind bars and thus not at liberty to commit violent crimes on the outside?). Of course, it does not have to be one or the other, since both effects may be operating. Society benefits from crime reduction regardless of why it occurs, but correctional scholars would like to know which of the processes (deterrence or incapacitation or an improved economy or some other variable) is most responsible for the decline.

Incapacitation: Incapacitation refers to the inability of criminals to victimize people outside prison walls while they are locked up. Its rationale is aptly summarized in James Q. Wilson's (1975) remark, "Wicked people exist. Nothing avails except to set them apart from innocent people" (p. 391). The incapacitation justification probably originated with *Enrico Ferri's* (1869–1929) concept of social defense. Ferri (1897/1917) was one of the early positivists who dismissed the classical ideas about human nature as myths. To determine punishment, notions of culpability, moral responsibility, and intent were to be secondary to an assessment of offenders' strength of resistance to criminal impulses, with the express purpose of averting future danger to society. He believed that moral insensibility and lack of foresight, underscored by low intelligence, were the criminal's most marked characteristics.

Ferri's (1897/1917) social defense asserts that the purpose of punishment is not to deter or to rehabilitate but to defend society from criminal predation. Ferri reasoned that the characteristics of criminals prevented them from basing their behavior on rational calculus principles, so how could their behavior be deterred? Given the assumptions of early positivism, the only reasonable rationale for punishing offenders is to incapacitate them for as long as possible so that they no longer pose a threat to the peace and security of society.

Obviously, incapacitation "works" while criminals are incarcerated. Elliot Currie (1999) uses robbery rates to illustrate this, stating that in 1995, there were 135,000 inmates in prison whose most serious crime was robbery, and that each robber on average commits five robberies per year. Had these robbers been left on the streets, they would have been responsible for an additional 135,000 × 5 or 675,000 robberies on top of the 580,000 actual robberies reported to the police in 1995. Similarly, Wright (1999) estimated that imprisonment averted almost 7 million offenses in 1990. The incapacitation effect is more starkly driven home by a study of the offenses of 39 convicted murderers committed *after* they had served their time for murder and had been released from prison. It was found that altogether they had 122 arrests for serious violent crimes (including 7 additional murders), 218 arrests for serious property crimes, and 863 "other" arrests between them (DeLisi, 2005, p. 165).

Our discussion of these 39 murderers brings up the idea of **selective incapacitation,** which refers to a punishment strategy that largely reserves prison for a select group of offenders. This select group should be composed primarily of violent repeat offenders but may also include other types of incorrigible offenders. Birth cohort studies (a *cohort* is a group composed of subjects having something in common, such as being born within a given time frame and/or in a particular place) from a number of different locations find that about 6% to 10% of offenders commit the majority of all crimes. For instance, in the 1945 birth cohort studies by Wolfgang, Figlio, and Sellin (1972), 6.3% of the 9,945 cohort members committed 71% of the murders, 73% of the rapes, and 82% of the robberies attributed to members of the cohort.

In saving prison space mostly for high-rate violent offenders, so the reasoning goes, we both better protect the community and save it money by incarcerating fewer nonviolent offenders. The problem with this strategy involves identifying high-rate violent offenders *before* they become high-rate violent offenders; identifying them after the fact, as in the above cohort studies, is easy. Generally speaking, individuals who begin committing predatory delinquent acts before they reach puberty are the ones who will continue to commit crimes across the life course (DeLisi, 2005; Moffitt & Walsh, 2003). Of course, although there are a number of excellent prediction scales in use today to assist us in estimating who will and who will not become a high-rate offender, the risk of too many false-positives (predicting someone will become a high-rate offender when in fact he or she will not) is always present (Piquero & Blumstein, 2007).

Rehabilitation: The term *rehabilitation* means to restore or return to constructive or healthy activity. Whereas deterrence and incapacitation are mainly justified philosophically on classical grounds, rehabilitation is primarily a positivist concept. The rehabilitative goal is based on a medical model that used to view criminal behavior as a moral sickness requiring treatment. Today, this model views criminality in terms of "faulty thinking" and criminals as in need of "programming" rather than "treatment." Although the goal is the same as that of deterrence, it is different in that the goal is to change offenders' attitudes so that they come to accept that their behavior was wrong, not to deter them by the threat of further punishment. Because we have a complete chapter (Chapter 14) devoted to correctional treatment and rehabilitation, we will defer our discussion of these concepts for the moment.

Reintegration: The goal of reintegration is to use the time criminals are under correctional supervision, either in institutions or in the community, to prepare them to reenter (or reintegrate with) the free community as well equipped to do so as possible. This goal is also known as *reentry* or *restoration*. In effect, reintegration is not much different from rehabilitation, but it is more pragmatic, focusing on concrete programs such as job training rather than attitude change. There are many challenges associated with this process, so much so that, like rehabilitation, it warrants a chapter to itself and will be discussed in detail in the chapter on parole (Chapter 8).

Table 1.1 is a summary of the key elements (justification, strategy, etc.) of the five punishment philosophies or perspectives discussed. The commonality that they all share to various extents is, of course, the prevention of crime.

Table 1.1

Summary of Key Elements of Different Correctional Perspectives

	Retribution	Deterrence	Incapacitation	Rehabilitation	Reintegration
Justification	Moral Just deserts	Prevention of further crime	Risk control Community protection	Offenders have correctable deficiencies.	Offenders have correctable deficiencies.
Strategy	None: Offenders simply deserve to be punished.	Make punishment more certain, swift, and severe.	Offenders cannot offend while in prison.	Treatment to reduce offenders' inclination to reoffend.	Concrete programming to make for successful reentry into society
Focus of perspective	The offense and just deserts	Actual and potential offenders	Actual offenders	Needs of offenders	Needs of offenders
Image of offenders	Free agents whose humanity we affirm by holding them accountable	Rational beings who engage in cost/benefit calculations	Not to be trusted but to be constrained	Good people who have gone astray. Will respond to treatment	Ordinary folk who require and will respond to concrete help

❖ Is the United States Hard or Soft on Crime?

A frequently heard criticism of the criminal justice system in the United States is that the nation is soft on crime. If we define hardness or softness in terms of incarceration rates, the figure indicating incarceration rates per 100,000 for countries belonging to the *Organisation for Economic Co-operation and Development* (OECD) in 2008–2009 shown below conveys the opposite message, as does the retention of the death penalty, which has been eschewed by other "civilized" nations. Only Russia (not a member of the OECD) with a rate of 532 per 100,000 comes close to the American incarceration rate, and the closest any modern Western nation comes to the U.S. rate is England and Wales, with a rate 5 times lower (see Figure 1.2). Unfortunately, comparisons among nations relative to this question are typically made using only Western democratic nations, which leads to the conclusion that the United States is hard on crime. But if we are to make valid comparisons, we cannot cherry pick our countries to arrive at a conclusion that fits our ideology. We must compare the United States with authoritarian as well as democratic nations.

If we define hardness/softness in terms of alternative punishments or the conditions of confinement, then the United States is "soft" on crime, although a better term would be *humane.* For instance, although China is listed by Mauer (2005) as having an incarceration rate more than 5 times lower than the United States, it is the world's leader in the proportion of its criminals it executes each year. Also, punishment in some fundamentalist Islamic countries such as Saudi Arabia and Afghanistan under the Taliban often includes barbaric corporal punishments for offenses considered relatively minor in the West. Drinking alcohol can get the drinker 60 lashes, robbers may have alternate hands and feet amputated, and

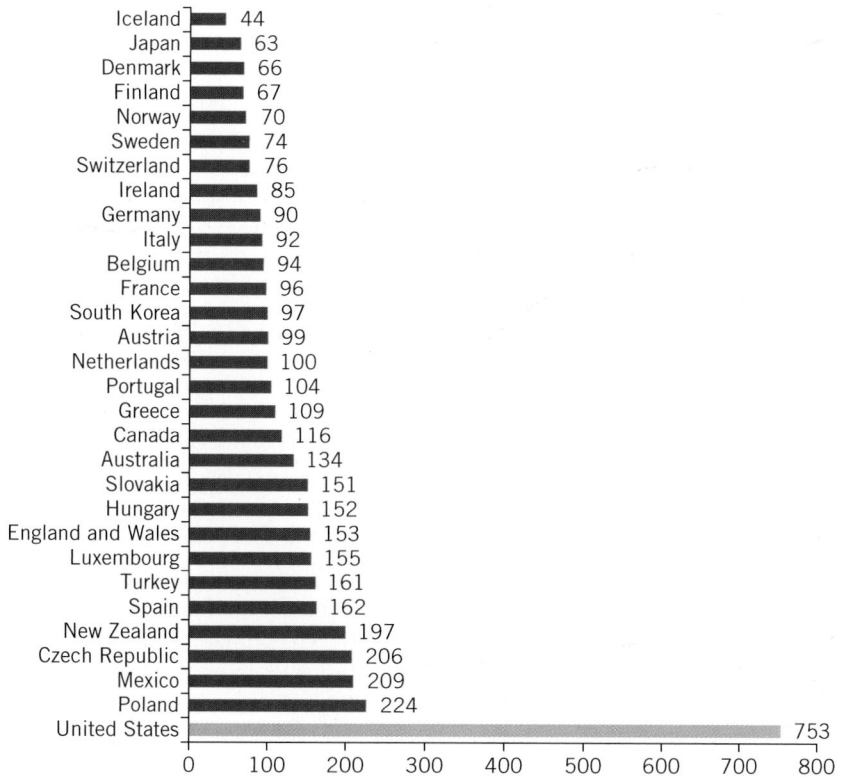

Figure 1.2

Incarceration Rate per 100,000 in OECD Countries (2008–2009)

Source: Schmitt, Warner, and Gupta (2010).

women accused of "wifely disobedience" may be subjected to corporal punishment (Walsh & Hemmens, 2011).

Another problem with assessing the hardness or softness of the American criminal justice system based on the rates is that they are calculated per 100,000 *citizens,* which is not the same as the rate per 100,000 *criminals.* If the United States has more criminals than these other countries, then perhaps the greater incarceration rate is justified. Of course, no one knows how many criminals any country has, but we can get a rough estimate from a country's crime rates. For instance, the U.S. homicide rate is about 5 times that of England and Wales, which roughly matches the United States' 5 times greater incarceration rate.

However, when it comes to property crimes, Americans are about in the middle of the pack of nations in terms of the probability of being victimized (less than in England and Wales, incidentally). This fact notwithstanding, burglars serve an average of 16.2 months in prison in the United States, compared with 6.8 months in Britain and 5.3 months in Canada (Mauer, 2005), which makes the United States harder on crime than its closest cultural relatives and suggests that we may be overusing incarceration to address our crime problem. (Alternatively, from a crime control perspective, these other nations can be seen as underutilizing incarceration at the expense of raising crime rates.)

So, is the United States softer or harder on crime than other countries? The answer obviously depends on how we conceptualize and measure the concepts of hardness and softness and with which countries we compare ourselves. Compared with countries that share our democratic ideals, we are tough (and because of our retention of the death penalty, some would even say barbaric) on crime; compared with countries most distant from Anglo/American ideals, we are extremely soft, and for that we should be grateful.

In the remainder of the book, we will discuss the various components and programs of corrections, their staff, and inmates. We will begin with two chapters (2 and 3) on the history of corrections, as much of what we do today to "correct" has been tried or explored by earlier generations: We have much to learn from those experiences. We then proceed through the criminal justice process (Chapters 4 through 8); what correctional work is like for staff (Chapter 9); and the unique experiences of women, minorities, and juveniles in corrections (Chapters 10 to 12). Chapters 13 and 14 are concerned with legal issues and treatment programming and research, as both affect the operation of correctional facilities and programs. Finally, the last chapter (Chapter 15) includes a brief overview of some emerging issues and influences that we expect will shape corrections for the next several years.

Summary

- Corrections is a social function designed to hold, punish, supervise, deter, and possibly rehabilitate the accused or convicted. It is also the study of these functions.

- Although it is natural to want to exact revenge ourselves when people do us wrong, the state has taken over this responsibility for punishment to prevent endless tit-for-tat feuds. Over social evolution, the state has moved to more restitutive forms of punishment that, while serving to tone down the community's moral outrage, tempers it with sympathy.

- Much of the credit for the shift away from retributive punishment must go to the great Classical School of criminology, which was imbued with the humanistic spirit of the Enlightenment. The view of human nature (hedonistic, rational, and possessing free will) held by thinkers of the time was that punishment should primarily be used for deterrent purposes, that it should only just exceed the gains of crime, and that it should apply equally to all who have committed the same crime regardless of any individual differences.

■ Opposing classical notions of punishment are those of the positivists who rose to prominence during the 19th century and who were influenced by the spirit of science. Positivists rejected the philosophical underpinnings regarding human nature of the classicists and declared that punishment should fit the offender rather than the crime.

■ The objectives of punishment are retribution, deterrence, incapacitation, rehabilitation, and reintegration, all of which have come into favor, gone out, and come back again over the years.

■ Retribution is simply just deserts—getting the punishment one deserves, with no other justification needed.

■ Deterrence is the assumption that people are prevented from committing crime by the threat of punishment.

■ Incapacitation means that the accused and convicted cannot commit further crimes (if they did so in the first place) against the innocent while incarcerated.

■ Rehabilitation centers around efforts to socialize offenders in prosocial directions while they are under correctional supervision so that they will not commit further crimes.

■ Reintegration refers to efforts to provide offenders with concrete skills they can use that will give them a stake in conformity.

■ The United States leads the world in the proportion of its citizens that it has in prison. Whether this is indicative of hardness (more prison time for more people) or softness (imprisonment as an alternative to execution or mutilation) depends on how we view hardness versus softness and with which countries we compare the United States.

Key Terms

Classical School

Corrections

Contrast effect

Deterrence

Discretion

Enlightenment

General deterrence

Hedonism

Hedonistic calculus

Incapacitation

Penology

Positivists

Principle of utility

Punishment

Recidivism

Rehabilitation

Reintegration

Restitutive justice

Retribution

Retributive justice

Selective incapacitation

Specific deterrence

Discussion Questions

1. Discuss the implications for a society that decides to eliminate all sorts of punishment in favor of forgiveness.

2. Why do we take pleasure in the punishment of wrongdoers? Is it a good or bad thing that we take pleasure in punishment? What evolutionary purpose does punishment serve?

3. Discuss the assumptions about human nature held by the classical thinkers. Are we rational, seekers of pleasure, and free moral agents? If so, does it make sense to try to rehabilitate criminals?

4. Discuss the assumptions underlying positivism in terms of the treatment of offenders. Do they support Garofalo or von Liszt in terms of the meaning these assumptions have for punishment.

5. Which justification for punishment do you favor? Is it the one that you think "works" best in terms of preventing crime, or do you favor it because it fits your ideology?

6. What is your position on the hardness/softness issue relating to the United States' stance on crime? We are tougher than other democracies. Is that okay with you? We are also softer than more authoritarian countries. Is that okay with you also? Why or why not?

CHAPTER 2

A History of Corrections

❖ Introduction: The Evolving Practice of Corrections

The history of corrections is riddled with the best of intentions and the worst of abuses. Correctional practices and facilities (e.g., galley slavery, transportation, jails and prisons, community corrections) were created, in part, to remove the riffraff—both poor and criminal—from urban streets or at least to control and shape them. Prisons and community corrections were also created to avert the use of more violent or coercive responses to such folk. In this chapter and the next, the focus is on exploring the history of the Western world's correctional operations and then American corrections, specifically, and the reoccurring themes that run through this history and define it.

It is somewhat ironic that one of the best early analyses of themes and practices in American prisons and jails was completed by two French visitors to the United States—*Gustave de*

Beaumont and *Alexis de Tocqueville*—while the country was in its relative infancy, in 1831, and experiencing the virtual birthing of prisons themselves (Beaumont & Tocqueville, 1833/1964). Tocqueville, as a 26-year-old French magistrate, brought along his friend Beaumont, supposedly to study America's newly minted prisons for 9 months. They ended up also observing the workings of its law, its government and political system, and its race relations, among other things (Damrosch, 2010; Tocqueville & Goldhammer, 1835/2004). The irony is that, as outsiders and social critics, Beaumont and Tocqueville could so clearly see what others, namely Americans, who were thought to have "invented prisons" and who worked in them, were blind to. In this chapter we will try to "see" what those early French visitors observed about Western and specifically American correctional operations.

Photo 2.1

In 1831, Tocqueville, as a 26-year-old French magistrate, brought along his friend Beaumont to study America's newly minted prisons.

Few visitors to the United States, or residents for that matter, explored or commented on the early correctional experience for women (*Dorothea Dix* being a notable exception—there will be more about her and her observations about the state of corrections in 1845 in Chapter 3). Yet some of the themes that run through the practice of corrections apply to women and girls as well, but with a twist. Women have always represented only a small fraction of the correctional population in both prisons and jails, and the history of their experience with incarceration, as shaped by societal expectations of and for them, can be wholly different from that of men. As literal outsiders to what was the "norm" for inmates of prisons and jails, and as a group whose rights and abilities were legally and socially controlled on the outside more than that of men and boys, women's experience in corrections history is worth studying and will be more fully explored in Chapter 10.

What is clear from the Western history of corrections is that what was *intended* when prisons, jails, and reformatories were conceived, and *how they actually operated,* then and now, were and are often two very different things (Rothman, 1980). As social critics ourselves, we can use the history of corrections to identify a series of "themes" that run through correctional practice, even up to today. Such themes will reinforce the tried, yet true, maxim, "Those who cannot remember the past are condemned to repeat it" (Santayana, 1905, p. 284). Too often we do not know or understand our history of corrections, and as a consequence, we are forever repeating it.

❖ Themes: Truths That Underlie Correctional Practice

There are some themes that have been almost eerily constant, vis-à-vis corrections, over the decades and even centuries. Some such themes are obvious, such as the influence that money, or its lack, exerts over virtually all correctional policy decisions. Political sentiments and the desire to make changes also have had tremendous influence over the shape of corrections in the past. Other themes are less apparent, but no less potent in their effect on correctional operation. For instance, there appears to be an evolving sense of compassion or humanity that, though not always clear in the short term, in practice, or in policy or statute, has underpinned reform-based decisions about corrections and its operation, at least in theory, throughout its history in the United States. The creation of the prison, with a philosophy of penitence (hence the *penitentiary*), was a grand reform itself, and as such it represented in theory, at least, a major improvement over the brutality of punishment that characterized early English and European law and practice (Orland, 1995).

Some social critics do note, however, that the prison and the expanded use of other such social institutions also served as a "social control" mechanism to remove punishment from public view, while making the state appear more just (Foucault, 1979; Welch, 2004). Therefore, this is not to argue that such grand reforms in their idealistic form, such as prisons, were not primarily constructed out of the need to control, but rather that there were philanthropic, religious, and other forces aligned that also influenced their creation and design, if not so much their eventual and practical operation (Hirsch, 1992). Also of note, the social control function becomes most apparent when less powerful populations like the poor, the minority, the young, or the female are involved, as will be discussed in the following chapters.

Other than the influence of money and politics and a sense of greater compassion/humanity in correctional operation, the following themes are also apparent in corrections history: the question of how to use labor and technology (which are hard to decouple from monetary considerations); a decided religious influence; the intersection of class, race, age, and gender in shaping one's experience in corrections; architecture as it is intermingled with supervision; methods of control; overcrowding; and finally the fact that good intentions do not always translate into effective practice. Though far from exhaustive, this list contains some of the most salient issues that become apparent streams of influence as one reviews the history of corrections. As was discussed in Chapter 1, some of the larger philosophical (and political) issues, such as conceptions of right and wrong and whether it is best to engage in retribution or rehabilitation (or both, or neither, along with incapacitation, deterrence, and reintegration) using correctional sanctions, are also obviously associated with correctional change and operation.

❖ Early Punishments in Westernized Countries

Human beings, throughout recorded history, have devised ingenious ways to "punish" their kind for real or perceived transgressions. Among tribal groups and even in more developed civilizations, such punishment might include, among other tortures, whipping, branding, mutilation, drowning, suffocation, executions, and banishment (which in remote areas was tantamount to a death sentence). The extent of the punishment often depended on the wealth and status of the offended party and the offender. Those accused or found guilty and who were richer were often allowed to make amends by recompensing the victim or his or her family, while those who were poorer and of lesser status were likely to suffer some sort of bodily punishment. But whatever the approach, and for whatever the reason, some sort of punishment was often called for as a means of balancing the scales of justice, whether to appease a god or gods or later Lady Justice.

As David Garland (1990) recounts, "ancient societies and 'primitive' social groups often invested the penal process with a wholly religious meaning, so that punishment was understood as a necessary sacrifice to an aggrieved deity" (p. 203). As urbanization took hold, however, and transgressions were less tolerated among an increasingly diverse people, the ancients and their governing bodies were more likely to designate a structure as appropriate for holding people. For the most part, such buildings or other means of confining people were often used to ensure that the accused was held over for "trial" or sometimes just for punishment (Orland, 1975, p. 13). Fines, mutilation, drawing and quartering, and capital punishment were popular ways to handle those accused or convicted of crimes (Harris, 1973; Orland, 1975).

> Although mutilation ultimately disappeared from English law, the brutality of Anglo-Saxon criminal punishment continued unabated into the eighteenth century. In the thirteenth century, offenders were commonly broken on the wheel for treason. A 1530 act authorized poisoners to be boiled alive. Burning was the penalty for high treason and heresy, as well as for murder of a husband by a wife or of a master by a servant. Unlike the punishment of boiling, that of burning remained lawful in England until 1790. In practice, and as a kindness, women were strangled before they were burned. The right hand was taken off for aggravated murder. Ordinary hangings were frequent, and drawing and quartering, where the hanged offender was publicly disemboweled and his still-beating heart held up to a cheering multitude, was not uncommon.
>
> In addition, until the mid-nineteenth century, English law permitted a variety of "summary" punishments. Both men and women (the latter until 1817) were flagellated in public for minor offenses. For more serious misdemeanors there was the pillory, which was not abolished in England until 1837. With his face protruding though its beams and his hands through the holes, the offender was helpless. Sometimes he was nailed through the ears to the framework of the pillory with the hair of his head and beard shaved; occasionally he was branded. Thereafter, some offenders were carried back to prison to endure additional tortures. (Orland, 1975, p. 15)

The First Jails

Jails were the first type of correctional facility to develop, and in some form they have existed for several thousand years. Whether pits or dungeons or caves were used, or the detained were tied to a tree, ancient people all had ways of holding people until a judgment was made or implemented (Irwin, 1985; Mattick, 1974; Zupan, 1991).

According to Johnston (2009), punishment is referenced in a work written in 2000 B.C. and edited by Confucius. The Old Testament of the Bible refers to the use of imprisonment from 2040–164 B.C. in Egypt and its use in ancient Assyria and Babylon. Ancient Greece and Rome reserved harsher physical punishments for slaves, whereas citizens might be subjected to fines, exile, imprisonment, or death, or some combination of these (Harris, 1973).

> Ancient Roman society was a slave system. To punish wrongdoers, *capitis deminutio maxima*—the forfeiture of citizenship—was used. Criminals became penal slaves. Doomed men were sent to hard labor in the Carrara marble quarries, metal mines, and sulphur pits. The most common punishment was whipping—and in the case of free men, it was accompanied by the shaving of the head, for the shorn head was the mark of the slave. (Harris, 1973, p. 14)

Early versions of *gaols* (or jails) and prisons existed in English castle keeps and dungeons and Catholic monasteries. These prisons and jails (not always distinguishable in form or function) held political adversaries and common folk, either as a way to punish them or incapacitate them or to hold them over for judgment by a secular or religious authority. Sometimes people might be held as a means of extorting a fine (Johnston, 2009). The use of these early forms of jails was reportedly widespread in England, even a thousand years ago. By the 9th century, Alfred the Great had legally mandated that imprisonment might be used to punish (Irwin, 1985). King Henry II in 1166 required that where no gaol existed in English counties, one should be built (Zupan, 1991) "[i]n walled towns and royal castles," but only for the purpose of holding the accused for trial (Orland, 1975, pp. 15–16). In Elizabethan England, innkeepers made a profit by using their facility as a gaol.

Such imprisonment in these or other gaols was paid for by the prisoners or through their work. Those who were wealthy could pay for more comfortable accommodations while incarcerated. "When the Marquis de Sade was confined in the Bastille, he brought his own furnishings and paintings, his library, a live-in valet, and two dogs. His wife brought him gourmet food" (Johnston, 2009, p. 12S). The Catholic Church maintained its own jails and prisonlike facilities across the European continent, administered by bishops or other church officials.

In fact, the Catholic Church's influence on the development of westernized corrections was intense in the Middle Ages (medieval Europe from the 5th to the 15th centuries) and might be felt even today. As a means of shoring up its power base vis-à-vis feudal and medieval lords and kings, the Catholic Church maintained not only its own forms of prisons and jails, but also its own ecclesiastical courts (D. Garland, 1990). Though proscribed from drawing blood, except during the Inquisition, the Church often turned its charges over to secular authorities for physical punishment. But while in their care and in their monasteries for punishment, the Catholic Church required "solitude, reduced diet, and reflection, sometimes for extended periods of time" (Johnston, 2009, p. 14S). Centuries later, the first prisons in the United States and Europe, then heavily influenced by Quakers and Protestant religions in the states, copied the Catholics' monastic emphasis on silence, placing prisoners in small austere rooms where one's penitence might be reflected upon—practices and architecture that, to some extent, still resonate today.

Galley Slavery

Another form of "corrections," **galley slavery,** was used sparingly by the ancient Greeks and Romans, but more regularly in the late Middle Ages in Europe and England, and stayed in use until roughly the 1700s. Under Elizabeth I, in 1602, a sentence to galley servitude was decreed as an alternative to the death sentence (Orland, 1975). Pope Pius VI (who was pope from 1775–1799) also reportedly employed it (Johnston, 2009, p. 12S). Galley slavery was used as a sentence for crimes or as a means of removing the poor from the streets. It also served the twin purpose of providing the requisite labor—rowing—needed to propel ships for seafaring nations interested in engagement in trade and warfare. For instance, these galley slaves were reportedly used by Columbus (Johnston, 2009). The "slaves" were required to row the boat until they collapsed from exhaustion, hunger, or disease; often they sat in their own excrement (Welch, 2004). Under Pope Pius, galley slaves were entitled to bread each day, and their sentences ranged from 3 years to life (Johnston, 2009). Though we do not have detailed records of how such a sentence was carried out, and we can be sure that its implementation varied to some degree from vessel to vessel, the reports that do exist indicate that galley slavery was essentially a sentence to death. Galley slavery ended when the labor was no longer needed on ships because of the technological development of sails.

Poverty and Bridewells, Debtors' Prisons, and Houses of Correction

However, galley slavery could only absorb a small number of the poor that began to congregate in towns and cities in the Middle Ages. Feudalism, and the order it imposed, was disintegrating; wars (particularly the Crusades prosecuted by the Catholic Church) and intermittent plagues did claim thousands of lives, but populations were stabilizing and increasing and there were not enough jobs, housing, or food for the poor. As the cities became more urbanized and as more and more poor people congregated in them, governmental entities responded in an increasingly severe fashion to the poor's demands for resources (Irwin, 1985). These responses were manifested in the harsh repression of dissent, increased use of death sentences and other punishments as deterrence and spectacle, the increased use of jailing to guarantee the appearance of the accused at trial, the development of poorhouses or bridewells and debtors' prisons, and the use of "transportation," discussed below (Foucault, 1979; Irwin, 1985).

Eighteenth-century England saw the number of crimes subject to capital punishment increase to as many as 225, for such offenses as rioting over wages or food (the Riot Act) or for "blacking" one's face so as to be camouflaged when killing deer in the king's or a lord's forest (the Black Act) (Ignatieff, 1978, p. 16). New laws regarding forgery resulted in two-thirds of those convicted of it being executed. Rather than impose the most serious sentence for many of these crimes, however, judges would often opt for the use of transportation, whipping, or branding. Juries would also balk at imposing the death sentence for a relatively minor offense and so would sometimes value property that was stolen at less than it was worth in order to ensure a lesser sentence for the defendant. In the latter part of the 1700s, a sentence of imprisonment might be used in lieu of, or in addition to, these other punishments.

Bridewells, or buildings constructed to hold and whip "beggars, prostitutes, and nightwalkers" and later as places of detention, filled this need; their use began in London in 1553 (Kerle, 2003; Orland, 1975, p. 16). The name came from the first such institution, which was developed at Bishop Ridley's place at St. Bridget's Well; all subsequent similar facilities were known as bridewells.

Bridewells were also workhouses, used as leverage to extract fines or repayment of debt or the labor to replace them. Such facilities did not separate people by gender or age or criminal and noncriminal status, nor were their inmates fed and clothed properly, and sanitary conditions were not maintained. As a consequence of these circumstances, bridewells were dangerous and diseased places where if one could not pay a "fee" for food, clothing, or release, the inmate, and possibly his or her family, might be doomed (Orland, 1975; Pugh, 1968). The use of bridewells spread throughout Europe and the British colonies, as it provided a means of removing the poor and displaced from the streets while also making a profit (Kerle, 2003). Such a profit was made by the wardens, keepers, and gaolers, the administrators of bridewells, houses of correction (each county in England was authorized to build one in 1609), and gaols, who, though unpaid, lobbied for the job as it was so lucrative. They made money by extracting it from their inmates. If an inmate could not pay, he or she might be left to starve in filth or be tortured or murdered by the keeper for nonpayment (Orland, 1975, p. 17).

Notably, being sent to "debtors' prison" was something that still occurred even after the American Revolution. In fact, James Wilson, a signer of the Constitution (and reportedly one of its main architects) and a Supreme Court justice, was imprisoned in such a place twice while serving on the court. He had speculated on land to the west and lost a fortune in the process (K. C. Davis, 2008).

Transportation

Yet another means of "corrections" that was in use by Europeans for roughly 350 years, from the founding of the Virginia Colony in 1607, was **transportation** (Feeley, 1991). Also used to rid cities and towns of the chronically poor or the criminally inclined, transportation, as with bridewells and gaols, involved a form of privatized corrections, whereby those sentenced to transportation were sold to a ship's captain. He would in turn sell their labor as indentured servants, usually to do agricultural work, to colonials in America (Maryland, Virginia, and Georgia were partially populated through this method) and to white settlers in Australia. Transportation ended in the American colonies with the Revolutionary War, but was practiced by France to populate Devil's Island in French Guiana until 1953 (Welch, 2004). Welch notes that transportation was a very popular sanction in Europe:

> Russia made use of Siberia; Spain deported prisoners to Hispaniola; Portugal exiled convicts to North Africa, Brazil and Cape Verde; Italy herded inmates to Sicily; Denmark relied on Greenland as a penal colony; Holland shipped convicts to the Dutch East Indies. (p. 29)

In America, transportation provided needed labor to colonies desperate for it. "Following a 1718 law in England, all felons with sentences of 3 years or more were eligible for transport to America. Some were given a choice between hanging or transport" (Johnston, 2009, p. 135).

It is believed that about 50,000 convicts were deposited on American shores from English gaols. If they survived their servitude, which ranged from 1 to 5 years, they became free and might be given tools or even land to make their way in the new world (Orland, 1975, p. 18). Once the American Revolution started, such prisoners from England were transported to Australia, and when settlers there protested the number of entering offenders, the prisoners were sent to penal colonies in that country as well as in New Zealand and Gibraltar (Johnston, 2009).

One of the most well-documented such penal colonies was **Norfolk Island,** 1,000 miles off the Australian coast. Established in 1788 as a place designated for prisoners from England and Australia, it was regarded as a brutal and violent island prison where inmates were poorly fed, clothed, and housed and were mistreated by staff and their fellow inmates (Morris, 2002). Morris, in his semi-fictional account of *Alexander Maconochie*'s effort to reform Norfolk, notes that Machonochie, an ex-naval captain, asked to be transferred to Norfolk, usually an undesirable placement, so that he could put into practice some ideas he had about prison reform. He served as the warden there from 1840–1844. What was true in this story was that, "In four years, Maconochie transformed what was one of the most brutal convict settlements in history into a controlled, stable, and productive environment that achieved such success that upon release his prisoners came to be called 'Maconochie's Gentlemen'" (Morris, 2002, book jacket). Maconochie's ideas included the belief that inmates should be rewarded for good behavior through a system of marks, which could lead to privileges and early release; that they should be treated with respect; and that they should be adequately fed and housed. Such revolutionary ideas, for their time, elicited alarm from Maconochie's superiors, and he was removed from his position after only 4 years. His ideas, however, were adopted decades later when the concepts of "good time" and parole were developed in Ireland and the United States. In addition, his ideas about adequately feeding and clothing inmates were held in common by such reformers, who came before him, as John Howard and William Penn and those who came after him, such as Dorothea Dix.

❖ Enlightenment—Paradigm Shift

Spock Falls in Love

As noted in Chapter 1, the Enlightenment period, lasting roughly from the 17th through the 18th century in England, Europe, and America, spelled major changes in thought about crime and corrections. But then, it was a time of paradigmatic shifts in many aspects of the Western experience as societies became more secular and open. Becoming a more secular culture meant that there was more focus on humans on earth, rather than in the afterlife, and, as a consequence, the arts, sciences, and philosophy flourished. In such periods of human history, creativity manifests itself in innovations in all areas of experience; the orthodoxy in thought and practice is often challenged and sometimes overthrown in favor of new ideas and even radical ways of doing things (K. C. Davis, 2008). Whether in the sciences with Englishman Isaac Newton (1643–1727), philosophy and rationality with the Englishwoman Anne Viscountess Conway (1631–1679), feminist philosophy with the Englishwoman Damaris Cudworth Masham (1659–1708), philosophy and history with the Scotsman David Hume (1711–1776), literature and philosophy with the Frenchman Voltaire (1694–1778), literature and philosophy with the Briton Mary Wollstonecraft (1759–1797) or the Founding Fathers of the United States (e.g., Samuel Adams, James Madison, Benjamin Franklin, Thomas Paine, and Thomas Jefferson), new ideas and beliefs were proposed and explored in every sphere of the intellectual enterprise (Duran, 1996; Frankel, 1996; Mackenzie, 1996). Certainly, the writings of *John Locke* (1632–1704) and his conception of liberty and human rights provided the philosophical underpinnings for the Declaration of Independence as penned by Thomas Jefferson. As a result of the Enlightenment, the French Revolution beginning in 1789 was also about rejecting one form of government—the absolute monarchy—for something that was to be more democratic and liberty based. (Notably, the French path to democracy was not straight and included a dalliance with other dictators such as Napoleon Bonaparte who came to power in 1799.)

Such changes in worldviews or paradigms, as Thomas Kuhn explained in his well-known work, *The Structure of Scientific Revolutions* (1962), when discussing the nonlinear shifts in scientific theory, come usually after evidence mounts and the holes in old ways of perceiving become all too apparent. The old theory simply cannot accommodate the new evidence. Such an event was illustrated on a micro, or individual, level in an episode of the original *Star Trek*

Photo 2.2

Philosopher John Locke's writings and his conception of liberty and human rights helped to provide the philosophical underpinnings for the Declaration of Independence.

television show when Spock (the logical, unemotional, and unattached second officer) fell in love with a woman for the first time after breathing in the spores of a magical flower on a mysterious planet. Those who experienced the Enlightenment period, much like reformers and activists of the Progressive (1880s to the 1920s) and Civil Rights (1960s and 1970s) Eras in the United States that were to follow centuries later, experienced a paradigm shift regarding crime and justice. Suddenly, as if magic spores had fundamentally reshaped thought and suffused it with kind regard, if not love for others, humans seemed to realize that change in crime policy and practice was called for, and they set about devising ways to accomplish it.

John Howard

John Howard (1726–1790) was one such person who acted as a change agent. As a Sheriff of Bedford in England and as a man who had personally experienced incarceration as a prisoner of war himself (held captive by French privateers), he was *enlightened* enough to "see" that gaols in England and Europe should be different, and he spent the remainder of his life trying to reform them (J. Howard, 1775/2000; Johnston, 2009). Howard's genius was his main insight regarding corrections: that corrections should not be privatized in the sense that jailers were "paid" by inmates (an inhumane and often illogical practice, as most who were incarcerated were desperately poor, a circumstance that explained the incarceration of many in the first place). Howard believed that the state or government had a responsibility to provide sanitary and separate conditions and decent food and water for those they incarcerate.

His humanity was apparent in that he promoted this idea in England and all over the European continent during his lifetime. His major written work, *The State of the Prisons in England and Wales, With Preliminary Observations, and an Account of Some Foreign Prisons* (1775/2000), detailed the horror that was experienced in the filthy and torturous gaols of England and Europe, noting that despite the fact that there were 200 crimes for which capital punishment might be prescribed, far more inmates died from diseases contracted while incarcerated (Note to reader: The Old English used by Howard in the following quote sometimes substitutes the letter "f" for the letter "s."):

Photo 2.3

John Howard (1726–1790) believed that the state or government had a responsibility to provide sanitary conditions and decent food and water for those they incarcerate.

I traveled again into the counties where I had been; and, indeed, into all the reſt; examining Houſes of Correction, City and Town-Gaols. I beheld in many of them, as well as in the County-Gaols, a complication of diftrefs: but my attention was principally fixed by the gaol-fever, and the ſmall-pox, which I ſaw prevailing to the deftruction of multitudes, not only of felons in their dungeons, but of debtors alfo. (p. 2)

Howard (1775/2000) found that gaol fever was widespread in all kinds of correctional institutions of the time: Bridewells, gaols, debtors' prisons, and houses of correction. Notably, in larger cities there were clear distinctions among these facilities and whom they held, but in smaller towns and counties there were not. In the neglect of inmates and the underfunding of the facilities, Howard found them all to be very alike. He noted that in some bridewells there was no provision at all made for feeding inmates. Though inmates of bridewells were to be sentenced to hard labor, he found that in many there was little work to do and no tools provided to do it: "The prifoners have neither tools, nor materials of any kind; but fpend their time in floth, profanenefs and debauchery, to a degree which, in fome of thofe houfes that I have feen, is extremely fhocking" (p. 8). He found that the allotment for food in county jails was not much better, remarking that in some there was none for debtors, the criminal, or the accused alike. He noted that these inmates, should they survive their suffering, would then enter communities or other facilities in rags, and spread disease wherever they went.

In his census of correctional facilities (including debtors' prisons, jails, and houses of correction or bridewells) in England and Wales, Howard (1775/2000) found that petty offenders comprised about 16% of inmates, about 60% were debtors, and about 24% were felons (which included those awaiting trial, those convicted and awaiting their execution or transportation, and those serving a sentence of imprisonment) (p. 25; Ignatieff, 1978). Ironically, Howard eventually died from typhus, also known as gaol fever, after touring several jails and prisons in Eastern Europe, specifically the prisons of Tsarist Russia.

In Focus 2.1

Modern-Day John Howard—Dr. Ken Kerle

The Corrections Section of the Academy of Criminal Justice Sciences (ACJS) established the "John Howard" Award in 2009 and gave the first one to a modern-day John Howard, Dr. Ken Kerle (retired Managing Editor of the *American Jails* magazine). Dr. Kerle has spent much of his adult life trying to improve jail standards both here in the United States and abroad. As part of that effort, he has visited hundreds of jails in this country and around the world. He has advised countless jail managers about how they might improve their operations. He has increased the transmission of information and the level of discussion between academicians and practitioners by encouraging the publication of scholars' work in *American Jails* magazine and their presentations at the American Jails Association meetings, and by urging practitioners to attend ACJS meetings. Kerle also published a book on jails titled *Exploring Jail Operations* (2003).

Bentham and Beccaria

As mentioned in Chapter 1, the philosophers and reformers *Jeremy Bentham* (1748–1832) in England and *Cesare Beccaria* (1738–1794) in Italy separately, but both during the Enlightenment period, decried the harsh punishment meted out for relatively minor offenses in their respective countries and, as a consequence, emphasized "certainty" over the severity and celerity components of the deterrence theory they independently developed. Beccaria, in his classic work *On Crimes and Punishments* (1764/1963) wrote,

> In order that punishment should not be an act of violence perpetrated by one or many upon a private citizen, it is essential that it should be public, speedy, necessary, the minimum possible in the given circumstances, proportionate to the crime, and determined by the law. (p. 113)

He argued that knowledge, as that provided by the sciences and enlightenment, was the only effective antidote to "foul-mouthed ignorance" (p. 105).

Bentham also proposed, in his *Plan of Construction of a Panopticon Penitentiary House* (1789/1969)—though the funding of it was not signed off on by King George III—the building of a special type of prison. As per Bentham, the building of a private "prison"-like structure—the **panopticon**, which he would operate—that ingeniously melded the ideas of improved supervision with architecture (because of its rounded, open, and unobstructed views) would greatly enhance supervision of inmates. Such a recognition of the benefits of some architectural styles as complementary to enhanced supervision was indeed prescient, as it presaged modern jail and prison architecture. His proposed panopticon would be circular, with two tiers of cells on the outside and a guard tower in its center, with the central area also topped by a large skylight. The skylight and the correct angling of the tower were to ensure that the guard was able to observe all inmate behavior in the cells, though owing to a difference of level and the use of blinds, the keeper would be invisible to the inmates. A chapel would also be located in the center of the rounded structure. The cells were to be airy and large enough to accommodate the whole life of the inmates in that the cells were to "[s]erve all purposes: work, sleep, meals, punishment, devotion" (Bentham, 1811/2003, p. 194). Somehow, Bentham notes in his plan without elaboration, the sexes were to be invisible to each other. He does not call for complete separation of all inmates, however, which becomes important when discussing the Pennsylvania and New York prisons in the following, but he does assert that the groups of inmates allowed to interact should be small, including only two to four persons (Bentham, 1811/2003, p. 195).

As an avowed admirer of John Howard, Bentham proposed that his Panopticon Penitentiary would include all of the reforms proposed by Howard and much more. Bentham (1811/2003) promised that inmates would be well fed, fully clothed, supplied with beds, supplied with warmth and light, kept from "strong or spirituous liquors," have their spiritual and medical needs fulfilled, be provided with opportunities for labor and education ("to convert the prison into a school") and to incentivize the labor so that they got to "share in the produce," be taught a trade so that they could survive once released, and be helped to save for old age (pp. 199–200). He would also personally pay a fine for every escape, insure inmates' lives to prevent their deaths, and submit regular reports to the "Court of the King's Bench" on the status of the prison's operation (pp. 199–200). Moreover, he proposed that the prison would be open in many respects not just to dignitaries, but to regular citizens, and daily, as a means of preventing abuse that might occur in secret. Bentham also recommended the construction of his prisons on a large scale across England, such that one would be built every 30 miles, or a good day's walk by a man. He planned, as he wrote in his 1830 diatribe against King George the Third, wryly titled "History of the War Between Jeremy Bentham and George the Third—By One of the Belligerents," that, "But for George the Third, all the prisoners in England would, years ago, have been under my management. But for George the Third, all the paupers in the country would, long ago, have been under my management" (Bentham, 1811/2003, p. 195).

Though his plan in theory was laudable and really visionary for his time, and ours, he hoped to make much coin as recompense for being a private prison manager—to the tune of 60 pounds sterling per prisoner, which when assigned to all inmates across England, was a considerable sum (Bentham, 1811/2003, p. 195). What stopped him, and the reason why he was so angry with his sovereign, was King George's unwillingness to sign the bill that would have authorized the funding and construction of the first panopticon. Bentham alleged that the king would not sign because the powerful Lord Spenser was concerned about the effect on the value of his property should a prison be located on or near it. Bentham's prison dream was dead, but eventually he was awarded 23,000 pounds for his efforts (p. 207). It was left to others to build panopticon prisons in both Europe and the states in the coming years.

William Penn

William Penn (1644–1718), a prominent Pennsylvania Colony governor and Quaker, was similarly influenced by Enlightenment thinking (though with the Quaker influence, his views were not so secular). Much like Bentham and Beccaria, Penn was not a fan of the harsh punishments, even executions, for relatively minor offenses, that were meted out during his lifetime. While in England, and as a result of his defense of religious freedom and practice, he was incarcerated in the local jails on more than one occasion, and even in the Tower of London in 1669, for his promotion of the Quaker religion and defiance of the English crown. He was freed only because of his wealth and connections (Penn, 1679/1981). As a consequence, when he had the power to change the law and its protections, and reduce its severity, he did so. Many years later (in 1682) in Pennsylvania, he proposed and instituted his **Great Law,** which was based on Quaker principles and de-emphasized the use of corporal and capital punishment for all crimes but the most serious (Clear, Cole, & Reisig, 2011; Johnston, 2009; Zupan, 1991). His reforms substituted fines and jail time for corporal punishment. He promoted Pennsylvania as a haven for Quakers who were persecuted in England and Europe generally, and for a number of other religious minorities (Penn, 1679/1981). His ideas about juries, civil liberties, religious freedom, and the necessity of amending constitutions so that they are adaptable to changing times, influenced a number of American revolutionaries, including Benjamin Franklin and Thomas Paine.

Many of Penn's contemporaries were not of the same frame of mind, however, and after his death, the Great Law was repealed and harsher punishments were again instituted in Pennsylvania, much as they existed in the rest of the colonies (Johnston, 2009; Welch, 2004). But the mark of his influence lived on in the development of some of America's first prisons.

Much like Howard and Bentham, Penn was interested in reforming corrections, but he was particularly influenced by his Quaker sentiments regarding nonviolence and the value of quiet contemplation. The early American prisons known as the **Pennsylvania model prisons**—the **Walnut Street Jail** (1790) in Philadelphia, the **Western Pennsylvania Prison** (1826) in Pittsburgh, and the **Eastern Pennsylvania Prison** (1829) in

Photo 2.4

William Penn proposed and instituted his Great Law, which was based on Quaker principles and deemphasized the use of corporal and capital punishment for all crimes but the most serious.

Philadelphia—incorporated these ideas (Johnston, 2009). Even the **New York model prisons, Auburn** and **Sing Sing Prisons,** often juxtaposed with Pennsylvania prisons based on popular depiction by historians (see Beaumont and Tocqueville, 1833/1964), included contemplation time for inmates and a plan for single cells for inmates that reflected the same belief in the need for some solitude.

❖ Colonial Jails and Prisons

The first jail in America was built in Jamestown, Virginia, soon after the colony's founding in 1606 (Burns, 1975; Zupan, 1991). Massachusetts built a jail in Boston in 1635, and Maryland built a jail for the colony in 1662 (Roberts, 1997). The oldest standing jail in the United States was built in the late 1600s and is located in Barnstable, Massachusetts (Library of Congress, 2010). It was used by the sheriff to hold both males and females, along with his family, in upstairs, basement, and barn rooms. Both men and women were held in this and other jails like it, mostly before they were tried for both serious and minor offenses, as punishment for offenses, or to ensure they were present for their own execution.

Such an arrangement as this—holding people in homes, inns, or other structures, that were not originally designated or constructed as "jails"—was not uncommon in early colonial towns (Goldfarb, 1975; Irwin, 1985; Kerle, 2003). As in England, inmates of these early and colonial jails were required to pay a "fee" for their upkeep (the same fee system that John Howard opposed). Those who were wealthier could more easily buy their way out of incarceration, or if that was not possible because of the nature of the offense, they could at least ensure that they had more luxurious accommodations (Zupan, 1991). Even when jailers were paid a certain amount to feed and clothe inmates, they might be disinclined to do so, being that what they saved by not taking care of their charges they were able to keep (Zupan, 1991). As a result, inmates of early American jails were sometimes malnourished or starving. Moreover, in the larger facilities they were crammed into unsanitary rooms, often without regard to separation by age, gender, or offense, conditions that also led to early death and disease. Though, Irwin (1985) does remark that generally Americans fared better in colonial jails than their English and European cousins did in their own, as the arrangements were less formal and restrictive in the American jails and were more like rooming houses. Relatedly, Goldfarb (1975) remarks,

> Jails that did exist in the eighteenth century were run on a household model with the jailer and his family residing on the premises. The inmates were free to dress as they liked, to walk around freely and to provide their own food and other necessities. (p. 9)

As white people migrated across the continent of North America, the early western jails were much like their earlier eastern and colonial cousins, with makeshift structures and cobbled together supervision serving as a means of holding the accused over for trial (Moynihan, 2002). In post–Civil War midwestern cities, disconnected outlaw gangs (such as the Jesse James Gang) were responded to in a harsh manner. Some communities even built *rotary jails,* which were like human squirrel cages. Inside a secure building, these rotating steel cages, segmented into small "pie-shaped cells" were secured to the floor and could be spun at will by the sheriff (Goldfarb, 1975, p. 11).

Of course, without prisons in existence per se (we will discuss the versions of such institutions that did exist shortly), most punishments for crimes constituted relatively short terms in jails, or public shaming (as in the stocks), or physical punishments such as flogging or the pillory, or banishment. Executions were also carried out, usually but not always for the most horrific of crimes such as murder or rape, though in colonial America, many more crimes qualified for this punishment (Zupan, 1991). As in Europe and England at this time, those who were poorer or enslaved were more likely to experience the harshest of punishments (Irwin, 1985; Zupan, 1991). Similar to Europe and England in this era, jails also held the mentally ill, along with debtors, drifters, transients, the inebriated, runaway slaves or servants, and the criminally involved (usually pretrial) (Cornelius, 2007).

Though the Walnut Street Jail, a portion of which was converted to a prison, is often cited as the "first" prison in the world, there were, as this recounting of history demonstrates, many precursors that were arguably "prisons" as well. One such facility, which also illustrates the "makeshift" nature of early prisons, was the **Newgate Prison in Simsbury, Connecticut** (named after the Newgate Prison in London). According to Phelps (1860/1996), this early colonial "prison" started as a copper mine, and during its 54 years of operation (from 1773 to 1827), some 800 inmates passed through its doors. The mine was originally worked in 1705, and one-third of the taxes it paid to the town of Simsbury at that time were used to support Yale College (p. 15). "Burglary, robbery, and counterfeiting were punished for the first offense with imprisonment not exceeding ten years; second offence for life" (p. 26). Later, those loyal to the English crown during the U.S. Revolutionary War, or Tories, were held at Newgate as well. Punishments by the "keeper of the prison" could range from shackles and fetters as restraints to "moderate whipping, not to exceed ten stripes" (p. 26). The inmates of Newgate Prison were held—stored, really—in the bowels of the mine during the evening (by themselves and with no supervision), and during the day were forced to work the mine or were allowed to come to the surface to labor around the facility and in the community. Over the course of the history of this facility, there were several escapes, a number of riots, and the burning of the topside buildings by its inmates. Early versions of prisons also existed in other countries.

Old Newgate Prison, near Hartford. Conn.

2563

Photo 2.5

Newgate Prison, a working copper mine, served as an early colonial prison.

Comparative Perspective: Early European and British Prisons

Some early European versions of prisons bucked the trend of harsh physical punishments even for minor offenses. Others, but only a few, even classified their inmates not just by economic and social status, but by gender, age, and criminal offense. For instance, in the Le Stinche Prison built in Florence, Italy, in the 1290s, the inmates were separated in this way (Roberts, 1997). Later, the Maison de Force Prison in Ghent, Belgium (1773), placed serious offenders in a different section of the prison from the less serious. A juvenile reformatory was even built in a separate wing of the Hospice of San Michele in Rome (1704) (Roberts, 1997). An architectural depiction of the Ghent prison shows an octagonal shape with a central court and then a partial view of separate living areas or courts for exercise, for women, vagrant men, and other men. Much like the American colonies and England, however, the early European prisons and jails classified inmates by their societal status and their ability to pay, with the concomitant amenities going to the wealthier.

> Incarcerated nobles who could pay the heftiest fees lived in comparative comfort with a modicum of privacy; less affluent prisoners were confined in large common rooms; the poorest inmates, and those who were considered the most dangerous, had to endure squalid dungeons. It was not unusual for men, women, and children, the sane and the mentally ill, felons and misdemeanants, all to be crowded indiscriminately in group cells. (Roberts, 1997, p. 5)

Another less enlightened type of prison existed in England in the form of the **"hulks,"** derelict naval vessels transformed into "prisons" for the overflowing inmates in England. Used in tandem with transportation and other forms of incarceration in the mid-1700s, and then increasing in use in the gap between the end of transportation to the American colonies with the Revolutionary War and the beginning of transportation of "criminals" to Australia, the last hulk was used on the coast of Gibraltar in 1875 (Roberts, 1997, p. 9). The English even confined some prisoners of war in a Hudson River hulk during the American Revolution. Inmates of these hulks were taken off to labor during the day for either public works or private contractors. The conditions of confinement were, predictably, horrible. "The hulks were filthy, crowded, unventilated, disease-ridden, and infested with vermin. The food was inadequate and the discipline was harsh" (Roberts, 1997, p. 11). Some inmates housed on the lower decks even drowned from water taken on by these broken-down ships.

Photo 2.6

Drawing of inmates in the hulk prison washroom

A major proponent of reform of English prisons, and also a Quaker, was *Elizabeth Gurney Fry* (1780–1845). She was an advocate for improved conditions, guidelines, training, and work skills for women inmates (Roberts, 1997). She provided the religious instruction herself to the women inmates.

Summary

- Human beings have been inventive in their development of punishments and ways in which to hold and keep people.
- Correctional history is riddled with efforts to improve means of coercion and reform.
- Those accused or convicted of crimes who had more means were less likely to be treated or punished severely.
- Sometimes the old worldviews (paradigms) are challenged by new evidence and ideas, and they are then discarded for new paradigms. The Enlightenment period in Europe was a time for rethinking old ideas and beliefs.
- Bentham, Beccaria, John Howard, and William Penn were all especially influential in changing our ideas about crime, punishment, and corrections.
- Correctional reforms, whether meant to increase the use of humane treatment of inmates or to increase their secure control, often lead to unintended consequences.
- Some early European and English versions of prisons and juvenile facilities were very close in mission and operation to America's earliest prisons.

Key Terms

Auburn Prison

Bridewells

Eastern Pennsylvania Prison

Galley slavery

Great Law

Hulks

New York model prisons

Newgate Prison in Simsbury, Connecticut

Norfolk Island

Panopticon

Pennsylvania model prisons

Sing Sing Prison

Transportation

Walnut Street Jail

Western Pennsylvania Prison

Discussion Questions

1. Identify examples of some themes that run throughout the history of corrections. What types of punishments tend to be used and for what types of crimes? What sorts of issues influence the choice of actions taken against offenders?

2. How were people of different social classes treated in early jails and bridewells?

3. We know that transportation ended because of the development of sails, which was an improvement in technology. Can you think of other types of correctional practices that have been developed, improved upon, or stopped because of advances in technology?

4. What role has religion played in the development of corrections in the past?

5. What types of things have remained the same in corrections over the years, and what types of things have changed? Why do you think things have changed or remained the same?

6. Several historical figures mentioned in this chapter advanced ideas that were viewed as radical for their day. Why do you think such ideas were eventually adopted? Can you think of similar sorts of seemingly "radical" ideas for reforming corrections that might be adopted in the future?

Useful Internet Sites

American Correctional Association: www.aca.org

American Jail Association: www.corrections.com/aja/

American Probation and Parole Association: www .appa-net.org.

Bureau of Justice Statistics (information available on all manner of criminal justice topics): http://bjs.ojp .usdoj.gov/

John Howard Society of Canada: www.johnhoward.ca

National Criminal Justice Reference Service: www .ncjrs.gov

Office of Justice Research (information available on all manner of criminal justice topics, specifically probation and parole here): www.ojp.usdoj.gov/bjs/ pub/pdf/ppus05.pdf

Pennsylvania Prison Society: www.prisonsociety .org

Vera Institute (information available on a number of corrections and other justice-related topics): www .vera.org

CHAPTER 3

Correctional History

Reforms and Themes

❖ Introduction: The Grand Reforms

In this chapter, we review the attributes of the seminal prison models of the early 1800s known as the **Pennsylvania prison system** (including the Walnut Street Jail and the Western and Eastern Pennsylvania prisons) and the **New York prison system** (including the Auburn and Sing Sing prisons). We include the eyewitness accounts of the operation of such systems in their early years, as these are provided by Beaumont, Tocqueville, and Dix.

Out of these two systems, the rampart for all American, and many European, prisons was constructed. As it became clear that neither prison model accomplished its multifaceted goals, and that its operation was so distorted and horrific for inmates, changes were gradually made as new reform efforts ensued. The Elmira Prison in New York was perhaps the most ambitious of these efforts, in the latter part of the 1800s, which in turn set the stage for the later development of correctional institutions. Though the implementation of the reform ideals at Elmira is much critiqued, it certainly was much more humane than the convict leasing system that operated at that time in the South. (More about these topics will be presented later in the chapter.)

What does become crystal clear from this two-chapter review of the history of corrections in the United States is that there are several themes that run through it. One such theme, of course, is the cyclical need for reform itself, but to what purpose it is not always clear.

❖ Early Modern Prisons and the Pennsylvania and New York Models

The Walnut Street Jail

The **Walnut Street Jail** was originally constructed in 1773 in Philadelphia, Pennsylvania, and

Photo 3.1

Drawing of the Walnut Street Jail (circa 1799)

operated as a typical local jail of the time: holding pretrial detainees and minor offenders; failing to separate by gender, age, or offense; using the fee system, which penalized the poor and led to the near starvation of some; and offering better accommodations, and even access to liquor and sex, to those who could pay for it (Zupan, 1991). It was remodeled, however, in 1790 and reconceptualized so that many correctional scholars, though not all, regard it as the "first" prison.

The remodeled cellhouse was of a frame construction and was built for the inmates of the "prison" section of the jail, with separate cells for each inmate. based on the reforms that John Howard (and later Bentham and Fry) had envisioned for English and European jails instituted in this prison: The fee system was dropped, inmates were adequately clothed and fed regardless of their ability to pay, and they were separated by gender and offense. Children were not incarcerated in the prison, and debtors were separated from convicted felons. Though inmates were to live in isolated cells (to avoid "contaminating" each other), some work requirements brought them together. In addition, medical care was provided and attendance at religious services was required. The availability of alcohol and access to members of the opposite sex and prostitutes was stopped.

The impetus for this philosophical change came from the reform efforts of the Philadelphia Society for Alleviating the Miseries of Public Prisons (or the Philadelphia Prison Society, currently known as the Pennsylvania Prison Society), led by *Dr. Benjamin Rush* who was a physician, reformer, statesman, and a signatory of the Declaration of Independence. Rush agitated for laws to improve the jail's conditions of confinement and a different belief about correctional institutions—namely, that they could be used to reform their inmates (Nagel, 1973; Roberts, 1997). Ideally, the Walnut Street Jail (prison) was to operate based on the religious beliefs of the Quakers and their emphasis on the reflective study of the Bible and an abhorrence for violence, which was so prevalent in other correctional entities. In 1789, the General Assembly of Pennsylvania enacted legislation based on these recommendations and the Pennsylvania system was born (Nagel, 1973).

The Walnut Street Jail, as a prison, was also an entity with a philosophy of penitence, which, it was hoped, would lead to reform and redemption. This philosophy was combined with an architectural arrangement shaped to facilitate it by ensuring that inmates were mostly in solitary cells. As Roberts (1997) aptly notes, the reason the Walnut Street Jail's new wing was the first real prison, as opposed to the other prisons such as Newgate of Connecticut that preceded it, or some of the early European prisons, was "[b]ecause it carried out incarceration as punishment, implemented a rudimentary classification system, featured individual cells, and was intended to provide a place for offenders to do penance—hence the term 'penitentiary'" (p. 26).

But in reality, the Walnut Street Jail soon became crowded, reportedly housing 4 times its capacity. As Johnston (2010) notes, "At one point 30 to 40 inmates were sleeping on blankets on the floor of rooms [which were] 18 feet square" (p. 13). Moreover, the institutional industry buildings that provided work for inmates burned down, leading to idleness, and the Walnut Street Jail (prison) by 1816 was little different from what it had been before the reforms (Harris, 1973; Zupan, 1991).

As Beaumont and Tocqueville (1833/1964) commented in 1831, after visiting and analyzing several prisons and jails in the United States, the implementation of the Walnut Street Jail had "[t]wo principal faults: it corrupted by contamination those who worked together. It corrupted by indolence, the individuals who were plunged into solitude" (p. 38).

Newgate Prison—New York City

Yet another **Newgate Prison, in New York City** (1797), was . . . was modeled after the Walnut Street Jail prison (as were early prisons in Trenton, New Jersey [1798]; Richmond, Virginia [1800]; Charlestown, Massachusetts [1805]; and Baltimore, Maryland [1811]), and even improved upon that model in some respects (Roberts, 1997). Thomas Eddy, the warden of Newgate, was a Quaker. The focus at the Newgate Prison in New York was on rehabilitation, religious redemption, work programs to support prison upkeep, and no corporal punishment. The builders of Newgate even constructed a prison hospital and school for the inmates. Unfortunately, because of crowding, single celling for any but the most violent inmates was not possible, and a number of outbreaks of violence erupted (such as a riot in 1802).

The Pennsylvania Prison Model (Separate System)

The **Western Pennsylvania Prison** (1821) was built in Pittsburgh, followed by the **Eastern Pennsylvania Prison** (1829) in Philadelphia, which was to replace the Walnut Street Jail prison (Nagel, 1973). The Western Pennsylvania Prison, built 8 years before Eastern, is little remarked upon or studied in comparison to Eastern. It was devised to operate in a solitary and separate fashion. Even labor was to be prohibited, as it was thought that this might interfere with the ability of the criminal to reflect and feel remorse for his or her crime (Hirsch, 1992). Despite the lessons learned from the Auburn prison, to be discussed shortly, namely, that complete separation without labor can be injurious to the person and expensive for the state to maintain, a point made by Tocqueville and Beaumont, the Western

Pennsylvania Prison was built to hold inmates in complete solitary confinement (hence the use of the term *separate system*), with no labor, for the full span of their sentence. However, as Beaumont and Tocqueville (1833/1964, p. 44) remark about the Western Pennsylvania prison, reducing all communication, and thus contamination in the authors' view, was almost impossible at this prison.

> Each one was shut up, day and night in a cell, in which no labor was allowed to him. This solitude, which in principle was to be absolute, was not such in fact. The construction of this penitentiary is so defective, that it is very easy to hear in one cell what is going on in another; so that each prisoner found in the communication with his neighbor a daily recreation, i.e. an opportunity of inevitable corruption. As these criminals did not work, we may say that their sole occupation consisted in mutual corruption. This prison, therefore, was worse than even that of Walnut Street, because, owing to the communication with each other, the prisoners at Pittsburgh (Western Pennsylvania Prison) were as little occupied with their reformation, as those at Walnut Street. And while the latter indemnified society in a degree by the produce of their labor, the others spent their whole time in idleness, injurious to themselves and burdensome to the public treasury.

As a consequence of these problems of architecture and operation, the Western Pennsylvania Prison as a model was abandoned and the Eastern Pennsylvania Prison came to epitomize the "Pennsylvania System," as opposed to the "New York System" of building and operating prisons. At the Eastern Pennsylvania Prison, known as "Cherry Hill" for much of its 150 years of operation, the idea that inmates could be contaminated or corrupted by their fellow inmates was officially embraced.

The Eastern Pennsylvania Prison was designed and built by the architect John Haviland, a relative newcomer from England. It cost three-quarters of a million dollars to build, which was an incredible expenditure for the time. It was the largest building in America in the 1820s (Alosi, 2008; Orland, 1975). The prison itself was huge, with seven massive stone spokes of cells radiating off of a central rotunda, as on a wheel. A 30-foot wall was constructed around the outside perimeter of the prison, thus physically and symbolically reinforcing the separation of the prison and its inhabitants from their community (Nagel, 1973). The cells were built large (15 by 7.5 feet with 12-foot ceilings), and those on the lowest tier had their own small outside exercise yard attached, so that inmates could do virtually everything in their

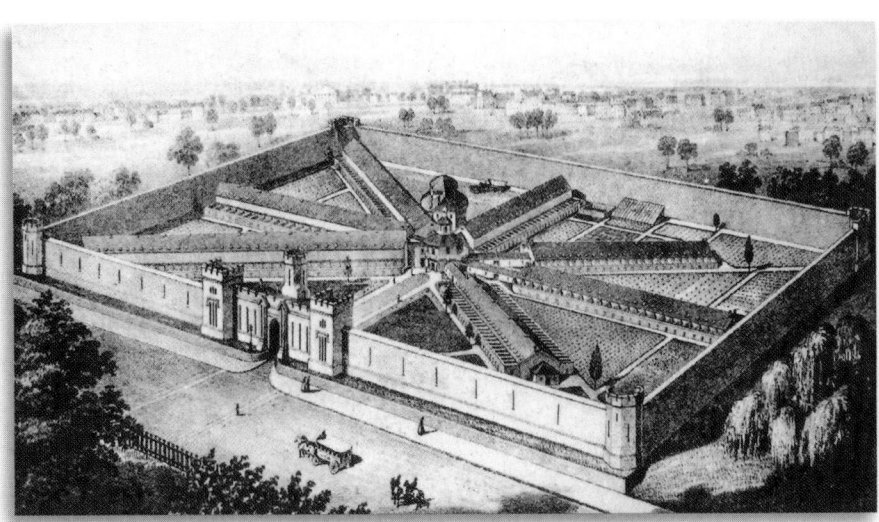

Photo 3.2

The Eastern Pennsylvania Prison was the largest building in America in the 1820s. (Lithograph, circa 1855.)

cells (Harris, 1973; Orland, 1975). The cells had both hot water and flush toilets, reportedly the first public building in the country to have such amenities. There were 400 solitary cells in this prison (Orland, 1975). At first, inmates were not to work, but that dictate was later changed and they were allowed to work in their cells (Harris, 1973). As Johnston (2010) explains,

> The solution to the problem of criminal contamination for the reformers was to be a regimen of near-total isolation and absolute separation of prisoners from one another, the use of numbers rather than names, and a program of work, vocational training, and religious instruction, all taking place within the inmate's individual cell. (p. 13)

The only contact inmates were to have with the outside was the clergy and some vocational teachers. "The reading of the Scriptures would furnish the offender with the moral guidance necessary for reform" (Nagel, 1973, p. 7). They had no access to visitors or letters or newspapers. Even their exercise yards were surrounded by a high stone fence. When they were brought into the prison, were taken for showers or to see the doctor, they had to wear a mask or a draped hood so as to maintain their anonymity and to prevent them from figuring out a way of escape (Alosi, 2008). As to how else they could occupy their time, "They made shoes, wove and dyed cloth products, caned chairs, and rolled cigars. Those products were sold to defray prison costs" (Roberts, 1997, p. 33).

The stated purpose of the solitary confinement was to achieve reform or rehabilitation. Quakers believed that God resides in everyone, and for a person to reach God, he or she must self-reflect. Silence is required for this self-reflection, the Quakers thought. The Quakers also believed that as God was in everyone, all were equal and were deserving of respect (Alosi, 2008).

Solitary confinement as a practical matter remained in existence at the Eastern Penitentiary until after the Civil War, but was not formally ended until 1913 (Alosi, 2008). When it was rigorously applied, there are indications that it drove inmates insane. In fact, and tellingly, most of the European countries who copied the Eastern Pennsylvania model and its architecture, did not isolate the inmates for this reason. Moreover, at a minimum it debilitated people by making them incapable of dealing with other people. For instance, the wardens' journals for Eastern in the early years indicate that it was not uncommon for an inmate to be released and then to ask to be reinstated at Eastern because he or she did not know how to live freely. Some inmates, once released, would actually sit on the curb outside the prison, as they said they no longer understood the outside world or how to function in it (Alosi, 2008).

Comparative Perspective: The Pentonville Prison

Though most American prisons of the 19th century ended up copying the Auburn and Sing Sing (New York model) prisons in design and operation, the Pentonville Prison, opened in London, England, in 1842, was modeled on the Pennsylvania prisons (Ignatieff, 1978). Inmates were confined in their rather large cells for most of the day, except for chapel (where they had their own solitary box) and recreation (where they had their own isolated yard). In their cells, inmates worked on crafts and at looms. As with the Pennsylvania prisons, at Pentonville silence was enforced and inmates were sent to solitary dungeon rooms for disobeying the rules. But also as with the Pennsylvania prisons, inmates devised ways to communicate between cells by tapping messages to each other at night. At first, as with the Pennsylvania prisons and initially the Auburn Prison, inmates were confined completely to their cells for 18 months. However, as inmates experienced the predictable mental illnesses associated with this isolation, such periods were eventually reduced to 9 months for new inmates (Ignatieff, 1978).

Though the separation of inmates under the Pennsylvania system was to be complete, there are indications that it was not. In testimony before a special investigation by a joint committee of the houses of the Pennsylvania Legislature in 1834 (before the whole prison was even completed), it was noted that a number of male and female inmates (there were a small number of female inmates housed separately at Eastern) were used for maintenance cleaning, and cooking at the facility and roamed freely around it, speaking and interacting with each other and staff (Johnston, 2010). Moreover, there were indications from this testimony that inmates were tortured to maintain discipline: One had died of blood loss from the iron gag put in his mouth, and another went insane after buckets of cold water were poured on his head repeatedly. It was alleged that food and supplies meant for inmates were given to guards or community members by the prison cook (who was a wife of one of the guards). There were also indications of the use and abuse of alcohol by staff and inmates and of sexual improprieties involving the warden and his clerk, some male inmates, and the female cook. Though ultimately, charges against the warden and his clerk related to these improprieties were dropped; only the cook was blamed, and the guards who testified about the scandal (the whistle-blowers) were fired.

In addition to these problems of implementation at Eastern, a debate raged among prison experts regarding the value of separation. As a result of the experiment with the Western Pennsylvania Prison, the early use of the Eastern prison, and the Auburn prison (which we will describe further on), the idea of "total separation" was under siege. As mentioned in the foregoing, it was observed that for those truly subjected to it, solitary confinement and separation caused serious psychological problems for some inmates. Despite these problems, about 300 prisons worldwide copied the Eastern Pennsylvania model, and tens of thousands of men and women did time there, including the 1920s gangster Al Capone. It was a famous prison worldwide, because of its philosophy, its architecture, and its huge size. It even became a tourist attraction in the 19th century to the extent that famous English author Charles Dickens noted it as one of the two sights he wanted to see when visiting the United States (the other was Niagara Falls) (Alosi, 2008). It turns out, after a visit of a few hours and talking to inmates, keepers, and the warden, Dickens was far from impressed with its operation (see In Focus 3.1).

Auburn, Sing Sing, and the New York (Congregate) System

The **New York model** for imprisonment was preferred over the Pennsylvania system, and copied extensively by American prison builders, in part because it disavowed the solitary confinement that Dickens and others lamented in the Pennsylvania prisons. Beaumont and Tocqueville (1833/1964) commented that the use of solitary confinement as normal practice for all inmates was ended at Auburn because it drove inmates insane. But it is not that the builders and planners for the Auburn Prison in New York learned from the Pennsylvania system; rather, that they learned from their own dalliance with solitary confinement. At first, the inmates of Auburn were housed in solitary confinement in their cells, a practice that was abandoned by 1822 because it led to mental anguish and insanity for inmates and it hampered the efficient production of goods that can only be done in the congregate. By 1822, a total of 5 prisoners had died, 1 had gone insane, and the remaining 26 were pardoned by the governor of New York as their mental faculties had deteriorated to such a great extent (Harris, 1973, p. 73). The governor ordered that inmates be allowed to leave their cells and work during the day, and a legislative committee in 1824 recommended the repeal of the solitary confinement laws (Harris, 1973).

In Focus 3.1

Charles Dickens's Impressions of the Eastern Pennsylvania Prison and the Silent System in 1842

Photo 3.3

Charles Dickens

In its intention, I am well convinced that it is kind, humane, and meant for reformation; but I am persuaded that those who devised this system of Prison Discipline, and those benevolent gentlemen who carry it into execution, do not know what it is that they are doing. I believe that very few men are capable of estimating the immense amount of torture and agony which this dreadful punishment, prolonged for years, inflicts upon the sufferers; and in guessing at it myself, and in reasoning from what I have seen written upon their faces, and what to my certain knowledge they feel within, I am only the more convinced that there is a depth of terrible endurance in it which none but the sufferers themselves can fathom, and which no man has a right to inflict upon his fellow-creature. I hold this slow and daily tampering with the mysteries of the brain, to be immeasurably worse than any torture of the body: and because its ghastly signs and tokens are not so palpable to the eye and sense of touch as scars upon the flesh; because its wounds are not upon the surface, and it extorts few cries that human ears can hear; therefore I the more denounce it, as a secret punishment which slumbering humanity is not roused up to stay.

I was accompanied to this prison by two gentlemen officially connected with its management, and passed the day in going from cell to cell, and talking with the inmates. Every facility was afforded me, that the utmost courtesy could suggest. Nothing was concealed or hidden from my view, and every piece of information that I sought, was openly and frankly given. The perfect order of the building cannot be praised too highly, and of the excellent motives of all who are immediately concerned in the administration of the system, there can be no kind of question. . . .

Standing at the central point, and looking down these dreary passages, the dull repose and quiet that prevails, is awful. Occasionally, there is a drowsy sound from some lone weaver's shuttle, or shoemaker's last, but it is stifled by the thick walls and heavy dungeon-door, and only serves to make the general stillness more profound. Over the head and face of every prisoner who comes into this melancholy house, a black hood is drawn; and in this dark shroud, an emblem of the curtain dropped between him and the living world, he is led to the cell from which he never again comes forth, until his whole term of imprisonment has expired. He never hears of wife and children; home or friends; the life or death of any single creature. He sees the prison-officers, but with that exception he never looks upon a human countenance, or hears a human voice. He is a man buried alive; to be dug out in the slow round of years; and in the mean time dead to everything but torturing anxieties and horrible despair. . . .

My firm conviction is that, independent of the mental anguish it occasions—an anguish so acute and so tremendous, that all imagination of it must fall far short of the reality—it wears the mind into a morbid state, which renders it unfit for the rough contact and busy action of the world. It is my fixed opinion that those who have undergone this punishment, MUST pass into society again morally unhealthy and diseased. (Dickens, 1842, n.p.)

Photo 3.4

Auburn Prison, built in 1816, is still in operation today though its name has changed to the Auburn Correctional Institution.

Beaumont and Tocqueville (1833/1964) supported the practice of maintaining the solitude of inmates at night and their silence during the day as they worked, as they believed, along with the Quakers of Pennsylvania, that solitude and silence led to reflection and reformation and also reduced cross-contamination of inmates. As to labor, they claimed "[i]t fatigues the body and relieves the soul," along with supplementing the income of the state to support the prison (p. 57).

The Auburn (New York) Prison cornerstone was laid in 1816, it received its first inmates in 1817, but it was not finished until 1819 (Harris, 1973). *Elam Lynds* (1784–1855), a strict disciplinarian and former army captain, was its first warden in 1821. Auburn has been in existence ever since (195 years at the time of this writing, in 2011), though its name changed to the Auburn Correctional Institution in 1970.

Auburn's cells were built back-to-back with corridors on each side. The prison has always had a gothic appearance, and its elaborate front and massive walls have been maintained up until today, with towers and a fortress facade.

The Auburn Prison has a storied history that spans the virtual beginning of prisons in the United States to the present day. As was already noted, Beaumont and Tocqueville visited it and recommended it over the Pennsylvania prisons. Auburn opened with a solitary confinement system, which was very quickly abandoned, and replaced it with the congregate, but silent system that formally lasted until the beginning of the 20th century. It

Photo 3.5

Sing Sing Prison, modeled after the Auburn Prison, was built by inmates from Auburn Prison in 1825.

was the progenitor of such widely adopted practices as the "lockstep" walk for inmates, the striped prison uniforms and the classification system that went hand in hand with it, and the well-known ball and chain. Warden Lynds believed in strict obedience on the part of inmates and the use of the whip by staff to ensure it (Clear et al., 2011). Under his regime, inmates were forbidden to talk or even to glance at each other during work or meals. Solitary confinement and flogging were used for punishing and controlling inmates. As noted in the foregoing, except for a few years at the beginning of Auburn's history, inmates were single-celled at night and the cells were quite small, even coffin-like (7 × 7 × 3.5), and during the day the inmates worked together, but silently, in factories and shops (Roberts, 1997).

The small cells like those at Auburn were cheaper to build, and prisons could house more inmates in the same amount of space than prisons with larger cells. Also, congregate work allowed for the more efficient production of more products, and thus more profit could be made (Roberts, 1997). However, putting all of these inmates together in one place presented some difficulties in terms of control and management. This is why the control techniques represented by the use of the lash, solitary confinement, marching in lockstep, and the requirement of silence came into play. As Roberts notes, "Ironically, whereas the penitentiary concept was developed as a humane alternative to corporal punishment, corporal punishment returned as a device to manage inmates in penitentiaries based on the Auburn System" (p. 44).

The Sing Sing Prison was modeled after Auburn, in some ways architecturally in that the cells were small and there were congregate areas for group work by inmates, but its cell blocks were tiered and very long. Inmate management and operations exactly mirrored the Auburn protocols. In fact, Sing Sing was built by Auburn inmates under the supervision of Auburn's Warden Lynds.

The prison was built on the Hudson River, near the town of Sing Sing and Mount Pleasant (and for many years the prison was referred to as Mount Pleasant), from locally quarried stone. Products produced at the prison could be transported to local towns via the river. Inmates sent there would refer to it as being sent "up the river," as it is 30 miles north of New York City (Conover, 2001). Its name derives from the Indian phrase *Sint Sinks*, which came from the older term *Ossine Ossine* and ironically means "stone upon stone" (Lawes, 1932, p. 68).

Warden Lynds picked 100 men from the Auburn Prison to build Sing Sing in 1825. The story of its construction in silence, as relayed by Lewis Lawes (1932), a later warden of Sing Sing, goes like this:

> Captain Lynds, then the foremost penologist of the day, was insistent, to the point of hysteria, on *silence* as the backbone of prison administration. "It is the duty of convicts to preserve an unbroken silence," was the first rule he laid down. "They are not to exchange a word with each other under any pretense whatever; not to communicate any intelligence to each other in writing. They are not to exchange looks, wink, laugh, or motion to each other. They must not sing, whistle, dance, run, jump, or do anything which has a tendency in the least degree to disturb the harmony or contravene to disturb the rules and regulations of the prison." . . . The sea gulls in the broad river, darting in large flocks here and there on the water, chirped raucously at these strange creatures sweating at their tasks in silence. Stone upon stone. (pp. 72–73)

Once constructed, it was noticed that with some effort inmates could communicate between the closely aligned cells, but nothing was done to rebuild the cells. Moreover, as the inmates from the old New York Newgate Prison were moved to Sing Sing right away and so were additional inmates from Auburn, the prison was full at 800 inmates by 1830 (Lawes, 1932).

Prison labor in the early years of prisons (before the Civil War) was contract labor and subject to abuse. Contractors would pay a set amount for inmates' labor and then would make sure they got the most work out of them, cutting costs where they could and bribing wardens and keepers when they needed to. Eventually, such contracts were ended as the cheap labor made prison-produced goods too competitive with products made by free workers (Conover, 2001).

When one thinks about old prisons, those castle-like fortress prisons, the image of Auburn and Sing Sing inmates and prisons come to mind, even unknowingly. So many United States prisons copied the New York design and operation of these prisons that even if one is not thinking of Auburn or Sing Sing per se, one is likely imagining a copy of them. By the time Beaumont and Tocqueville (1833/1964) visited the states in 1831, they reported that the Auburn Prison had already been copied in prisons built in Massachusetts, Maryland, Tennessee, Kentucky, Maine, and Vermont.

It was not just the physical structure or the silent, but congregate, inmate management that was copied, however, from Auburn and Sing Sing, but the inmate discipline system as well. As Orland (1975) summarized regarding the Connecticut prison regulations of the 1830s, which were borrowed from the New York model,

> Inmates were exhorted to be "industrious, submissive, and obedient"; to "labor diligently in silence"; they were forbidden to "write or receive a letter" or to communicate in any manner "with or to persons" without the warden's permission; they were prohibited from engaging in conversation "with another prisoner" without permission or to "speak to, or look at, visitors." (p. 26)

In Focus 3.2

Lewis E. Lawes's Observations About Sing Sing History and Discipline

In 1920, Lawes began his tenure as Sing Sing warden and later commented on how the severity of prison discipline had waxed and waned at this prison over the years. At first it was very severe with the use of the "cat-o-nine-tails" whip. "It was made of long strips of leather, attached to a stout wooden handle, and was not infrequently wired at the tips. The 'cat' preferred its victim barebacked" (Lawes, 1932, p. 74–75). Under a warden in 1840, however, the cat was retired and inmates could have a few visits and letters. A Sunday school and library were constructed and the warden walked among the men. Within a few years, though, a new warden was appointed with a new political party in power and all of the reforms were abandoned and the cat was resurrected. A few years later, when a reportedly insane inmate was literally whipped to death, the public was outraged and the use of the lash declined for men and was prohibited for women. The prison discipline was consequently softened, and this cycle continued for the rest of the 1800s from severe to soft discipline. Lawes maintained, after reviewing all of the wardens' reports since the opening of Sing Sing, that escapes were highest during times of severe punishment, despite the risks inmates took should they be caught.

He also observed that the prison had problems with management and control in other ways, noting that by 1845 an outside accountant found that the prison held 20 fewer female and 33 fewer male inmates in the facility than it had officially on the books, that $32,000 was missing, and that there was no explanation as to where these people were or where the money had gone (Lawes, 1932, p. 82). The wardens' and other official reports indicated that inmates were poorly fed and that diseases were rampant at Sing Sing. By 1859, some of Sing Sing's small cells became doubles to accommodate the overcrowding, and the punishments got worse. By 1904, the official report was that the prison was in a disgraceful condition. "Such was the Sing Sing of the Nineteenth Century. A hopeless, oppressive, barren spot. Escapes were frequent, attempts at escape almost daily occurrences. Suicides were common" (p. 88).

❖ Early Prisons and Jails Not Reformed

Lest one be left with the impression that all prisons and jails in the early 1800s in America were reformed, we should note the fact that they were not. Beaumont and Tocqueville (1833/1964) commented, for instance, on the fact that the New Jersey prisons, right across the river from the reformist New York system, were vice ridden, and that Ohio prisons, though ruled by a humanitarian law, were "barbarous," with half the inmates in irons and "[t]he rest plunged into an infected dungeon" (p. 49). But in New Orleans, they found the worst, with inmates incarcerated with hogs. "In locking up criminals, nobody thinks of rendering them better, but only of taming their malice; they are put in chains like ferocious beasts; and instead of being corrected, they are rendered brutal" (p. 49).

As to jails, Beaumont and Tocqueville (1833/1964) noticed no reforms at all. Inmates who were presumed innocent, or if guilty had generally committed a much less serious offense than those sent to prison, were incarcerated in facilities far worse in construction and operation than prisons, even in states where prison reform had occurred. In colonial times, inmates in American jails were kept in house-like facilities and were allowed much more freedom, albeit with few amenities that they did not pay for themselves. Dix (1843/1967) described many jails, particularly those that did not separate inmates, as a "free school of vice." However, as the institutionalization movement began for prisons, jails copied their large, locked-up and controlled atmosphere, without any philosophy of reform to guide their construction or operation (Goldfarb, 1975). By mid-century, some jails had employed the silent or separate systems popular in prisons, but most were merely congregate and poorly managed holding facilities (Dix, 1843/1967). Such facilities on the East Coast by the latter quarter of the 1800s were old, crowded, and full of the "corruptions" that the new prisons were trying to prevent (Goldfarb, 1975). In the end, Beaumont and Tocqueville (1833/1964) blamed the lack of reform of prisons in some states and the failure to reform jails hardly at all, on the fact that there were independent state and local governments who handled crime and criminals differently. "These shocking contradictions proceed chiefly from the want of unison in the various parts of government in the United States (p. 49).

Prisons: "The Shame of Another Generation"

The creation of prisons was a grand reform, promoted by principled people who were appalled at the brutality of discipline wielded against those in their communities. Prisons were an exciting development supported by Enlightenment ideals of humanity and the promise of reformation. They were developed over centuries, in fits and starts, and had their genesis in other modes of depriving people of liberty (e.g., galley slavery, transportation, jails, bridewells, houses of corrections, early versions of prisons), but they were meant to be much better, so much better, than these.

It is not clear whether the problems arose for prisons in their implementation or in their basic conceptualization. In societies where the poor and dispossessed exist among institutions where law and practice serve to maintain their status, is it any wonder that prisons, as a social institution that reflects the values and beliefs of that society, would serve to reinforce this status. All indications are that most prisons, even those that were lauded as the most progressive in an earlier age of reform, were by the mid-19th century regarded as violent and degrading places for their inmates and staff.

Dorothea Dix's Evaluation of Prisons and Jails

Dorothea Dix was a humanitarian, a teacher, and a penal and insane asylum reformer who, after 4 years of study of prisons, jails, and almshouses in northeast and midwestern states, wrote the book *Remarks on Prisons and Prison Discipline in the United States,* in 1843 (reprinted in 1845 and 1967). The data for her book were assembled from multiple observations at prisons; conversations and correspondence with staff, wardens, and inmates in prisons; and a review of prisons' annual reports.

Photo 3.6

Dorothea Dix was a humanitarian, a teacher, and a penal and insane asylum reformer who, after 4 years of study of prisons, jails, and almshouses in northeast and midwestern states, wrote the book *Remarks on Prisons and Prison Discipline in the United States,* in 1843.

Dix tended to prefer the Pennsylvania model over the New York because she thought inmates benefited from separation from others. However, she forcefully argued that both prison models that had promised so much in terms of reform for inmates, were in fact abject failures in that regard. She found these and most prisons to be understaffed, overcrowded, and with inept leadership that changed much too often. She noted that at Sing Sing about 1,200 lashes, using the cat-of-nine-tails, were administered every month to about 200 men, an amount she thought too severe, though she believed the use of the lash, especially in understaffed and overcrowded prisons like Sing Sing and Auburn, was necessary to maintain order (Dix, 1843/1967). In contrast, at the Eastern Penitentiary she commented that punishments included mostly solitary confinement in darkened cells, which to her appeared to lead to changed behavior of recalcitrant inmates. Dix argued as far as inmate discipline goes that "Man is not made better by being degraded; he is seldom restrained from crime by harsh measures" (p. 4).

Thus she argued against the long sentences for minor offenses that she found in prisons of the day (e.g., Richmond, Virginia; Columbus, Ohio; Concord, Massachusetts; Providence, Rhode Island) and the disparity in sentencing from place to place. She thought such sentences were not only unjust, but that they led to insubordination by inmates and staff who recognized the arbitrary nature of the justice system. On the other hand, in her study of prisons she found that the pardoning power was used too often, and this again led, she thought, to less trust in the just and fair nature of the system and to insubordination of its inmates.

Dix also remarked on the quality and availability of food and water for inmates in early correctional facilities. She found the food to be adequate in most places, except Sing Sing where there was no place to dine at the time of the second edition of her book (1845), and the water inadequate in most places except the Pennsylvania prisons where it was piped into all of the cells. Her comments on the health, heating, clothing, cleanliness, and sanity of inmates were detailed, also, by institution, and indicated that though there were recurrent problems with these issues in prisons of the time, some prisons (i.e., Eastern Penitentiary) did more than others to alleviate miseries by changing the diet, providing adequate clothing, and making warm water for washing available to inmates.

She did not find that more inmates were deemed "insane" in Pennsylvania-modeled prisons based on her data, or at least not more than one might expect, even in the Pennsylvania prisons. Given the history of the separate system being linked to insanity, she was sensitive to this topic. However, by 1845, when she published the second edition of her book, inmates at Eastern were not as "separate" from others as they had been, both formally and informally, and this might explain the relative paucity of insanity cases in her data. By this time, inmates were allowed to speak to their keepers (guards) and attend church and school.

Dix also explored the moral and religious instruction provided at the several state prisons and county prisons (jails) that she visited. Except for the Eastern Penitentiary, she found them all deficient in this respect and that the provisions of such services were severely lacking in the jails.

Dix studied a peculiar practice of the early prisons, that of allowing visitors to pay to be spectators at the prisons. Adults were generally charged 25 cents and children were half price in some facilities. In Auburn in 1842, the prison made $1,692.75 from visitors; in Columbus, Ohio, in 1844, the prison made $1,038.78; and Dix documented five other prisons that allowed the same practice, a practice she thought should be "dispensed with" as it "would not aid the moral and reforming influences of the prisons" (p. 43). Of course, this fascination with watching inmates continues today with reality-based television shows filmed in prisons and jails.

Finally, Dix tried to explore the idea of recidivism, or as she termed it, reform. She wanted to know how many inmates leave their prison and are reformed or "betake themselves to industrious habits, and an honest calling; who, in place of vices, practice virtues; who, instead of being addicted to crime, are observed to govern their passions, and abstain from all injury to others" (p. 66). When she asked about how many inmates were reformed, she found that none of the prisons had records in that regard, but that wardens tended to think most inmates were reformed because they did not return to that particular prison. She was doubtful that this failure to return to that prison should be regarded as an indication of low recidivism as she suspected that inmates got out and changed their names and dispersed across the countryside, where they may return to crime undetected. In most respects in all of these areas, she concluded from her study of several prisons that the Eastern Penitentiary was far superior to most prisons and that the Sing Sing Prison was far inferior, but she thought even the Eastern Penitentiary was far from perfect. Rather she called for more focus on the morals and education of the young and on preventing crime as a means of improving prisons and reducing their use—a call that sounds very familiar today.

The Failure of Reform Is Noted

Dix's writings foretold the difficulties of implementing real change, even if the proposal is well-intentioned. Simply put, prisons in the latter half of the 19th century were no longer regarded as places of reform. As Rothman (1980) states,

> Every observer of American prisons and asylums in the closing decades of the nineteenth century recognized that the pride of one generation had become the shame of another. The institutions that had been intended to exemplify the

humanitarian advances of republican government were not merely inadequate to the ideal, but were actually an embarrassment and a rebuke. Failure to do good was one thing; a proclivity to do harm quite another—and yet the evidence was incontrovertible that brutality and corruption were endemic to the institutions. (p. 17)

Newspapers and state investigatory commissions by the mid-19th century were documenting the deficiencies of state prisons. Instead of the relatively controlled atmosphere of the Pennsylvania or Auburn prisons of the 1830s, there was a great deal of laxity and brutality (Rothman, 1980). Prisons were overcrowded and understaffed, and the presence of prison contractors led to corruption such as paying off wardens to look the other way as inmate labor was exploited or, alternatively, the wardens and staff used inmates and their labor for their own illegal ends. At Sing Sing, an 1870 investigation found that inmates were largely unsupervised, had access to contraband, and were unoccupied for most of their sentence (Rothman, 1980). A lack of space and staff were likely to lead to more severe punishments to deter misbehavior. Therefore, when their keepers did pay attention to inmates, it might be to administer brutal medieval-style punishments, which might include the "pulley" where inmates were hung by their wrists or by their thumbs, or they were forced to wear the iron cap or cage on the head that weighed 6 to 8 pounds, or they were tied up by their hands so high behind them that they had to "stand" on their toes, or the guards resorted to the "lash and paddle" (Rothman, 1980, p. 19). Solitary confinement in a dark dungeon-like cell was popular too, especially with little sustenance (water and bread) and a bucket. A Kansas prison employed the "water crib":

> The inmate was placed in a coffin-like box, six and one-half feet long, thirty inches wide, and three feet deep, his face down and his hands handcuffed behind his back. A water hose was then turned on, slowly filling the crib. The resulting effect of this procedure was the sensation of slowly drowning, with the inmate struggling to keep his head up above the rising water line. (Rothman, 1980, p. 20)

❖ The Renewed Promise of Reform

The 1870 American Prison Congress

The first major prison reform came about approximately 50 years after the first New York and Pennsylvania prisons were built, and doubtless as a result of all of those calls for change. The 1870 American Prison Congress was held in Cincinnati, Ohio, with the express purpose of trying to recapture some of the idealism promised with the creation of prisons (Rothman, 1980). Despite their promises of reform and attempts at preventing "contamination," the early prisons had become, by the 1860s, warehouses without hope or resources. All of the themes mentioned at the beginning of Chapter 2—save the desire for reform, and that was remedied with the next round of reforms to follow the Congress—applied to the operation of the 19th-century prisons: They were overcrowded, underfunded, brutal facilities where too many inmates would spend time doing little that was productive or likely to prepare them to reintegrate into the larger community.

Appropriately enough, then, the Declaration of Principles that emerged from the Prison Congress of 1870 was nothing short of revolutionary at the time and provided a blueprint for prisons we see today (Rothman, 1980). Some of those principles were concerned with the grand purposes of prisons—to achieve reform—while others were related to their day-to-day operation (e.g., training of staff, eliminating contract labor, the treatment of the insane) (American Correctional Association, 1983).

Elmira

As a result of these principles, a spirit of reform in corrections again spurred action, and the **Elmira Reformatory** was founded in 1876 in New York (Rothman, 1980). The reformatory would encompass all of the rehabilitation focus and graduated reward system (termed the **marks system,** as in if one behaves, it is possible to earn marks that in turn entitle one to privileges). The marks system, as mentioned in Chapter 2, was practiced by Machonochie, and later by Crofton in Irish prisons, and was promoted by reformers. Elmira was supposed to hire an educated and trained staff and to maintain uncrowded facilities (Orland, 1975).

Zebulon Brockway was appointed to head the reformatory, and he was intent on using the ideas of Machonochie and Crofton to create a "model" prison (Harris, 1973, p. 85). He persuaded the New York legislature to pass a bill creating the indeterminate sentence, which would be administered by a "board" rather than the courts. He planned on the reformatory handling only younger men (ages 16 to 30), as he expected that they might be more amenable to change. He planned to create a college at Elmira that would educate inmates from elementary school through college. He also sought to create an industrial training school that would equip inmates with technical abilities. In addition, he focused on the physical training of inmates, including much marching, but also the use of massages and steam baths (Harris, 1973). The marks system had a three-pronged purpose: to discipline, to encourage reform, and to be used to justify "good time" to reduce the sentence of the offender. Brockway did not want to resort to the use of the lash.

Much lauded around the world and visited by dignitaries, the Elmira Reformatory, and Brockway's management of it, led to the creation of good time, the indeterminate sentence (defined in Chapter 4), a focus on programming to address inmate deficiencies, and the promotion of probation and parole. "After Brockway, specialized treatment, classification of prisoners, social rehabilitation and self-government of one sort or another were introduced into every level of the corrections system" (Harris, 1973, pp. 86–87).

Unfortunately, and as before, this attempt at reform was thwarted when the funding was not always forthcoming, and the inmates did not conform as they were expected to. The staff, who were not the educated and trained professionals that Brockway had envisioned, soon resorted to violence to keep control. In fact, Brockway administered the lash himself on many occasions (Rothman, 1980). It should not be forgotten, however, that even on its worst day, the Elmira prison was likely no worse, and probably much more humane, than were the old Auburn or Sing Sing prisons.

The Creation of Probation and Parole

As indicated in Chapters 6 and 8, probation and parole were developed in the first half of the 19th century, and their use spread widely across the United States in the early 20th century. The idea behind both was that programming and assistance in the community, while supervising the offender, could effectuate the reduction in the use of incarceration and help the offender to transition more smoothly back into the community. Doubtless, the intent was good, but the execution of this reform was less than satisfactory; however, it did represent an improvement over the correctional practices that preceded it (Rothman, 1980). For a full discussion of the history and current operation of probation and parole, or community corrections, see Chapters 6 and 8.

❖ Southern and Northern Prisons and the Contract and Lease Systems, and Industrial Prisons

Southern prisons, because of the institution of slavery, developed on a different trajectory from that of other prisons. As indicated by Young's (2001) research, prisons were little used before the Civil War. In agriculturally based societies, labor is prized and needed in the fields, particularly slave labor, and that served as a rampart for the Southern economy. Once slavery was abolished with the Thirteenth Amendment to the Constitution, Southern states in the Reconstruction period following the Civil War began incarcerating more people, particularly ex-slaves, and recreating a slave society in the corrections system. As Oshinsky (1996) documents for Mississippi prisons, blacks were picked up and imprisoned for relatively minor offenses and forced to work like slaves on prison plantations or on plantations of Southern farmers.

The North and Midwest, and later the West, built prisons somewhat on the Auburn model, but for the most part abandoned the attempt to completely silence inmates. It was no longer emphasized, as maintaining such silence required an excess of staff and constant vigilance, which were usually not available in these understaffed and overcrowded facilities (J. B. Jacobs, 1977). Inmates in such prisons worked in larger groups under private or public employers, and order was maintained with the lash or other innovations in discipline as discussed in Chapter 2 (see also Lawes [1932] regarding the management of Sing Sing). Though there was no pretense of high-minded reform going on in these prisons, their conditions and the accommodations of inmates were thought to be far superior to those provided in Southern prisons of the time. Conditions under both the contract and lease systems could be horrible, but were likely worse under the Southern lease system where contractors were often responsible for both housing and feeding inmates. Such contractors had little incentive for feeding or keeping inmates in good condition, as the supply of labor from the prison was almost inexhaustible.

Industrial Prisons

The contract system morphed into industrial prisons in the latter part of the 19th and first few decades of the 20th century in several states. Inmates were employed either by outside contractors or by the state to engage in the large-scale production of goods for sale in the open market or to produce goods for the state itself. Eventually, as the strength of unions increased, and particularly as the Depression struck in 1930, the sale of cheap prison-made goods was restricted by several state and federal laws, limiting the production of goods in prisons to just products that the state or nonprofits might be able to use.

❖ Correctional Institutions or Warehouse Prisons?

In James Jacobs's classic book *Stateville: The Penitentiary in Mass Society* (1977), the author describes the operation of and environmental influences on the **Stateville Prison** in Illinois. It was built as a panopticon in 1925 in reaction to the deplorable conditions of the old Joliet,

Illinois, prison built in 1860. Joliet was overcrowded, and the Stateville Prison was also built to relieve that overcrowding, but ironically, by 1935 Stateville itself was full at 4,000 inmates and the population at Joliet had not been reduced at all.

In a reformist state such as Illinois at the time (juvenile court reform began here, and it was one of the first states to initiate civil service reforms), Stateville was conceived as a place where inmates would be carefully classified into treatment programs that would address their needs and perceived deficiencies, where inmates could earn good time and eventual parole. Inmates were believed to be "sick," and a treatment regimen provided by the prison would address that sickness and hopefully "cure" them so that they might become productive members of society. Thus, correctional institutions would use the "**medical model**" to treat inmates. Even though it was built as a maximum security prison, Stateville's conception fit the definition of a correctional institution, where inmates were not to be merely warehoused, but to be corrected and treated. However, though inmates in the Illinois system were classified and good time was available for those who adhered to the rules, there was little programming available, the prison was crowded, it was understaffed, and the staff who were employed were ill trained (J. B. Jacobs, 1977). Moreover, the first 10 years of operation were filled with disorganized management and violent attacks on staff and inmates in a prison controlled by Irish and Italian gangs.

In essence, and despite the intent to create a correctional institution, Stateville became what is termed a **Big House prison.** These, according to Irwin (2005), are fortress stone or concrete prisons, usually maximum security, whose attributes include "isolation, routine, and monotony" (p. 32). Strict security and rule enforcement, at least formally, and a regimentation in schedule are other hallmarks of such facilities. The **convict code** or the rules that inmates live by vis-à-vis the institution and staff are clear-cut: "1. Do not inform; 2. Do not openly interact or cooperate with the guards or the administration; 3. Do your own time" (p. 33).

The next 25 years of the Stateville Prison (1936–1961) were marked by the authoritarian control of one warden (Ragen), the isolation of staff and inmates from the larger world, strict formal rule enforcement, and informal corruption of those rules. Some of the trappings of a correctional institution were present (i.e., good time for good behavior and parole), but inmates for the most part were merely warehoused, double and triple celled. Those inmates who were favored by staff and the warden were given better housing and a whole array of privileges. Corruption seethed under the surface with the relaxation of rules for tougher inmates, black market trade by both staff and inmates, and the warden turning his head when beatings of inmates by staff occurred. By the mid-1950s, Ragen, who had been appointed director of corrections for the state in 1941, was redefining its purpose as one of rehabilitation (J. B. Jacobs, 1977). So that his prisons would appear to be at the forefront of the move to a rehabilitative focus, the numbers of inmates in school and in vocational programming did increase, though staff, under the guise of providing vocational training, were able to use the inmate labor to repair their appliances and cars for free.

By the 1960s, the Stateville and other Illinois prisons, much like the rest of the country, were under pressure internally by more career-oriented professionals interested in work in management of prisons, and externally by greater racial consciousness and an emerging inmates' rights movement. Eventually, such prisons had to open their doors to other ideas and perspectives, and sometimes the press, as well as court-mandated legal review of their practices (J. B. Jacobs, 1977).

The 1960s through the 1990s saw a boom in prison building across the country, most of the medium- and minimum security variety, which were more likely to classify inmates according to both security and treatment needs, institute rehabilitative programming (although the amount and value of this have varied from state to state and by time period), and employ the use of good time and parole (except in those states that abolished it as part of a determinate sentencing schema; see Chapter 4). Thus, by the 1960s and 1970s, the ideal of a correctional institution had been more fully realized in many parts of the country and in some prisons. However, the extent to which it truly was realized is in doubt. Staff hired to work in these prisons, other than the few treatment staff, tended to have only a GED or high school diploma and were not paid a professional wage. The prisons were understaffed. They were also often crowded, and educational and other treatment programs, even work programs, were limited. Good time was usually a given, though you could lose it. You did not really earn it; rather, you did time and got it. Parole was typically poorly supervised and by the 1970s and through the 1980s and early 1990s, several states and the federal government eliminated it as they moved to determinate sentencing (see Chapter 4).

By the mid-1970s, a conservative mood regarding crime had gripped the country and a real skepticism had developed about the value of rehabilitative programming. The media and political actors played on the fear of crime, and despite the fact that overall street crime has been decreasing since the early 1980s in the United States and violent crime since the mid-1990s, a prison building boom ensued (Irwin, 2005). Prisons of the 1980s, 1990s, and into the 2000s reflect all of these earlier trends and influences. The maximum- and super-maximum security prisons of today (and possibly some medium- and minimum security prisons) are merely **warehouse prisons** where inmates' lives and movement are severely restricted and rule bound. There is no pretense of rehabilitation in warehouse prisons; punishment and incapacitation are the only justification for such places. The more hardened and dangerous prisoners are supposed to be sent there, and their severe punishment is to serve as a deterrent to others in lesser security prisons.

These lesser security prisons, the medium and minimum security prisons, which comprise roughly two-thirds of all prisons, do still have the trappings of rehabilitation programming, though it is limited in scope and funding, and they usually afford good time and even parole (most states still have a version of these). They, too, are often crowded and understaffed, and their staff are not as educated or well paid as one might wish. However, such prisons do approximate the original ideal of a correctional institution.

The rest of this book will be primarily focused on the correctional institution model as it is often imperfectly implemented in the United States. There are some who argue (e.g., Irwin, 2005) that the rehabilitative ideal is not realized in prisons, and instead that programming is too often used to control inmates rather than to help locate another life path for them that does not involve crime. Correctional institutions intended to rehabilitate instead end up warehousing the "dangerous classes" (Irwin, 2005) or the poor and the minority. And, of course, our history of corrections would lead us to be skeptical of any easy claims to rehabilitative change. (For a fuller discussion of rehabilitative programming, see Chapter 14.) As will be explored in this book, too often a plan, though well intentioned, is inadequately conceived and executed, and as a result nothing changes, or worse, we achieve precisely the opposite results.

❖ Themes That Prevail in Correctional History

As was mentioned in Chapter 2, there are several themes that appear to underlie the history, and current operation, of corrections in the United States. The overriding one, of course, has been money. Operating a correctional institution or a program is a costly undertaking and from the first, those engaged in this business have had to concern themselves with how to fund it. Of course, the availability of funding for correctional initiatives is shaped by the political sentiments of the time. Not surprisingly, schemes to fund correctional operations often included ways to utilize inmate labor. Complementary themes that have shaped how money might be made and spent and how inmates or clients might be treated have included a move to a greater compassion and humanity in correctional operation; the influence that the demographics of inmates themselves have played (e.g., race, class, gender); religious sentiments about punishment and justice; architecture as it aligns with supervision; the pressure that crowding places on correctional programs and institutions; and the fact that though reforms might be well-intentioned, they do not always lead to effective or just practice. Again, this list of themes is not exhaustive, but it does include some of the prevailing influences that span correctional history in the United States and that require the attention of each successive generation.

Summary

- Howard, Beaumont and Tocqueville, and Dix all conducted studies of corrections in their day and judged the relative benefits of some practices and institutions over others, based on their data.
- The Pennsylvania and the New York early prisons were the models for most American prisons of the 19th century.
- The ideal conception of prisons was rarely achieved in reality.
- The Elmira Prison arose out of a prison reform movement that occurred roughly 50 years after the Auburn Prison was built.
- The Southern and Northern versions of prisons that followed the Civil War were not like Elmira and were instead focused on utilizing inmate labor for the production of goods for private contractors.
- The Stateville Prison, though conceived as a correctional institution and all that that implies, for the most part became a Big House prison.
- Correctional institutions, as a type of prison, do exist in a less-than-perfect form in the United States.
- Correctional institutions, as that term has been expanded to apply generically to jails, prisons, and some forms of community corrections, have been shaped by several themes throughout their history. These themes, though apparently constant, are

a product of the times. For instance, the Eastern Penitentiary would not be built today as a *general use prison* because it would be considered cruel to isolate inmates from other human contact. Yet this kind of isolation, sometimes even with the tiny cells, is seen as beneficial by those who today build and operate super-max prisons for "special uses" to control incorrigible inmates (Kluger, 2007).

- In the following chapters, we will see such themes and the history of corrections as detailed here, dealt with again and again. However, although we continue to repeat both the mistakes and successes of the past, that does not mean we cannot, and have not, made any progress in corrections. There is no question that on the whole the vast majority of jails and prisons in this country are much better than were those for much of the last 200 years, though the unprecedented use of correctional sanctions in the United States would be regarded by some as overly harsh and thus a regressive trend. These themes presented here merely represent conundrums (e.g., how much money or compassion or religious influence, is the "right" amount), and as such we are constantly called upon to address them.

Key Terms

Big House prison	Eastern Pennsylvania Prison	Marks system
Convict code	Elmira Reformatory	Medical model

New York model

New York prison system

Newgate Prison in New York City

Pennsylvania prison system

Stateville Prison

Walnut Street Jail

Warehouse prison

Western Pennsylvania Prison

Discussion Questions

1. Discuss the relative benefits and drawbacks of the Pennsylvania versus the New York models of early prisons. What did Beaumont and Tocqueville and Dix think of them and why? Which type of prison would you rather work in, or be incarcerated in, and why?

2. What role did Penn, Bentham, Beccaria, and Howard play in reforming the prisons and jails of their time? Are the concerns they raised still valid today?

3. How has the question of inmate labor shaped the development of prisons over time?

4. Note why there is often a disconnect between the intentions of reformers and the ultimate operation of their reforms. Why is it difficult for theory to be put into practice? How might we ensure that there is a truer implementation of reforms?

5. How are the "themes" that run through the history of corrections represented in current practices today? Why do these themes continue to have relevance for correctional operation over the centuries?

Useful Internet Sites

American Correctional Association: www.aca.org

American Jail Association: www.corrections.com/aja/

American Probation and Parole Association: www.appa-net.org

Auburn Prison, New York: www.correctionhistory.org

Bureau of Justice Statistics (information available on all manner of criminal justice topics): http://bjs.ojp.usdoj.gov/

John Howard Society (Canada): www.johnhoward.ca

National Criminal Justice Reference Service: www.ncjrs.gov.

Office of Justice Research (information available on all manner of criminal justice topics, specifically probation and parole here): http://nij.gov/

Pennsylvania Prison Society: www.prisonsociety.org

Vera Institute (information available on a number of corrections and other justice-related topics): www.vera.org

CHAPTER 4

Sentencing

The Application of Punishment

❖ Introduction: The Scope of Sentencing

Sentencing refers to a post-conviction stage of the criminal justice process. A **sentence** is the punitive penalty ordered by the court after a defendant has been convicted of a crime either by a jury, in a bench trial by a judge, or in a plea bargain. Sentencing typically occurs about 30 days after conviction. The goals of sentencing are to implement one or more of the punishment philosophies discussed in Chapter 1—retribution, deterrence, incapacitation, rehabilitation, or reintegration. In some states, juries may be entitled to pronounce sentence, but in most states, and in federal court, sentencing is performed by a judge except in death penalty cases, where it is the jury's responsibility. The penalties meted out at sentencing can range from various forms of probation coupled with fines and restitution orders and/or treatment orders to house arrest/ electronic monitoring, work release, jail time, prison time, or the death penalty, all of which are discussed elsewhere in this book. The severity of the penalty depends on the crime or crimes the defendant is convicted of and the extent of his or her criminal history, although other factors both legitimate and illegitimate may also come into play.

It is a major concern of the American criminal justice system that punishments received by defendants at sentencing should be consistent with justice. **Justice** is a moral concept that is difficult to define, but in essence it means treating people in ways consistent with norms of fairness and in accordance with what they justly deserve by virtue of their behavior. Perhaps the best definition was provided by the Greek philosopher Aristotle many centuries ago: "Justice consists of treating equals equally and unequals unequally according to relevant differences" (quoted in Walsh & Stohr, 2010, p. 133). In terms of sentencing, this means that those who have committed the same crime and have similar criminal histories are considered legal "equals" and should be treated equally. Those who have committed different crimes and have different criminal histories are considered legal "unequals" and should therefore be treated unequally; that is, those in this group should be treated either more leniently or more harshly than others, depending on their crime and criminal background.

You may ask what these "relevant differences" are and who defines them. Strictly speaking, the relevant differences in sentencing should be limited to legally relevant factors (crime seriousness and prior record), but *extralegal* factors are often also brought into play such as gang affiliation and a history of substance abuse. Depending on what these factors are, justice is either served or not served by adding them. A judge who sentences a remorseful mother to probation, whose children would become wards of the state if she were sent to prison, is probably acting justly. This may be so even if the same judge sentences an unremorseful single male to prison who has committed the same crime and has a criminal record identical to the mother's, and thus the judge is treating legal equals unequally. On the other hand, if the judge sentences legal equals unequally only because one defendant is a woman and the other a man, or only because one defendant is black and the other white, he or she is not acting justly.

In this chapter, we will keep this concept of justice, or injustice, in mind as we review the types of sentencing in use and sentencing disparity. We will also focus on drug courts and their emerging importance as an alternative sentence in corrections. Finally, we will review the presentencing paperwork and decision making that precedes a sentencing determination.

❖ Types of Sentences: Indeterminate, Determinate, and Mandatory

A prison sentence an individual receives can be indeterminate or determinate. An **indeterminate sentence** is one in which the actual number of years a person may serve is not fixed but is rather a range of years such as, the person "shall be imprisoned for not less than 2 years to a maximum of 10 years." More serious crimes increase both minimum and maximum time periods. Indeterminate sentences were previously much more common than they are today, but a number of states still retain this system. Indeterminate sentences fit the positivist's rehabilitation philosophy of punishment because they allow for offenders to be released after they have served their minimum sentence if they demonstrate to the parole board's satisfaction that they have made efforts to turn their life around. Such sentences are tailored to the offender and aimed at rehabilitation rather than tailored to the crime and designed to be strictly punitive.

The indeterminate sentencing model prevailed most strongly under the so-called medical model whereby offenders were considered "sick" and in need of a cure. Because some criminals may be "sicker" than others, the time made available for the "cure" must be flexible. Offenders that behaved themselves in prison and could demonstrate that they were "reformed" could be rather quickly released; ill-behaved and stubborn offenders might have to serve the upper boundary (10 years in the above example) and then be released, "rehabilitated" or not. It has been precisely because of its flexibility that indeterminate sentencing has been accused of contributing to sentencing disparity. For instance, even if two offenders receive the same "2 to 10 years," one may serve only 2 years because she can keep out of trouble and knows how to play the rehabilitation/parole game while the other, who is more rowdy and does not play the game as well, may serve 2 or 3 additional years. Supporters of the model, however, will reply that it is not the judiciary that is at fault (after all, both were sentenced identically by judges), but rather it was the inmates themselves who caused the discrepancy by their different behaviors while incarcerated.

Prisoners released from state prisons in 1996 served an average of only 44% of their sentences under predominantly indeterminate sentencing structures (Ditton & Wilson, 1999). Rising crime rates in the 1980s and early 1990s (for violent crimes only) saw a groundswell of opposition to what many saw as "mollycoddling" criminals, and there were many calls for longer sentences. In response to public demands, most states enacted **truth-in-sentencing laws.** These laws require that there be a truthful, realistic connection between the custodial sentence imposed on offenders and the time they actually serve, and they mandate that inmates serve at least 85% of their sentences before becoming eligible for release. In addition, many states restricted good-time credit and/or parole eligibility.

Determinate sentences became more prevalent after the enactment of truth-in-sentencing laws. A **determinate sentence** means that convicted individuals are given a fixed number of years they must serve rather than a range. Under a determinate sentencing structure, the maximum prison time for a given crime is set by the state legislature in state statutes. This structure is more in tune with the classical notion that the purpose of punishment is to deter and that all who commit the same crime must receive a fixed sentence. This does not mean that everyone convicted of the same crime receives the same set penalty. For instance, the maximum time for burglary may be set at 15 years, and a repeat offender may be sentenced to the full 15. Another person who is a young first offender may only receive 5 years. Whatever the sentence, offenders know under this sentencing structure how much time they will have to serve. Longer and more determinate sentences satisfy the urge for greater punishment for offenders, and they also serve an incapacitation function. However, time off for good behavior is still granted.

Another type of sentencing is mandatory sentencing, or sometimes known as mandatory minimum sentencing. A **mandatory sentence** can exist in the context of both determinate and indeterminate sentencing structures and simply means that probation is not an option for some crimes and that the minimum time to be served is set by law. It is set by law because legislative bodies in various states have decided that some crimes are just too serious for probation consideration (certain violent crimes), or they have decided that there is a particular problem, such as drug trafficking or the use of a gun during the commission of a crime, that requires mandatory imprisonment as a deterrent.

Prison sentences imposed for two separate crimes, whether they occurred during the same incident (say, robbery and aggravated assault) or in different incidents (say, two separate burglaries), can be ordered to be served concurrently or consecutively. A **concurrent sentence** is one in which two separate sentences are served at the same time.

If the robbery and aggravated assault crimes both carried sentences of 10 years, for instance, the offender's release date would be calculated on the basis of 10 rather than 20 years. A **consecutive sentence** is one in which two or more sentences must be served sequentially (one at a time). If our robber/aggravated assaulter receives two 10-year sentences to be served consecutively, his or her release date would be based on 20 rather than 10 years. Consecutive sentences therefore increase the time a person spends in prison. The judge's decision to impose concurrent or consecutive sentences for persons convicted to two crimes may rest mainly on factors such as the seriousness of the crime, criminal history, plea bargain arrangements, and offender cooperation. Some have suggested that judges may actually impose harsher sentences on those offenders with the audacity to demand a trial rather than accept a plea bargain because it makes extra work for the judge. This philosophy has been expressed as the judge's warning—"You take some of my time and I'll take some of yours" (Neubauer, 2008).

Photo 4.1

A defendant listens as his sentence is announced by the judge.

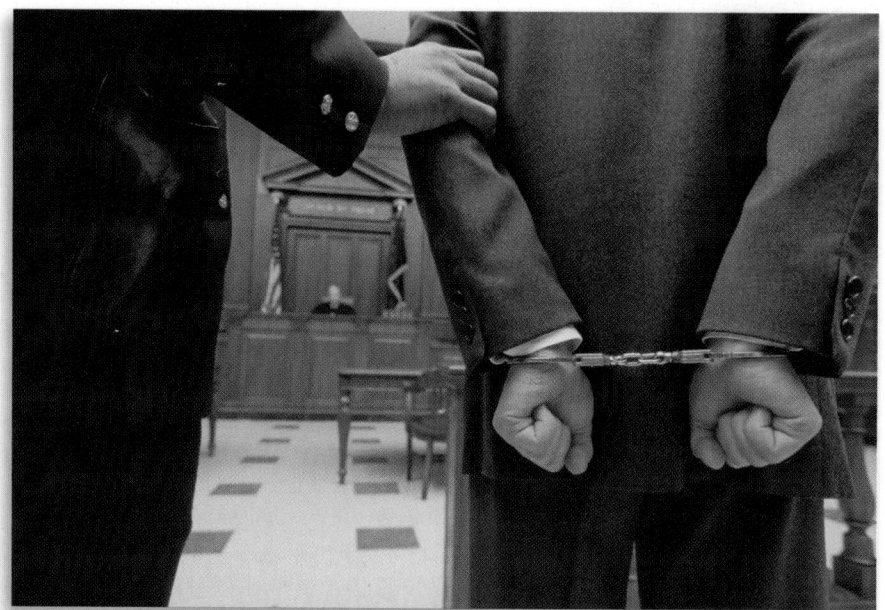

❖ Habitual Offender Statutes

Habitual offender (or "three strikes and you're out") **statutes** are derived from the same punitive atmosphere that led to truth-in-sentencing laws. These statutes essentially mean that offenders with a third felony conviction may be sentenced to life imprisonment regardless of the nature of the third felony. This is a way of selectively incapacitating felons only after they have demonstrated the inability to live by society's rules. This all sounds very well until we factor in the financial costs of these sentences. Few of us would be against the lifetime incarceration of seriously violent offenders, but many states include relatively minor nonviolent crimes in their habitual offender statutes. For instance, the United States Supreme Court upheld the life sentence of a felon under Texas's habitual offender statute even though the underlying felonies involved nothing more serious than obtaining a total of less than $230 over a 15-year period by false pretenses (fraudulent use of a credit card and writing bad checks) in three separate incidents (*Rummel v. Estelle*, 1980). Very few of us would consider this a just sentence, and apart from the disproportionate nature of the sentence, the cost to the taxpayers of Texas of keeping Rummel in custody is many thousand of times greater than the $230 he fraudulently obtained.

A life sentence still carries with it the possibility of parole, but some life sentences are imposed as **life without parole (LWOP)**. Such sentences may seem popular with the public at large until they get the bill. According to Nellis (2010), in 2008 there were 140,095 prisoners serving LWOP sentences in the United States, a 400% increase from 1984. LWOP sentences are usually imposed on those convicted of murder, but habitual property offenders have also been given such sentences. Long-term incapacitation of violent or habitual offenders may be sound policy, but how much time is enough? In one large-scale study, only one-fifth of lifers who are released after long stays (15–30 years) in prison were rearrested within 3 years versus two-thirds of non-lifers who were released (Mauer, King, & Young, 2004). Old age is the best "cure" for criminal behavior that we have, so perhaps releasing "lifers" after 20 to 30 years of imprisonment is both humane and fiscally responsible. Given the ever-increasing medical needs of people as they age, elderly inmates add a highly disproportionate financial burden on the taxpayer.

❖ Other Types of Sentences: Shock, Split, and Non-Custodial Sentences

Judges have many sentencing options open to them besides straight imprisonment. The fact is that over 90% of sentences imposed in our criminal courts do not involve imprisonment (Neubauer, 2008). One type of sentence that does include imprisonment is shock incarceration, also called **shock probation**. This type of sentence is used to literally shock offenders into going straight by exposing them to the reality of prison life for a short period, typically no more than 30 days, followed by probation. Shock probation is typically reserved for young, first-time offenders who have committed a relatively serious felony but who are considered redeemable.

Split sentences are sentences that require felons to serve brief periods of confinement in a county jail prior to probation placement. Jail time may have to be served all at once or be spread over a certain period, such as every weekend in jail for the first year of probation placement. This is designed to show offenders that jail is a place to stay away from and thus to convince them that it would be a good idea to abide by all the conditions imposed by the court. Another form of split sentence is work release whereby a person is consigned to a special part of the jail on weekends and nights, but released to go to work during the day. Thus, these mainly non-custodial sentences typically mean a probation sentence coupled with certain conditions that must be followed in order to remain in the community. The conditions may involve such things as paying fines, paying restitution, attending drug or alcohol treatment, doing community service, remaining gainfully employed or actively looking for work, and any number of other more specific conditions. These different non-custodial sentences and probation conditions will be discussed more fully in the chapters on probation, parole, and treatment.

Drug Courts

A sentence to a **drug court** requires participants to be involved in an intensive treatment program that lasts about one year. Participants typically have pled guilty to a nonviolent drug-related felony charge. In response to the growing drug problem, the first drug court was established in Miami-Dade County, Florida, in 1989. Twenty years later, there were 2,037 drug courts active in all 50 states, a growth that suggests there is much that is positive about drug courts (Mackin, Lucas, & Lambarth, 2010). Under the supervision of the judge, probation officers, and other caseworkers, participants attend counseling groups and 12-step meetings, regularly appear before a judge, and must submit to random urine testing. If a participant successfully completes the program, in almost all jurisdictions the criminal charges will be dismissed. The U.S. Department of Justice (Ashcroft, Daniels, & Herraiz, 1997) provided the 10-component model presented below for state and county agencies implementing their drug court systems.

1. Drug Courts integrate alcohol and other drug treatment services with justice system case processing.

2. Using a non-adversarial approach, prosecution and defense counsel promote public safety while protecting participants' due process rights.

3. Eligible participants are identified early and promptly placed in the Drug Court program.

4. Drug Courts provide access to a continuum of alcohol, drug and other related treatment and rehabilitation services.

5. Abstinence is monitored by frequent alcohol and other drug testing.

6. A coordinated strategy governs Drug Court responses to participants' compliance.

7. Ongoing judicial interaction with each Drug Court participant is essential.

8. Monitoring and evaluation measure the achievement of program goals and gauge effectiveness.

9. Continuing interdisciplinary education promotes effective Drug Court planning, implementation, and operations.

10. Forging partnerships among Drug Courts, public agencies, and community-based organizations generates local support and enhances Drug Court effectiveness.

Photo 4.2

An offender listens to the judge in drug court.

Note the strong emphasis on interagency cooperation, the provision of services to participants, and the strict monitoring of their behavior. In addition to saving the states many millions of dollars in jail and prison costs, drug courts appear to be quite successful in reducing recidivism. For instance, the Baltimore County Juvenile Drug

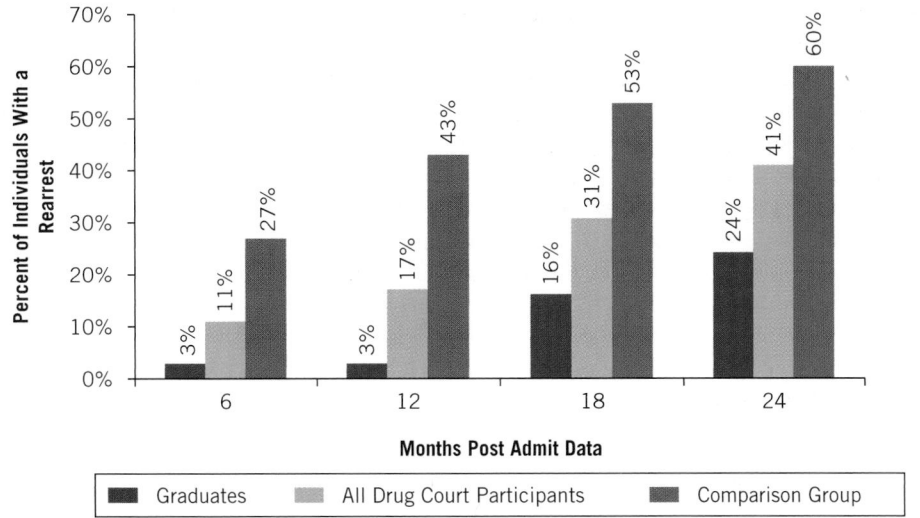

Figure 4.1

Comparison of Rearrest Rates for Juvenile Drug Court Participants and Nonparticipants at 6 Through 24 Months Post Admittance

Source: Mackin et al. (2010).

Court outcome analysis (Mackin et al., 2010) estimated that the program saved the county $8,762 per participant over 24 months because of lower recidivism rates and savings from not incarcerating participants. Figure 4.1 provides an illustration of recidivism outcomes for drug court graduates, participants who did not graduate, and the control group consisting of juveniles who fit the criteria for participation but did not take part. Note that while the likelihood of rearrest increased for all groups over time, the graduates had a lower arrest rate at 24 months than the control group did at 6 months.

❖ Sentencing Disparity, Legitimate and Illegitimate

Sentencing disparity occurs when there is wide variation in sentences received by different offenders. This disparity is legitimate if it is based on considerations such as crime seriousness or prior record, but illegitimate, or discriminatory, if it is not. We think of sentencing disparity as discriminatory if there are differences in punishment in cases in which no rational justification can be found for it. The biggest concern is racial discrimination. There is no doubt that the American criminal justice system has a dark history of racial discrimination, but does this indictment still apply?

African Americans receive harsher sentences on average than white or Asian American offenders, a fact often seen as racist. Sentencing variation is reasonable and just if the members of the group being more harshly punished commit more crimes than the individual members of other groups, but is discriminatory and unjust if they do not. All data sources show that African Americans as a group commit more crime, especially violent crime, than whites or Asians (Federal Bureau of Investigation [FBI], 2010). But the question is whether this racial disproportionality in offending is sufficient to account for the disparity in sentencing. One sentencing scholar concluded that it was not: "[R]acial bias continues to pervade the U.S. criminal justice sentencing system [although] the effects of this bias are somewhat hidden . . . or may even have less to do with the race of the defendant than with the race of the victim" (Kansal, 2005, p. 17). Another scholar concluded the opposite: "Although critics of American race relations may think otherwise, research on sentencing has failed to show a definitive pattern of racial discrimination" (Siegel, 2006, p. 578). Different researchers thus arrive at different conclusions, but there is widespread agreement on one point: The more stringent researchers are in controlling for the effects of legally relevant variables (crime seriousness and criminal record), the less likely they are to find racial discrimination (Siegel, 2006). One study of over 46,000 federal defendants in 23 states found no evidence of racial bias after controlling for legally relevant variables (Wang, Mears, Spohn, & Dario, 2009). Some critics argue, however, that racial bias occurs much earlier in the system, at the arrest stage where racial profiling or targeting of minority neighborhoods for police sweeps occurs, or even earlier as minority juveniles, particularly those who are poor, are introduced into the juvenile justice system rather than being diverted to family members or social services (Welch, 2011; see also Chapter 11 of this volume for a more complete discussion of this topic).

Sentencing research is thus complicated and can be misleading. Researchers typically report average effects, among which are multiple interacting variables hiding specifics. Table 4.1 shows average length in months of felony sentences in state courts in 2006 broken down by race and gender (Durose, Farole, & Rosenmerkel, 2010). You can see that African American males typically have longer sentences than whites, and males have longer sentences than females. Also note that for violent offenses, white females receive longer sentences than black females. However, no conclusions about racial or gender bias can be drawn from the table because it tells us nothing about how serious each of the crimes was (some robberies, sexual assaults, and assaults are more vicious than others, for instance) or the criminal histories of the men and women represented in the table.

One of the biggest concerns in the sentencing disparity literature is the huge differences in sentencing received for crack possession versus sentences imposed for possession of powder cocaine. Of particular concern has been the difference in sentencing imposed on those who used or sold the cheaper, but pharmacologically similar, crack cocaine, which tended to be minority group offenders, particularly African Americans, versus sentences for using or selling powder cocaine, which tends to be more expensive and more likely used and trafficked by white offenders. In 1986, Congress passed the Anti-Drug Abuse Act, which established a 100-to-1 quantity ratio differential between powder and crack cocaine. The act also specified that simple possession of crack cocaine was to be treated more seriously than the simple possession of other illegal drugs. According to a U.S. Sentencing Commission Report to Congress in 1995, in 1986, Congress was reacting to media hype about how addictive crack was, with congressional members claiming crack use was at "epidemic" levels, crack babies were severely impaired, and that crime related to crack use was out of control in some cities. At the time of this 1995 report, the Commission knew that "88.3 percent of the offenders convicted in federal court for crack cocaine distribution in 1993 were Black and 7.1 percent were Hispanic," and critics were concerned that instead of fair and evenhanded sentences for all, the effect of the Anti-Drug Abuse Act was to be unfair and harsh in sentencing of racial minorities (U.S. Sentencing Commission, 1995, p. 1).

Criticisms of the different treatment of people convicted of possession of pharmacologically identical drugs resulting in the increased incarceration of minorities for longer periods of time mounted to the point where Congress had to do something. In 2009, a **Fair Sentencing Act** was introduced and passed by Congress, and was signed into law by President Obama on August 3, 2010. Under the act, the amount of crack cocaine subject to the 5-year minimum sentence is increased from 5 grams to 28 grams, thus reducing the 100-to-1 ratio to an 18-to-1 ratio (28 grams of crack gets as much time as 500 grams of powder cocaine). Although it has been reduced, there is still a large sentencing differential between possessors of crack versus powder cocaine. However, this ratio probably reflects lawmakers'

Table 4.1

Average Length of Felony Sentences in 2006 by Offense, Race, and Gender

Mean maximum sentence length for persons who were—

Most serious conviction offense	White		Black	
	Male	Female	Male	Female
Sentenced to incarceration				
All offenses	40 mo.	25 mo.	45 mo.	25 mo.
Violent offenses	75 mo.	52 mo.	88 mo.	41 mo.
Murder/Non-negligent manslaughter	265	225	266	175
Sexual assault	115	72	125	32
Robbery	89	61	101	54
Aggravated assault	42	30	48	29
Other violent offenses	43	55	41	17
Property offenses	31 mo.	22 mo.	35 mo.	23 mo.
Burglary	41	29	50	34
Larceny	24	17	23	19
Fraud/Forgery	27	22	27	23
Drug offenses	31 mo.	22 mo.	36 mo.	22 mo.
Possession	21	17	25	15
Trafficking	39	26	40	27
Weapon offenses	34 mo.	24 mo.	34 mo.	24 mo.

Source: Durose et al. (2010).

perceptions that crack is more intimately related to violence (in territorial battles) and to a higher probability of addiction than the powder variety, though it is not clear there is scientific evidence to support this last point (Leigey & Bachman, 2007).

❖ Structuring Sentencing: The Presentence Investigation Report

To assist judges in deciding on a sentence, a **presentence investigation report (PSI)** is commonly used. There are few documents as important to the defendant as the PSI. It is used for many other purposes besides sentencing, such as treatment planning, classification to supervision levels in probation/parole departments and prisons, and parole decisions (Walsh & Stohr, 2010). A PSI is usually completed in 30 days or less so that the convicted individual can be sentenced in a timely manner. PSIs are usually written by probation officers informing the judge of various aspects of the offense for which the defendant is being sentenced as well as facts about the defendant's background (educational, family, and employment history), gang ties, substance abuse, character, and criminal history. On the basis of this information, officers make recommendations to the court regarding the sentence the offender should receive. Because probation officers enjoy considerable discretionary power relating to how their reports are crafted to be favorable or unfavorable to offenders, many scholars view them as the agents who really determine the sentences that offenders receive (Champion, 2005). Other researchers, however, suggest that the high rate of judicial agreement with officer recommendations reflects an anticipatory effect whereby officers become adept at "second-guessing" a judge's likely sentence for a given case and recommend accordingly (Durnescu, 2008).

In Focus 4.1 below is an example of a (fictional) PSI containing the usual required information. PSIs come in a variety of sizes, the smallest being a 1- or 2-page short-form report used in misdemeanor cases or less serious "run-of-the-mill" felony cases. For serious or complicated cases, we may see 10- to 15-page reports. The report given here is an example of a mid-range report used for relatively serious crimes, although the trend is for smaller, more concise reports focusing primarily on legally relevant variables.

The following is a fictionalized version of case.

In Focus 4.1

Example of a PSI

The following is a fictionalized version of a PSI.

GEM COUNTY ADULT PROBATION DEPARTMENT

Williamstown, Iowa 74812

Name:	Joan Place	Judge:	Franklin Riley
Indictment No.:	CR 6742	Probation Officer:	James Smith
Age:	28	Attorney:	William Paley
Race:	white		
Sex:	female	Offense:	forgery
Marital Status:	divorced	Conduct (IRC #2908) two counts	

Circumstances of the Offense

On 8/10/94, Mr. John Smith, security operative for the Omaha Trust Company (OTC), reported to the police that the defendant cashed forged checks in the amount of $917.00 at various OTC branch banks. These checks were drawn against the account of one Mrs. Patricia DeValera, 4561 Black St. The defendant stole Mrs. DeValera's checks while employed by her as a nurse's aide. Mr. Smith also indicated that the defendant cashed forged checks in the amount of $575.00 on the account of Mr. Richard Blane, a former boyfriend of the defendant. The total loss to the Omaha Trust bank is $1,492.00.

Defendant's Version

The defendant's written statement is reproduced verbatim as follows:

Took checks filled it out in amount I needed for drugs and signed it, forged a name and cashed the checks in Aug. of 1994. No, I did not pay back the person. I was so drug dependent that I took my boyfriends checks, Mrs. Devaleras checks too. All I could live for at the time was heroin and alcohol (mainly beer). I'm sorry I did these things, normally I wouldn't of forged the checks if I wouldn't of needed drugs. All I could do was live for drugs. I've been threw the withdrawals of drugs when I put myself in the treatment center on Wilson St.

It is noted that the defendant places the blame for her criminal activity on her craving for heroin and alcohol. Her statements of remorse ring rather hollow in light of her new forgery arrest while undergoing presentence investigation. She was arrested on this new charge on 2/22/95 and released on $1,000 bond (10% allowed). Upon learning of this new arrest, I rearrested her on 2/24/95 and placed her in the county jail, where she has been ever since.

Prior Record

Juvenile: None known

Adult:

8/18/94 WPD Forgery, 5 counts, amended to 1 count: Present Offense.

11/6/94 WPD Forgery, 3 counts, pending under CR841234.

2/5/95 W-PD Forgery, pending.

Above record reflects juvenile, OPD, BCI, and FBI record checks.

Present Family Status

The defendant is the fourth of five children born to Ann and Frank Place. Her father passed away in January of 1991. On 6/17/85, the defendant married one James Fillpot. Mr. Fillpot was described as a heavy drug abuser, and is now serving a life sentence on an aggravated murder charge (he was convicted of murdering the defendant's alleged lover). The defendant divorced Mr. Fillpot shortly after his 1989 conviction, and shortly thereafter (6/89) she married one Ralph Burke. Mr. Burke is an alcoholic with an extensive criminal record. After an extremely abusive 2 years of marriage, the defendant's second marriage ended in divorce on 5/12/92. No children were born to either of these marriages. At the time of her arrest, the defendant was living with her mother at the above listed address.

Present Employment or Support

The defendant is unemployed at the present time, and was existing on $76.00 in food stamps at the time of her arrest. She receives no general relief monies. Her last period of employment was as a nurse's aide for Mrs. DeValera, one of the victims of the present offense. This employment was for the period encompassing March through August, 1993. The defendant's longest period of employment was with the Red Barron Restaurant, 3957 Laskar Rd, as a waitress from March, 1988 through November, 1991. This employer has not responded to our request for information as yet.

(Continued)

(Continued)

Health

Physical

The defendant describes her current physical health as "O.K." She relates no significant hospitalizations, diseases, or current health problems with the exception of her substance abuse. Her substance abuse is quite extensive. She claims that she has used anywhere from $20 to $300 per day of heroin. Needle marks on her arms attest to her frequent usage. She also relates that she likes to consume 6 to 12 beers per night, which she claims that she receives free from boyfriends. Her substance abuse goes back to her first marriage 13 years ago when Mr. Fillpot introduced her to heroin. Her second husband, Mr. Burke, got her heavily involved in alcohol. She admits that she has experimented with many other drugs, but states that heroin and beer are her drugs of choice. This officer contacted the Wilson Street Drug Rehabilitation Facility regarding the defendant's claimed attendance there. It appears that she did voluntarily admit herself there, but left after the first 15-day phase. I also made an effort to get her into the ROAD drug rehabilitation program. However, after two interviews with ROAD personnel, the defendant was denied admission because they thought that her only motive for seeking admission was her current legal difficulties.

Mental

The defendant is a graduate of Borah High School. She graduated 287th out of a class of 348. She attained a cumulative GPA of 1.64 on a 4.0 scale. Although no IQ information is available, the defendant impresses as functioning well within the average range of intelligence as gauged by her written and verbal statements. She did indicate that she was easily led, and that she does not think much of herself. Her choice of marriage partners (both very abusive to her) and her current boyfriends give the impression that she is attracted to men who will verify her low opinion of herself.

Statutory Penalty

N.R.C. 2913.31

Forgery "shall be imprisoned for a period of 6 months, 1 year, or 1 and one-half years and/or fined up to $2,500."

Evaluative Summary

Before the court is a 28-year-old woman facing her first felony conviction. However, she has numerous other forgery charges pending at this time. There would seem to be little doubt that the genesis of her criminal activity is her severe abuse of alcohol and drugs. She also appears to posses a low concept of herself as indicated by her very poor choice of marriage partners, both of whom were serious substance abusers, and both of whom were physically abusive to her. She appears to be intimately involved in the drug subculture. I initiated the procedure to get the defendant admitted to the Port of Hope (POH) residential drug treatment center. However, after conducting two interviews with the defendant, personnel from the POH decided that her motivation for seeking treatment was her current legal difficulties. Consequently, her application for treatment there was denied. They did indicate, however, that they would reconsider her application after the disposition of the present offense. Therefore, I recommend that the defendant be placed on probation, ordered to pay complete restitution, and to rigorously pursue entry into the POH. It is also recommended that she remain in the county jail after sentencing to reinitiate her application with HOP.

Approved	Respectfully Submitted
_____	_____
James F. Collins	Megan G. Mann
Unit Supervisor	Probation Officer

PSI Controversies

Although the PSI has generally been considered a positive aid to individualized justice, it is not without its problems. Because the future of a defendant depends to a great extent on the content of the report, the information contained therein should be reliable and objective. All pertinent information must be verified by cross-checking with more than one source, and those sources should be reliable. The officer must be careful in the terms he or she uses to describe the offender. The use of phrases such as "morally bankrupt" or "sweet young lady" may reveal more about the officer's attitudes and values rather than the defendant's character.

If you were the subject of a PSI, would you not like to see what was in it so that you could challenge any erroneous information harmful to you that may be contained in it? There have been a number of arguments for and against allowing defendants and their attorneys access to the PSI. It is feared that if victims and other informants from whom the investigating officer has sought information were to know that the offender will see their comments, they may refuse to offer their information, and thus the judge will not have complete information on which to make the sentencing decision. However, 16 states currently require full disclosure; other states require disclosure but omit information that may lead to retaliation, such as the officer's recommendation or negative comments from informants. Despite objections and real concerns about confidentiality, the trend is to allow defendants access to their PSI reports. For instance, in the federal system, Section 3552 of the U.S. Code requires that

> The court shall assure that a report filed pursuant to this section is disclosed to the defendant, the counsel for the defendant, and the attorney for the Government at least ten days prior to the date set for sentencing, unless this minimum period is waived by the defendant. The court shall provide a copy of the presentence report to the attorney for the Government to use in collecting an assessment, criminal fine, forfeiture or restitution imposed.

In the federal system and in some state systems, probation/parole officers no longer write PSI reports. Rather, they merely complete sentencing guidelines and certain other assessment tools and calculate the presumed sentence (Abadinsky, 2009). This and a number of other factors may be signaling a move away from individualized justice (the idea that punishment should be tailored to the individual and be consistent with rehabilitation) and back to the classical idea discussed in Chapter 1 that the punishment should fit the crime and serve as a deterrent.

❖ Structured Sentencing: Sentencing Guidelines

Prior to 1984, federal judges enjoyed almost unlimited sentencing discretion as long as they stayed within the statutory maximum penalties. This led to much criticism regarding sentencing disparities and moved Congress to establish the **United States Sentencing Commission**. The commission was charged with the task of creating mandatory sentencing guidelines to rein in judicial discretion (Reynolds, 2009). **Sentencing guidelines** are forms containing scales that come with a set of rules for numerically computing sentences that offenders should receive based on the crime they committed and on their criminal records. Guidelines are typically devised by federal or state sentencing commissions and provide classifications of suggested punishments based on an offender's scores on those scales. Such guidelines are in use in the federal system and in a number of state systems. Because the sentencing guidelines are a set of rules and principles that a judge follows to decide a defendant's sentence, they curtailed the discretionary powers of judges, as was intended by Congress. Most people viewed this as a good thing since unbridled discretion can lead to wide sentencing disparities based only on a judge's subjective evaluations and whims. At one end of the scale we might get "hanging" judges, and at the other end "bleeding heart" judges, and so a defendant's fate may depend largely on the temperament or ideology of the sentencing judge by whom he or she has the good luck or bad fortune to be sentenced. The guidelines used by the federal government and some states are limited to crime seriousness and prior record, while others are more comprehensive and assign points not only for the statutory degree of seriousness of the offense and prior record, but also the amount of harm done; whether the offender was on bail, probation, or parole at the time; prior periods of incarceration; and any number of other factors. These numbers are then applied to a grid at the point at which they intersect, which contains the appropriate sentence. The example in Figure 4.2 is a more comprehensive guideline that includes many other pieces of information deemed important.

In some jurisdictions, adherence to the guidelines is presumptive; in others, it is merely advisory. Although they were presumptive (mandatory) in the federal system for many years after their creation, at present they are only advisory. By *presumptive* we mean that the sentence indicated by the guideline must be imposed unless there are compelling reasons for not following it. *Advisory* guidelines are used simply to guide the judges' decisions by providing a uniform set of standards for them to consult if they wish. Recall from Chapter 1 that a major concern of the Classical School of criminal justice was to make the law more fair and equal by removing a great deal of judicial discretion and providing standards set by the legislature for making punishment for equal crimes the same. According to Lubitz and Ross (2001), sentencing guidelines have achieved a number of outcomes consistent with this classical ideal and with Aristotle's definition of justice. These outcomes include

1. A reduction in sentencing disparity.

2. More uniform and consistent sentencing.

3. A more open and understandable sentencing process.

4. Decreased punishment for certain categories of offenses and offenders and increased punishment for others.

5. Aid in prioritizing and allocating correctional resources.

6. Provision of a rational basis for sentencing and increased judicial accountability.

Figure 4.2 is a sentencing guideline from Ohio. As you can see, it takes into consideration many more factors than the seriousness of the offense and prior record and leaves quite a bit of room for subjective judgment, especially in the culpability, mitigation, and credits section. How would you complete this guideline for the Joan Place case outlined in our example PSI? What degree of culpability/mitigation would you assign her, and what credits would you give her?

❖ The Future of Sentencing Guidelines

As useful as guidelines have proven to be for reducing sentencing disparity and curtailing judicial discretion, their future format and function is by no means assured. As

Defendant's Name _____   _____   Case No. _____

Figure 4.2

Example of a Comprehensive Sentencing Guideline

OFFENSE RATING

1. Degree of Offense

Assess points for the one most serious offense or its equivalent for which offender is being sentenced, as follows: 1st felony = 4 points; 2nd felony = 3 points; 3rd felony = 2 points; 4th felony = 1 point.   ____

2. Multiple Offenses

Assess 2 points if one or more of the following applies: (A) Offender is being sentenced for two or more offenses committed in different incidents; (B) offender is currently under a misdemeanor or felony sentence imposed by any court; or (C) present offense was committed while offender on probation or parole.   ____

3. Actual or Potential Harm

Assess 2 points if one or more of the following applies: (A) Serious physical harm to a person was caused; (B) property damage or loss of $300 or more was caused; (C) there was a high risk of any such harm, damage or loss, though not caused; (D) the gain or potential gain from theft offense(s) was $300 or more, or (E) dangerous ordnance or a deadly weapon was actually used in the incident, or its use was attempted or threatened.   ____

4. Culpability

Assess 2 points if one or more of the following applies: (A) Offender was engaging in continuing criminal activity as a source of income or livelihood; (B) offense was party, or (C) offense included shocking and deliberate cruelty in which offender participated or acquiesced.   ____

5. Mitigation

Deduct 1 point for each of the following as applicable: (A) There was substantial provocation, justification, or excuse for offense; (B) victim induced or facilitated offense; (C) offense was committed in the heat of anger; and (D) the property damaged, lost, or stolen was restored or recovered without significant cost to the victim.   ____

NET TOTAL = OFFENSE RATING   ____

OFFENDER RATING

1. Prior Convictions

Assess 2 points for each verified prior felony conviction, any jurisdiction. Count adjudications of delinquency for felony as convictions.   ____

Assess 1 point for each verified prior misdemeanor conviction, any jurisdiction. Court adjudications of delinquency for misdemeanor as convictions. Do not count traffic or intoxication offenses, or disorderly conduct, disturbing the peace, or equivalent offenses.   ____

2. Repeat Offenses

Assess 2 points if present offense is offense of violence, sex offense, theft offense, or drug abuse offense, and offender has one or more prior convictions for same type of offense.   ____

3. Prison Commitments

Assess 2 points if offender was committed on one or more occasions to a penitentiary, reformatory, or equivalent institution in any jurisdiction. Count commitments to Ohio Youth Commission or similar commitments in other jurisdictions.   ____

4. Parole and Similar Violations

Assess 2 points if one or more of the following applies: (A) Offender has previously had probation or parole for misdemeanor or felony revoked; (B) present offense committed while offender on probation or parole; (C) present offense committed while offender free on bail; or (D) present offense committed while offender in custody.   ____

5. Credits

Deduct 1 point for each of the following as applicable: (A) Offender has voluntarily made bona fide, realistic arrangements for at least partial restitution; (B) offender was age 25 or older at time of first felony conviction; (C) offender has been substantially law abiding for at least 3 years; and (D) offender lives with his or her spouse or minor children or both, and is either a breadwinner for the family or, if there are minor children, a housewife.   ____

NET TOTAL = OFFENDER RATING   ____

mentioned above, the federal guidelines are now only "advisory," meaning that judges can consult them and follow them or not, which has opened the door once again to unwarranted sentencing discrepancies that guidelines were supposed to rein in. The about-turn began with the separation of responsibilities of the trial judge and the trial jury. The role of judges is to be finders of law; the role of juries is to be finders of facts. A famous case based on these principles came before the U.S. Supreme Court in 2005 (*United States v. Booker,* 2005).

The circumstances of the case are that Freddie Booker was arrested in 2003 in possession of 92.5 grams of crack cocaine. He also admitted to police that he had sold an additional 566 grams. A jury found Booker guilty of possession with intent to sell at least 50 grams, for which the possible penalty ranged from 10 years to life. At sentencing, the judge used additional information (the additional 566 grams and the fact the Booker had obstructed justice) to sentence Booker to 30 years. Booker's sentence would have been 21 years and 10 months based on the facts presented to the jury and proved beyond a reasonable doubt. This was permissible under the guidelines, and 30 years was the minimum guideline sentence Booker could have received. Booker appealed his sentence, arguing that his Sixth Amendment rights had been violated by the judge "finding facts" when this is the proper role of the jury (An earlier federal appeals court had ruled that the facts of prior convictions are the only ones judges can "find" as justification for increasing sentencing.). In other words, anything other than prior record that is used to increase a criminal penalty beyond what the guidelines call for must be submitted to a jury, and proved beyond a reasonable doubt. The Supreme Court agreed with Booker that his sentence violated the Sixth Amendment and sent the case back to District Court with instructions either to sentence Booker within the sentencing range supported by the jury's findings or to hold a sentencing hearing before a jury (Bissonnette, 2006).

The remedial portion of the Court's opinion (What can be done to prevent this happening again?) is much more controversial. The Supreme Court held that the guidelines were to be advisory only, and therefore no longer binding on judges. However, the Court did require them to "consult" the guidelines and take them into consideration, but there is no way of ensuring that judges do just that. John Ashcroft, the U.S. attorney general at the time, called the decision "a retreat from justice," and Congressman Tom Feeney decried "the extraordinary power to sentence" now afforded federal judges who are accountable to no one, and said that the decision "flies in the face of the clear will of Congress" (Bissonnette, 2006, p. 1499). In fact, Booker was resentenced by the same judge to the same 30-year sentence that he originally received. Because the sentencing guidelines had by then become merely advisory, the judge did not have to further justify his sentence since it was within the range of the statutorily defined penalty. The Court's ruling on guidelines only applies to the federal system at present.

❖ The American Correctional Association's Statement on Sentencing

Because of changing sentencing policies (determinate, mandatory minimums, and particularly the policies driven by the "War on Drugs"), there has been a huge increase in the prison population in the United States. According to the American Correctional Association (ACA), sentencing policies should be aimed at controlling crime at the lowest cost to taxpayers, and offenders should be placed in the least restrictive environment consistent with public safety. The ACA strongly promotes and supports any policies that render sentencing fair and rational, and has issued their 2009 official statement on sentencing policy, reproduced below.

In Focus 4.2

Policy Statement of the American Correctional Association Regarding Sentencing

The American Correctional Association actively promotes the development of sentencing policies that should:

A. Be based on the principle of proportionality. The sentence imposed should be commensurate with the seriousness of the crime and the harm done;

B. Be impartial with regard to race, ethnicity, and economic status as to the discretion exercised in sentencing;

C. Include a broad range of options for custody, supervision and rehabilitation of offenders;

D. Be purpose-driven. Policies must be based on clearly articulated purposes. They should be grounded in knowledge of the relative effectiveness of the various sanctions imposed in attempts to achieve these purposes;

E. Encourage the evaluation of sentencing policy on an ongoing basis. The various sanctions should be monitored to determine their relative effectiveness based on the purpose(s) they are intended to have. Likewise, monitoring should take place to ensure that the sanctions are not applied based on race, ethnicity, or economic status;

F. Recognize that the criminal sentence must be based on multiple criteria, including the harm done to the victim, past criminal history, the need to protect the public and the opportunity to provide programs for offenders as a means of reducing the risk for future crime;

G. Provide the framework to guide and control discretion according to established criteria and within appropriate limits and allow for recognition of individual needs;

H. Have as a major purpose restorative justice—righting the harm done to the victim and the community. The restorative focus should be both process and substantively oriented. The victim or his or her representative should be included in the "justice" process. The sentencing procedure should address the needs of the victim, including his or her need to be heard and, as much as possible, to be and feel restored to whole again;

I. Promote the use of community-based programs whenever consistent with public safety; and

J. Be linked to the resources needed to implement the policy. The consequential cost of various sanctions should be assessed. Sentencing policy should not be enacted without the benefit of a fiscal-impact analysis. Resource allocations should be linked to sentencing policy so as to ensure adequate funding of all sanctions, including total confinement and the broad range of intermediate sanction and community-based programs needed to implement those policies.

Source: Quoted in Walsh & Stohr (2010). This Public Correctional Policy was unanimously ratified by the American Correctional Association in 1994, and was reviewed and amended in 1999, 2004, and 2009. Published with permission of the American Correctional Association, Alexandria, Virginia.

Comparative Perspective: Sentencing in China

China is a one-party state with a socialist legal system with few procedural protections. The idea behind the lack of individual rights and procedural limitations is that they only matter when the state and the individual are distinct entities and at odds with one another. In a socialist society, the individual and the state are supposed to be one and the same, and

(Continued)

(Continued)

thus a person does not need protection from the self. According to Lu and Miethe (2002), a defendant mounting a strong defense is not seen as remorseful, so "the stronger the defense the more severe the punishment is likely to result" (p. 271). Penal sanctions in China are characterized by the use of "reeducation" and forced labor camps with the stated function of resocializing inmates to rid them of "politically incorrect" thoughts and behaviors.

Chinese sentences are classified as control, criminal detention, fixed-term imprisonment, life imprisonment, and the death penalty. Control is imposed for minor offenses and is analogous to probation. Offenders under control continue to work but are continually under surveillance by the police and the informal control of neighborhood committees. Public surveillance allows members of "people's mediation committees," which is essentially every adult in China, to monitor the probationer's behavior and report infractions to the police (Walsh & Hemmens, 2011). In traditional societies such as China and Japan, public shaming is a very difficult cross to bear, and sometimes results in suicide.

Criminal detention is analogous to a jail sentence in the United States, where people are deprived of their freedom for a short time for having committed a relatively minor crime. Offenders may be granted permission to go home 1 or 2 days each month and may be paid for work, which makes such a sentence almost like a work release sentence in the United States. The next step up in severity is fixed-term imprisonment, which may range from 6 months to 15 years, and the step after that is life imprisonment. Individuals sentenced to fixed-term sentences or to life imprisonment are subjected to long periods of hard labor as long as they are physically able. Hard labor is thought to be the best way to rehabilitate criminals (Walsh & Hemmens, 2011).

The most serious sentence is one of death. As we have seen, the death penalty is applied quite indiscriminately in China, and can be applied for about 70 different offenses, including murder, rape, economic crimes committed by high-level officials, and "hooliganism." According to Amnesty International (2005), there were at least 3,400 confirmed executions in China in 2004, but sources inside China put the true figure at around 10,000. Adjusting for population size differences, the official figure (3,400) is still over 13 times more executions than occurred in the United State in the same year. There are two types of death sentence: immediate and delayed. A delayed sentence is a 2-year suspension of sentence during which defendants must show that they are reformed. If a person is considered rehabilitated, the sentence is usually changed to a long period of incarceration; if not, he or she is executed. An immediate sentence is carried out within 7 days of imposition of the penalty. Such a sentence is imposed when in the court's opinion the defendant is beyond rehabilitation. Execution is by a single gunshot at the base of the skull.

With a rather arbitrary set of procedures and almost unlimited judicial discretion, it is not surprising that one of the biggest sentencing issues in China is sentencing disparity. With this in mind, the Chinese courts introduced sentencing guidelines in 2008 with the intention of introducing uniformity in sentencing. These guidelines contain the same criteria for determining sentences as American guidelines do, and are presumptive (judges must follow them). Curiously, for the harshest of sentences—life imprisonment and death—judges retain full sentencing discretion (Chen, 2010).

Summary

- Sentencing is a post-conviction process in which the courts implement one or more of the punitive philosophies: retribution, deterrence, incapacitation, or rehabilitation. Sentencing decisions should be in accordance with justice.

- There are three major sentencing models: indeterminate (a range of possible years), determinate (a specific number of years), and mandatory (can exist under either of the above models but means that the person must be sent to prison; probation is not an option). Sentencing to a drug court is becoming increasingly popular in the United States.

- Truth-in-sentencing laws have led to longer sentences, a stronger move to determinate and mandatory sentencing, and to statutes such as habitual offender statutes.

- Sentencing disparity—sentences not accounted for by legally relevant variables—is a major concern in the criminal justice system. A big concern is whether African Americans' more severe sentences are accounted for by their greater involvement in crime or by racism. The sentences imposed for crack versus powder cocaine possession have been a contentious issue because of racial differentials.

■ Efforts have been made to "individualize" justice by providing judges with presentence reports, written by probation officers, which contain many factors about the person the judge is to sentence. A major controversy involving these reports is whether the defense should be able to view them.

■ Sentencing guidelines are designed to eliminate sentencing disparity by submitting a person's crime seriousness and prior record (in some states, with additional information included) to a scoring system. The person is then supposed to be sentenced the same way as every other person who receives the same score.

■ Certain legal problems with sentencing under guidelines moved the U.S. Supreme Court to rule that the federal guidelines, which were previously mandatory, are now merely advisory. This opens up the door once again for wide levels of judicial discretion, and thus for sentencing disparity.

Key Terms

Concurrent sentence

Consecutive sentence

Determinate sentence

Drug court

Fair Sentencing Act

Habitual offender statutes

Indeterminate sentence

Justice

Life without parole (LWOP)

Mandatory sentence

Presentence investigation report (PSI)

Sentence

Sentencing disparity

Sentencing guidelines

Shock probation

Split sentences

Truth-in-sentencing laws

United States Sentencing Commission

Discussion Questions

1. Is it ever just, right, and moral to sentence equals in terms of legally relevant variables unequally? Give an example.

2. If you are being sentenced for a felony, would you prefer to know when your date for parole consideration is to come, or would you prefer an indeterminate sentence where you could possibly "work your way out" and get released earlier?

3. What is your opinion of habitual offender statutes that lock people up for life if convicted of a third felony?

4. What research strategy is required to asses the racial sentencing disparity issue?

5. What are the pros and cons of allowing the defense access to the presentence investigation report? Where do you stand on the issue?

6. Sentencing guidelines were designed to rein in excessive judicial sentencing discretion, and most criminal justice personnel consider this a very good thing. So why did the U.S. Supreme Court throw a wrench into the works by making the federal guidelines advisory only?

Useful Internet Sites

The Sentencing Project: www.sentencingproject.org

United States Sentencing Commission: www.ussc.gov

Vera Institute of Justice: www.vera.org

CHAPTER 5

Jails

❖ Introduction: The Community Institution

The American jail is a derivative of various modes of holding people for trial that have existed in Western countries for centuries. Whether fashioned from caves or mines or old houses or as separate buildings, jails were developed originally as a primary means of holding the accused for trial, for execution, or in lieu of a fine. As was noted in Chapter 2, jails were called gaols in the England of the Middle Ages and were operated by the *shire reeve*, or sheriff, and his minions.

Jails have been in existence much longer than prisons and their mission is much more diverse, especially now: These days, jails are usually local and community institutions that hold people who are presumed innocent before trial; they hold convicted offenders before they are sentenced; they hold more minor offenders who are sentenced for terms that are usually less than a year; they hold juveniles (usually in their own jails or separated from adults or before transport to juvenile facilities); they hold women (usually separated from men and sometimes in their own jails); they hold people for the state or federal authorities; and, depending on the particular jail population being served and the capacity of any given facility, they serve to incapacitate, deter, rehabilitate, punish, and reintegrate.

Though described as correctional afterthoughts by scholars, and despite their multifaceted and critical role in communities, jails have often received short shrift in terms of monetary support and professional regard (Kerle, 1991, 2003; Thompson & Mays, 1991; Zupan, 1991). The vast majority of jails are operated by county sheriffs whose primary focus has been law enforcement rather than corrections. As a result, jail facilities have often been neglected, resulting in dilapidated structures, and jail staffs with less training and pay than probation and parole officers or correctional staffs working at the state or federal level in prisons. Jail staffs also often receive less pay and training than deputy sheriffs working in the same organization (sheriff's office) as the jail. The late comic Rodney Dangerfield's perennial lament "I (they) don't get no respect" surely applies to jails more than perhaps any other social institution.

In this chapter, we discuss how this forgotten social institution fulfills a vital community role, one that includes all of the functions described in the preceding paragraph, as well as serving as a repository for people who are only nominally criminal and have nowhere else to go (e.g., the homeless or the mentally ill). The role of jails also includes the holding of some state or federal inmates, as prisons are too full. In some larger counties, the holding of longer-term sentenced inmates or those who have numerous physical, mental, and substance abuse problems—not to mention educational deficits—has led to more programming and treatment in jails. Part and parcel of this interest in treatment is the emergence of community reentry programs, as will also be discussed in the chapter on parole (Chapter 8), as a means of preventing crime and addressing the multifaceted needs of jail ex-inmates. In this chapter, these emerging trends will be explored, as will the challenges jails face, but first we will discuss the types of institutions that constitute jails.

❖ Jail Types

The typical jail is operated by the sheriff of a county. However, some cities, states, and the federal government operate jails, and sometimes multiple jurisdictions combine resources to administer a jail that serves a region. Some American Indian tribes have their own jails, and many police departments have short-term lockup facilities to hold suspects or those accused of crimes. Currently, there are about 2,900 jails in the United States, and 68 jails are operated by American Indian tribes (Minton, 2007; Sabol & Minton, 2008). When a state or the federal government, or another governmental entity, has inmates for a jail but no facility of its own in a given vicinity, the entity will typically ask the county to hold that inmate. Counties are usually more than willing to do this, as they are paid a fee that often exceeds the cost of holding inmates, which makes holding inmates for other jurisdictions a money-making enterprise.

Most jails are composed of one or two buildings in close proximity to each other. They are usually operated somewhat close to a city or town center, except when located on reservations or at military facilities.

Many jails have adopted technological changes that have greatly enhanced their ability to supervise and control inmates. The use of cameras, voice and visual check–operated doors by a control center, electronic fingerprint machines, and even video visiting are revolutionizing the jail experience. Certainly these changes are making the facility more secure, but also, in the case of video visiting, they may make it easier to maintain contact with the outside.

Photo 5.1

Exterior of the Travis County Jail, Texas, a typical large urban jail

Photo 5.2

Interior of a typical jail cell

❖ Jail Inmates and Their Processing

Jails operate 7 days a week, 24 hours a day, as crime does not take a holiday. They hold all kinds of inmates, from the serious convicted offender awaiting transport to a state or federal prison, down to the accused misdemeanant who cannot make bail. About 60% of jail inmates have not been convicted of the crime for which they are being held (Minton, 2010). Jails receive inmates from local, state, federal, and tribal police officers. They process about 13 million inmates every year, with most inmates in and out within a few days or a week, some within hours, though others might be held for more than a year if they are sentenced state or federal inmates (Minton, 2010). Because of their complicated and diverse role, and as a means of keeping track of the inmates they are responsible for holding, jails will often follow a set procedure that is prescribed by both tradition and practice.

The first part of the typical processing of an inmate at a county or city jail is the delivery of the arrestee to the facility by a law enforcement officer. As is discussed in the following, many arrestees may be stressed, upset, mentally disturbed, or intoxicated. In the latter case, the officer may choose to administer a breathalyzer test at the jail. If the arrestee is injured, the jail booking staff may require that the arrestee be taken by law enforcement to the hospital to be checked out before he or she is admitted to the jail.

If not injured, the law enforcement officer will fill out the paperwork for admittance of the arrestee to the facility. Usually the arrestee is still with the officer when this is occurring and often still in handcuffs. Once the required paperwork and processing are completed, the jail will accept the arrestee, search him or her, and begin its own paperwork for admitting the arrestee. At this juncture, and depending on the alleged offense, the arrestee may be allowed to contact family and friends and/or a bail bondsman. The arrestee might be released directly into the community if the alleged offense is minor. However, if the alleged offense is serious enough, the arrestee will need to await arraignment by a judge to determine bail and in the interim is booked into the jail.

During the booking process, jails will strip search arrestees (now inmates), take their property, and issue clothing and other essentials. If the new inmate is intoxicated or belligerent, booking staff may place him or her in a special holding cell. In the latter case, this might involve a padded room or a restraint chair. Once the inmate is sober and calm, he or she is then classified and moved to a more permanent housing area in the jail. Larger jails often keep new inmates in a separate area or cell before they place them in a general housing unit so that they can be observed and classified (based on the inmate's alleged offense, alleged criminal coconspirators, criminal history, gang involvements, health and other needs, etc.).

❖ Overcrowding

As indicated in other sections of this book, jails have to deal with the same kinds of overcrowding issues that have afflicted prisons. **Overcrowding** occurs when the number of inmates exceeds the physical capacity (i.e., the beds and space) available. Each year, and over the last several decades, the number of jail beds needed by jurisdictions has increased, and they have been filled almost as soon as they have been built (Minton, 2010). Between 2008 and 2009, there was an unprecedented decrease in jail inmates of 1.1%. As of midyear 2009 (the latest data available when this book was published), on average, jails were operating at 90% of their capacity and the highest capacity for the decade (2000–2009) was achieved in 2006 and 2007 at 96% (Minton, 2010, p. 5). This percentage use of capacity is actually better than in past years when jails of the 1980s and 1990s were operating at well over their rated capacity (Cox & Osterhoff, 1991; Gilliard & Beck, 1997; Klofas, 1991). Also, and notably, even an average of 90% for 2009 means that many of the jails in the United States are operating at over that average.

The percentages of capacity can be misleading when one considers overcrowding. Certain sections of jails are designated for specific types of offenders that cannot or do not mix well (e.g., males and females, but also juveniles, trustees, inmates with medical problems, etc.). The percentage capacity may indicate that the jail is not completely full, but any given section might be overwhelmed with inmates.

Photo 5.3

Inmates in the Reception Housing Area of a California State Prison. California, like many states has suffered from severe overcrowding in recent decades.

Such overcrowding limits the ability of the jail to fulfill its multifaceted mission: Less programming can be provided, health and maintenance systems are overtaxed, and staff are stressed by the increased demands on their time and the inability to meet all inmate needs. From the inmates' perspective, their health, security, and privacy are more likely to be threatened when the numbers of inmates in their living units increase and the amount of space, and possibly the number of staff, does not. The jail staff also lose their ability to effectively classify and sometimes control inmates; they may be unable to keep the serious convicted offenders away from the presumed innocent unconvicted, or more minor-offending inmates. Judges and jail managers will struggle over how to keep the jail population down to acceptable limits and as a result, even serious offenders may be let loose into communities as a means of reducing the crowding. Therefore, though the "get tough" laws in many states were passed with the explicit intent of incarcerating more people for longer, their actual unintended effect in some jails may be to incarcerate serious offenders less (as there is no room) and all offenders in less safe and secure facilities.

Though suits by jail inmates are usually not successful, some are. Welsh (1995) found in his study of lawsuits involving California jails that the issue that courts gave greatest credence to was overcrowding. Perhaps this is because overcrowding is clearly quantifiable (the rated capacity is clear and the inmate count is obvious), but it is likely that it was regarded as so important by courts because it can lead to a number of other seemingly intractable problems, such as those just mentioned.

❖ Gender, Juveniles, Race, and Ethnicity

As indicated from the data supplied in Table 5.1, most jail inmates are adult minority males, though the number of whites represents the largest racial grouping of the men and the number of whites as a proportion of the total men increased slightly from 2000 to 2009. Women comprised over 12.2% of jail inmates in 2009, which is more than in 2000 (11.4%), but less than in 2006 and 2007 (12.9%). The reason often cited for the overall increases in incarceration in jails and prisons and the increases for women and minorities in jails and prisons, in particular, has been the prosecution of the drug war since the 1980s and 1990s. The "get tough" policies, which have led to longer periods of incarceration in prisons, have also led to a greater propensity to catch and keep low-level drug offenders in jails (Irwin, 2005; Owen, 2005; Welch, 2005; Whitman, 2003). The focus of arrests in the drug war has often been on the low-level sellers, rather than the buyers or the drug kingpins, and that has netted more minorities and women into the system. Mandatory sentences, juvenile waivers, and sentence enhancements for certain offenses have collectively led to longer sentences for most offenders and backed up numbers of offenders in some jails either awaiting transfer to state or federal prisons or doing their time in the jails rather than in the overcrowded prisons.

It is not clear why there have been recent declines in the numbers of women and minorities, vis-à-vis men and whites, incarcerated in jails, a particularly notable phenomenon in large city jails

Photo 5.4
A female jail inmate awaits her cell assignment. Women comprised over 12.2% of jail inmates in 2009.

(Minton, 2010). It could just be a minor shift, which will not become a trend, or it could signal a longer-term change in the use of jails due to the recession of 2007–2010, a rethinking in the prosecution of the drug war, or some other variable not yet identified by researchers. Longer-term trends do indicate that the number of adult males in jail from 1990 to 2006 almost doubled, while the numbers of adult females and juveniles almost tripled. Percentage increases for women and juveniles are also large: In 1990, women represented only about 9% of jail populations and juveniles about 0.6%, whereas by 2000, women comprised 12.2% and juveniles 1.0% of jail populations (Bureau of Justice Statistics, 1998, 2007; see also Table 5.1).

Across the two largest racial groupings (whites and African Americans) and the largest ethnic grouping (Hispanics), there have been significant increases in jail incarceration. The raw number of whites has increased from 1990 (when there were fewer whites incarcerated in jails than African Americans). Proportionate to their representation in the population, however, African Americans are much more likely to be incarcerated in American jails than are whites or Hispanics. As reported by the Bureau of Justice Statistics for 2006 (2008), "Blacks were almost three times more likely than Hispanics and five times more likely than whites to be in jail" (p. 2). Again, this higher proportional rate of incarceration for African Americans in particular can likely be attributed to their greater concentration in impoverished neighborhoods and the focus of the drug war that has tended to target such living areas and the selling and use of crack cocaine (see the discussion of enhanced sentences for crack cocaine in Chapter 4).

❖ The Poor and the Mentally Ill

The late corrections scholar John Irwin (1985) once referred to the types of people who are managed in jails as the "rabble," by which he meant "[d]isorganized and disorderly,

-Detached
-of disrepute

Table 5.1

Percentage of inmates in local jails, by characteristics, midyear 2000 and 2005–2009

Characteristic		2000	2005	2006	2007	2008	2009
Sex							
	Male	88.6%	87.3%	87.1%	87.1%	87.3%	87.8%
	Female	11.4	12.7	12.9	12.9	12.7	12.2
Adults		98.8%	99.1%	99.2%	99.1%	99%	99.1%
	Male	87.4	86.5	86.3	86.3	86.4	86.9
	Female	11.3	12.6	12.9	12.8	12.6	12.1
Juveniles[a]		1.2%	0.9%	0.8%	0.9%	1%	0.9%
	Held as adults[b]	1.0	0.8	0.6	0.7	0.8	0.8
	Held as juveniles	0.2	0.1	0.2	0.2	0.2	0.2
Race/Hispanic origin[c]							
	White[d]	41.9%	44.3%	43.9%	43.3%	42.5%	42.5%
	Black/African American[d]	41.3	38.9	38.6	38.7	39.2	39.2
	Hispanic/Latino	15.2	15	15.6	16.1	16.4	16.2
	Other[d,e]	1.6	1.7	1.8	1.8	1.8	1.9
	Two or more races[d]	—	0.1	0.1	0.1	0.2	0.2
Conviction status/[b]							
	Convicted	44%	38%	37.9%	38%	37.1%	37.8%
	Male	39	33.2	32.8	32.9	32.3	33
	Female	5	4.9	5	5.2	4.8	4.8
	Unconvicted	56	62	62.1	62	62.9	62.2
	Male	50	54.2	54.3	54.3	55.2	54.8
	Female	6	7.7	7.8	7.7	7.8	7.4

Note: Detail may not sum to total due to rounding.

[a]Persons under age 18 at midyear.

[b]Includes juveniles who were tried or awaiting trial as adults.

[c]Estimates based on reported data and adjusted for nonresponse.

[d]Excludes persons of Hispanic or Latino origin.

[e]Includes American Indians, Alaska Natives, Asians, Native Hawaiians, and other Pacific Islanders.

— = Data not collected.

Source: Minton (2010).

the lowest class of people" (p. 2). These were not just the undereducated, the under- or unemployed, or even the poor and mentally ill. He meant to include all those descriptors as they related to the state of being disorganized and disorderly and as those designations

*Greater risk
for police
conduct*

might lead to permanent residence in a lower class, but he also meant that jail inmates tend to be "detached" and of "disrepute" in the sense that they offend others by committing mostly minor crimes in public places.

Certainly, the fact of being homeless puts that person at a greater risk for negative contact with the police; lacking a home, private matters are more likely to be subject to public viewing in public spaces, and this disturbs or offends some community members, which leads to police involvement. Those who are mentally ill are more likely to be homeless, as they are unable to manage the daily challenges that employment and keeping a roof over one's head and food in one's mouth require (McNiel, Binder, & Robinson, 2005; Severson, 2004).

Jails in the United States are full of the mentally ill, the homeless, and the poor. Recent data from the Bureau of Justice Statistics (BJS) (based on interviews of local jail inmates in 2002) indicates that about 64% of jail inmates (75% of females and 63% of males) have a mental health problem (as compared to 56% of state prisoners and 45% of federal prisoners) (James & Glaze, 2006, p. 1). In contrast, about 10.6% of the United States population has symptoms of mental illness. Moreover, for virtually every manifestation of mental illness, more jail inmates than state or federal prisoners were likely to exhibit symptoms, including 50% more delusions and twice as many hallucinations (James & Glaze, 2006, p. 2). Of those jail inmates with a mental health problem, the specific diagnosis included mania (54%), major depression (30%), and a psychotic disorder (24%). The specific identification of a mental illness for each inmate by the BJS research team was based on a recent clinical diagnosis or symptoms that fit the criteria of the *Diagnostic and Statistical Manual of Mental Disorders* (DSM-IV).

A whole host of problems have been found to be associated with mental illness including homelessness, greater criminal engagement, prior abuse, and substance use (McNiel et al., 2005). Among the findings from this BJS study of jails was that those with a mental illness were almost twice as likely to be homeless as those jail inmates without a mental illness designation (17% as opposed to 9%) (James & Glaze, 2006, pp. 1–2). More inmates with a mental health problem had prior incarcerations than those without such a problem (one-quarter as opposed to one-fifth). About 3 times as many jail inmates with a mental health problem had a history of physical or sexual abuse than those without such a problem (24% as opposed to 8%). Almost three-quarters of the inmates with a mental health problem were dependent on, or abused, alcohol or illegal substances (74% as opposed to 53% of those without a mental health problem). In short, mental illness, along with poverty, was entangled in a whole array of societal issues for jail inmates.

Further evidence for this supposition was found by McNiel et al. (2005) in their study in San Francisco County. They found that mental illness, substance abuse, and jail incarcerations were inextricably connected as life events. Those who were mentally ill and homeless were also more likely to have a substance abuse problem, and it was also likely for this population that jail incarcerations were part of their existence as well.

❖ Medical Problems

One of the social issues that is particularly problematic for jail inmates, and the people who manage them, is the relatively poor health of people incarcerated in jails (2007). According to the same 2002 study of jail inmates by the Bureau of Justice Statistics, more than a third of jail inmates, or 229,000 people, reported a medical problem more serious than a cold or the flu (cited in Maruschak, 2006, p. 1). Most of these medical maladies preceded placement in jail and included (in order of prevalence) arthritis, hypertension, asthma, heart problems, cancer, paralysis, stroke, diabetes, kidney problems, liver problems, hepatitis, sexually transmitted diseases, tuberculosis, and HIV. A small percentage of inmates (2%) were so medically impaired that they needed to use a cane or a walker or a wheelchair.

As one might expect, the elderly are much more prone to some of these medical maladies than would be younger inmates. In the BJS study, 61% of those over 45 reported a medical problem (Maruschak, 2006, p. 1). With the exception of asthma and HIV, which tended to be more prevalent among younger inmates, the older inmates were much more likely to have the other medical problems tallied in this report; which means that older inmates are more costly to manage in jails because of their greater need for medical care.

Like the older inmates, women were much more likely to report medical problems to the BJS researchers (53% for women as opposed to 35% for men) (Maruschak, 2006, p. 2). They reported a rate of cancer that was almost 8 times that of men (831 per 10,000 inmates, compared to 108 per 10,000 inmates), with the most common type of cancer being cervical for women and skin for men. In fact, of every medical problem documented in the study, the women reported more prevalence than the men, with the exception of paralysis (where they were even with men) and tuberculosis where a slightly greater percentage of men reported more (4.3% for men as opposed to 4.0% for women) (Maruschak, 2006, p. 2).

Incarcerated youth have their own set of potentially debilitating health problems that also present an immediate health risk to communities. In a study of adolescents in a juvenile detention center in Chicago, about 5% of the teens had contracted gonorrhea and almost 15% had chlamydia (Broussard et al., 2002, p. 8). Girls were over 3 times more likely to have one of these diseases than were boys in this study.

According to the 1976 Supreme Court *Estelle v. Gamble* case, inmates have a constitutional right to reasonable medical care. The court held that to be deliberately indifferent to the medical needs of inmates would violate the Eighth Amendment prohibition against cruel and unusual punishment. Needless to say, treating such problems requires that a jail of any size have budgetary coverage for the salaries of nurses; a contract with a local doctor, mental health provider, and dentist; and an arrangement with local hospitals. Moreover, regular staff need basic training in CPR and other medical knowledge (e.g., to know when someone is exhibiting the symptoms of a heart attack or stroke or the symptoms of mental illness), so that when problems arise, they recognize how serious it might be and know how to address it or whom to call (Kerle, 2003; Rigby, 2007).

Some jails are addressing these issues by contracting with private companies to provide medical services or using telemedicine as a means of delivering some services. The National Commission on Correctional Health Care recommends that should jails go the route of private provision of services, they make sure that such programs are properly accredited so that the services provided meet national standards (Kerle, 2003). When such matters as obtaining/maintaining quality care are not attended to, as is sometimes the case in jails and prisons (Vaughn & Carroll, 1998; Vaughn & Smith, 1999)**,** jail inmates are likely to suffer the consequences in terms of continued poor health (Sturgess & Macher, 2005). In addition, jails may be sued for failure to provide care, and communities might be exposed to contagious diseases, along with the legal bills (Clark, 1991; Macher, 2007; Rigby, 2007). Clearly, the provision of decent health care to incarcerated persons is important not just because the Supreme Court mandates it, or because it is the moral thing to do for people who are not free to access health care on their own, but because the vast majority of jail inmates return to the community, most within a week or two (Kerle, 2003). Therefore, to prevent the spread of diseases and to save lives both inside and outside of jails, basic medical care would appear to be called for. Some jails are evidently expending energy to address this area of incarceration, as 4 in 10 of the inmates in the 2002 BJS study reported that they had had a medical exam since their admission (Maruschak, 2006, p. 1).

❖ Substance Abuse and Jails

It is one of those oft-cited assumptions that people in prisons and jails have substance abuse problems, but this is one area of social commentary that actually fits social reality. According to a 2002 BJS study of jail inmates (the latest available data at the time of writing), fully 68% of jail inmates reported substance abuse or dependence problems (Karberg & James. 2005, p. 1). In fact, half of convicted inmates reported being under the influence at the time they committed their offense, and 16% said they committed the crime to get money for drugs. Female and white inmates were both more likely to report usage at the time of the offense (Karberg & James, 2005, p. 5). For convicted offenders who used at the time of offense, alcohol was more likely to be in their system than drugs (33.3% for alcohol as opposed to 28.8% for drugs). The drugs of choice for abusers and users varied and included by prevalence of use marijuana, cocaine or crack, hallucinogens, stimulants (including methamphetamines), and inhalants (Karberg & James, 2005, p. 6). Not surprisingly, those who reported a substance abuse problem were also more likely to have a criminal record and to have been homeless before incarceration. M. D. White, Goldkamp, and Campbell (2006) found in their study conducted in New Mexico that many people who are arrested and subsequently come into contact with the local jail, have "co-occurring disorders" such as mental illness and substance abuse problems (p. 303).

Photo 5.5

Inmates at Metro-Davidson County Detention Facility (Nashville, TN) participate in the LifeLine Substance abuse program. Prisoners pass a stuffed animal around—giving it to someone who has helped him.

Violent offenders were more likely to use alcohol than other substances at the time of the offense. But violent offenders were also least likely, with the exception of public order offenders, to report being on drugs or alcohol at the time of the offense (Karberg & James, 2005, p. 6).

Fully 63% of those with a substance abuse problem had been in a treatment program before (Karberg & James, 2005, p. 1). Most such programs were of the self-help variety such as Alcoholics Anonymous or Narcotics Anonymous. However, 44% of these people had actually been in a residential treatment program or a detoxification program or had received professional counseling or had been put on a maintenance drug (Karberg & James, 2005, p. 8). Treatment for convicted offenders in jails, as of 2002, was at 6%. Notably, provision of treatment in jails is difficult because most inmates are out of the facility within a week and about 60% are unconvicted, so as people who are "presumed innocent," they cannot be coerced into getting treatment. Therefore, treatment programs are usually focused on those who meet all of the following criteria: They have a substance abuse problem, they are convicted, and they are longer-term inmates. Even having said this, the amount of treatment programming in jails does not fit the obvious need (Kerle, 2003).

❖ Suicides and Sexual Violence in Jails

Suicides

As indicated from the data presented above, those incarcerated in jails often enter them at some level of intoxication. Moreover, many have a mental disability, and if this is their first experience with jail, it might be exacerbated by the shock of incarceration. Most who are booked into jails are impoverished. Also being booked itself may represent both the mental and physical lowest point of their lives. Such a combination of conditions may predispose some jail inmates to not just contemplate suicide, but to attempt it (Winfree & Wooldredge, 1991; Winter, 2003).

In 1986, the National Center on Institutions and Alternatives (NCIA) did a study of suicides in jails. Twenty years later, in 2006, the National Institute of Corrections funded another NCIA study of the status of jail suicides. Based on 464 suicides that occurred in 2005 and 2006, the NCIA published the following findings regarding suicide victims in jails and characteristics of the suicides (Hayes, 2010, p. xi):

- Sixty-seven percent were white.
- Ninety-three percent were male.
- The average age was 35.
- Forty-two percent were single.
- Forty-three percent were held on a personal and/or violent charge.
- Forty-seven percent had a history of substance abuse.
- Twenty-eight percent had a history of medical problems.
- Thirty-eight percent had a history of mental illness.
- Twenty percent had a history of taking psychotropic medication.
- Thirty-four percent had a history of suicidal behavior.
- Deaths were evenly distributed throughout the year; certain seasons and/or holidays did not account for more suicides.

- Thirty-two percent occurred between 3:01 p.m. and 9 p.m.
- Twenty-three percent occurred within the first 24 hours, 27% between 2 and 14 days, and 20% between 1 and 4 months after incarceration.

These data indicate that the profile of the suicide-prone inmate in jail is that of someone who is male, white, younger (though the BJS data indicate both younger and older inmates are prone to committing suicide), in jail on a violent offense charge, with a history of substance abuse, and at the beginning of his or her jail incarceration. Other data, from the BJS and other sources, flesh out and contextualize these findings.

Data obtained by the Bureau of Justice Statistics in a 2-year study (2000–2002) of deaths while in custody also suggest that age, gender, and race are important variables in predicting suicide, along with jail size (Mumola, 2005). White males under 18 and over 35 and those inmates with a more violent commitment history were more likely to commit suicide than African American males or those in other age groups and who were not incarcerated for a violent offense (Mumola, 2005; Winter, 2003). Winter found in her study of 10 years of suicide data from jails in one midwestern state that those who committed suicide tended to be younger, were arrested for a violent offense, had no history of mental or physical illness, did not necessarily "exhibit suicidal tendencies," and were more likely to be intoxicated with alcohol when admitted (p. 138).

Moreover, according to the authors of the BJS study, the suicide rate at large, primarily urban jails, which tend to hold fewer whites, was about half that of the smaller jails (cited in Mumola, 2005). Similarly, in a study by Tartaro and Ruddell (2006), the researchers also found that smaller jails (with less than a 100-bed capacity) had a 2 to 5 times greater prevalence of attempted and completed suicides than larger jails did (p. 81). In this study, crowded jails and those with "special-needs and long-term inmates" were also more likely to have a higher suicide completion. The shock of incarceration may be one explanation for jail suicide rates, although why this shock might be greater for those in smaller jails is not entirely clear. The BJS and NCIA data do indicate that about half of the suicides occur within the first 9 days—for women it was 4 days—and in the cell of the person committing the suicide (Mumola, 2005).

Larger jails, with their greater resources and higher level of training for staff, may be better equipped than their smaller counterparts to monitor and prevent suicides in their facilities. For instance, if the younger inmates are fearful of being housed with, and possibly abused by, adults, some less crowded and perhaps larger jails may have the luxury of segregating young men from older men and thereby lessening the fear that might precipitate some suicides. Winter's (2003) conclusion, after studying 18 years worth of administrative data on suicides in a midwestern jail, is that keeping and accessing more complete records regarding suicides is critical to preventing them. It is possible that larger, more urban jails are better able to handle this responsibility. Relatedly, in their comparison study of rural and urban jails, Applegate and Sitren (2008) remarked on the greater capacity of urban jails, relative to rural jails, to provide services to inmates, which one assumes would directly and indirectly affect the rate of suicides in these jails.

We do know that the rate of suicide among inmates in jails, despite its marked decrease over the last 9 years, is still twice as high as would be true for a comparable group of free citizens (Mumola, 2005). Jails have 3 times the rate of suicides that prisons do, though their homicide rates are comparable (Mumola, 2005). The good news, however, is that jail and prison deaths due to suicide (and homicide) have declined precipitously from 1983 to 2002, with the rate of prison suicides declining by half during this time period and jail suicides by almost two-thirds (Mumola, 2005, p. 2). In 1983, jail suicides accounted for the major cause of death for inmates, but by 2002, illness had replaced suicide as the primary reason for death.

Sexual Violence

The **Prison Rape Elimination Act of 2003** mandated that the BJS collect data on sexual assaults in adult and juvenile jails and prisons and that it identify facilities with high levels of victimization. According to the National Inmate Survey for 2008 and 2009 (which included 286 local jails and was conducted by BJS researchers), 3.1% of jail inmates (as opposed to 4.4% of prison inmates) reported experiencing sexual victimization perpetrated by other jail inmates or staff in the previous 12 months (Beck & Harrison, 2010, p. 1). Extrapolating these sample findings to the national population of jail and prison inmates, the BJS researchers estimate that fully 88,500 inmates in prisons and jails experienced sexual victimization during this time period (Beck & Harrison, 2010, p. 2). For jails, it was estimated that there were 24,000 victims, 11,600 inmate on inmate (6,000 nonconsensual) and 15,800 staff on inmate (11,400 "unwilling," as sexual contact between staff and inmates is legally nonconsensual) (Beck & Harrison, 2010, p. 2).

Female inmates in jails (as well as in prisons) were more than twice as likely as male inmates to experience sexual victimization perpetrated by another inmate (3.1% for females vs. 1.3% for males) (Beck & Harrison, 2010, p. 1). Male inmates in jails and prisons were the more likely victims of staff perpetrators and most of those staff were females. There was higher victimization of whites by other inmates and blacks by staff, and generally younger inmates were targeted by both inmates and staff for sexual victimization, though this was not necessarily true for inmate-on-inmate victimization in prisons. Lesbian, gay, and bisexual inmates were much more likely to be victimized by both staff and inmates in both prisons and jails. Among those victimized in the jails, 19% of the men, but only 4% of the women, were victimized in the first 24 hours (Beck & Harrison, 2010, p. 1). Most of the victimization occurred after 6 p.m.

In an earlier study by BJS researchers of sexual victimization in correctional institutions, the data, summarized here, were collected in different, but in many instances comparable, ways for 2004, 2005, and 2006 (Beck, Harrison, & Adams, 2007). Data were obtained from administrative records, surveys, and interviews with current and former inmates and from all state departments of corrections, the Federal Bureau of Prisons, and a sample of jails.

Based on just these 3 years of data, which is really not enough to establish a trend, the amount of allegations of sexual violence in all adult correctional institutions increased to almost 3 (2.91) per 1,000 inmates in 2006 from 2.46 per 1,000 inmates in 2004 (Beck et al., 2007, p. 3). Most of these alleged incidents occurred at night in the victim's cell and involved the use or threat of force. But notably, most of these allegations were not substantiated, nor were they investigated or found to be supported by evidence, by prison or jail officials. Having said this, however, we should recognize that in most such instances of sexual violence it would be very difficult to find evidence, as it is in the free world, particularly if the one perpetrating the victimization was a staff member, which, of course, is why the inmate survey data presented first in this section becomes so important.

In the above study, prison inmates were more likely to report sexual violence allegations than were jail inmates (Beck et al., 2007). Moreover, about 49% of the time, inmates alleged that staff perpetrated the sexual violence against inmates (36%) or sexually harassed the inmate (13%) (Beck et al., 2007, p. 4). Conversely, it was alleged that inmates were involved in the sexual victimization of other inmates roughly 51% of the time.

A total of 967 incidents of sexual violence in all correctional institutions were substantiated in 2006, whereas 885 were in 2005 (Beck et al., 2007, p. 4). Rates of substantiated incidents were lowest in federal and privately operated prisons. This finding might mean that there was less victimization in the federal and private prisons, or it might mean that they were less vigorous in investigating and thus substantiating it. The Inmate

Survey (Beck & Harrison, 2010), however, validated this finding of less victimization in federal prisons. Female staff were more likely perpetrators in prisons, though we know from research on Texas prisons that when the offense was actual sexual battery, the staff offender was more likely to be male (Marquart, Barnhill, & Balshaw-Biddle, 2001). Male staff were more likely the perpetrators of sexual violence in jails. For instance, in a recent (2007) case involving the Yuma County, Arizona, jail, three male officers were charged with unlawful sexual conduct with three female inmates (Reuffer, 2007).

Notably, based on the earlier BJS data, in 57% of the substantiated incidents in 2006, the sexual relationship "appeared to be willing" (Beck et al., 2007, p. 6). Some states and localities maintain, however, that if there is sexual involvement between staff and inmates in a correctional institution, it is by definition "unwilling," as incarcerated people are powerless and vulnerable, compared to their keepers, and so do not have the ability to say "no" or to give consent. The BJS researchers report the following findings regarding victims and perpetrators of sexual violence in correctional institutions (Beck et al., 2007, pp. 7–8):

- In State and Federal prisons, 65% of inmate victims of staff sexual misconduct and harassment were male, while 58% of staff perpetrators were female.

- In local jails, 80% of victims were female, while 79% of perpetrators were male.

- 49% of staff perpetrators in prisons were age 40 or older, while 65% of victims were under age 35.

- 56% of staff perpetrators in jails were age 40 or older, while 86% of victims were under age 35.

- Among staff perpetrators in prisons and jails, 71% were white; 20%, black; and 7%, Hispanic. Among inmate victims, 66% were white; 23%, black; and 8%, Hispanic.

- A correctional officer was identified as the perpetrator in 54% of incidents in prisons, and in 98% of incidents in jails.

- Three-quarters of staff perpetrators in 2006 lost their jobs; 56% were arrested or referred for prosecution.

- Half of inmates involved in staff sexual misconduct were transferred or placed in segregation.

- In most incidents of staff sexual misconduct or harassment (76%), victims received no medical follow-up, counseling or mental health treatment.

❖ Innovations in Jails: New Generation/Podular Direct Supervision Jails, Community Jails, Co-Equal Staffing, Reentry Programs for Jails

New Generation/Podular Direct Supervision Jails

In the 1980s, a new kind of jail was under construction in the United States, then called a **new generation jail** and now known as a **podular direct supervision jail**. Its two key components included a rounded or "podular" architecture for living units and the "direct," as opposed to indirect or intermittent, supervision of inmates by staff; in other words, staff were to be in the living units full time (Applegate & Paoline, 2007; Farbstein & Associates, 1989; Gettinger, 1984; Zupan, 1991). It was believed that the architecture would complement the ability to supervise, and the presence of staff in the living unit would negate the ability of inmates to control those units. Other important facets of these jails are the provision of more goods and services in the living unit (e.g., access to telephones, visiting booths, recreation, library books) and the more enriched leadership and communication roles for staff.

Not surprisingly, several scholars recognized that the role for the correctional officer in a podular direct supervision jail would have to change. Zupan (1991), building on the work of Gettinger (1984), identified seven critical dimensions of new generation jail officer behavior: (1) proactive leadership and conflict resolution skills; (2) building a respectful relationship with inmates; (3) uniform, and predictable, enforcement of all rules; (4) active observation of all inmate doings and occurrences in the living unit; (5) attending to inmate requests with respect and dignity; (6) disciplining inmates in a fair and consistent manner; and (7) being organized and in the open with the supervisory style. Whether officers in podular direct supervision jails are always adequately selected and trained to fit these dimensions of their role is as yet an open research question (Applegate & Paoline, 2007; Nelson & Davis, 1995; Wener, 2006).

Photo 5.6

A female correctional officer operates a new generation jail control pod.

New generation jails, though hardly "new" anymore, became wildly popular in the United States by the late 1980s and through the 1990s (Kerle, 2003; Wener, 2005). Reportedly, in the 21st century, about one-fifth of medium and larger jails are said to be new generation facilities (Tartaro, 2002). Their architecture, though not all features of such jails, can be seen in most new jails and prisons built these days, whether or not they include direct supervision.

It is widely acknowledged by correctional scholars and practitioners that though podular direct supervision jails or prisons are not necessarily a panacea for all that ails corrections today (i.e., crowding, few resources, etc.), they often do represent a significant improvement over more traditional jails (Kerle, 2003; Perroncello, 2002; Zupan, 1991). If operated correctly, and including all of the most important elements, they are believed to be less costly in the long run (due to fewer lawsuits), be safer for both staff and inmates, provide a more developed and enriched role for staff, and include more amenities for inmates. This is a big *if*, however, and some research has called these claims of a better environment for inmates and staff and a more enriched role for staff into question, as the implementation of the new generation model has sometimes faltered or been incomplete in many facilities (Applegate & Paoline, 2007; Stohr, Lovrich, & Wilson, 1994; Tartaro, 2002, 2006). Clearly, more research on new generation jails is called for to determine their success (or failure) in revolutionizing the jail environment for staff and inmates.

Community Jails

Another promising innovation in jails has been the development of **community jails** (Barlow, Hight, & Hight, 2006; Kerle, 2003; Lightfoot, Zupan, & Stohr, 1991). Community jails are devised so that programming provided on the outside does not end at the jailhouse door, as the needs such programming was addressing have not gone away and will still be there when the inmate transitions back into the community. Therefore, in a community jail, those engaged in educational, drug or alcohol counseling, or mental health programming will seamlessly receive such services while incarcerated and again as they transition out of the facility (Barlow et al., 2006; Bookman, Lightfoot, & Scott, 2005; National Institute of Corrections, 2008). So whether one is in and out of the facility within a few days or a few months, needs are met and services provided so that the reintegration into the community is smoother for the inmate and the community in question.

Managers of community jails also recognize that they cannot staff or resource the jail sufficiently to address every need of their inmates. Rather, community experts who are regularly engaged in the provision of such services are the appropriate persons to provide them, whether the inmate is in a jail or free in the community; in both instances, it is argued, he or she is a community member and entitled to such services (Barlow et al., 2006; Lightfoot et al., 1991).

Obviously, the development of community jails requires that some resources (particularly space) be devoted to the accommodation of community experts who provide for inmates' needs. Unfortunately, it is the rare jail that has the luxury of excess space for allocation to such programming. Therefore, the solution may lie in inclusion of such space in jail architectural plans, though this certainly is not optimal given the immediacy of inmate needs discussed in the foregoing section.

The second problem that faces jail managers interested in creating "community jails" is convincing local service providers, and lawmakers if need be, that people in jails have a right to and a continued need for services and that the continued provision of such services, by community experts, benefits both those inmates and the larger community. Needless to say, making this case, as reasonable as it might sound, can be a "hard sell" to those social service agencies that already have scarce resources and to policy makers concerned that more tax dollars might be required to fund such resource provision in jails. For these reasons, larger jails and communities, with their economies of scale and a greater proportion of their populations in need of social services, might be better situated to operate community jails and thus achieve their purported benefits of less crime, from the continuous provision of services in jails (Kerle, 2003; Lightfoot et al., 1991).

Co-Equal Staffing

Another promising innovation in jails that has occurred in the last couple of decades, in some sheriff's departments, has been the development of **co-equal staffing** programs that provide comparable pay and benefits to those who work in the jail with those who work on the streets as law enforcement (Kerle, 2003). Historically, jails have not been a dumping ground (to use Irwin's [1985] terminology) just for inmates, but for staff as well. If a sheriff deemed that a staff person could not "make it" on the streets as law enforcement, he or she was given a job in the jail where apparently the individual's lack of skills and ability was not seen as a problem. Moreover, jail staff were (and often still are) paid less and received less training than their counterparts working on the streets (Stohr & Collins, 2009). As a result, jails do find it difficult to attract and keep the best personnel, or even if they can attract the more talented applicants, jail jobs were and are used as "stepping stones" to better-paying, and higher-status, jobs on the law enforcement side of sheriffs' agencies (Kerle, 2003).

Since the 1980s, however, many sheriff's departments, though far from a majority, have recognized the problems created by according this second-tier status to those who work in jails (Kerle, 2003). Consequently, they have instituted programs whereby staff who work in the jails, who often are given deputy status, are trained and paid similarly to those who work in the free communities. Some anecdotal evidence from sheriff's departments indicates that this change has had a phenomenal effect on the professional operation of jails (as they are better staffed) and on the morale of those who labor in them (Kerle, 2003).

Reentry Programs for Jails

Perhaps the newest "thing" in jails these days (and in prisons, too) is a rethinking about how to keep people out of them! (Reentry will be discussed in greater detail in Chapter 9.) Rather than focusing on deterrence or incapacitation so much (that is *so* 1980s and 1990s), jail practitioners are studying how to make the transition from jails to the community smoother and more successful so that people do not commit more crime and return (Bookman et al., 2005; Freudenberg, 2006; McLean, Robarge, & Sherman, 2006; Osher, 2007). As is indicated by the discussion in the foregoing material of all of the medical, psychological, social, and not to mention criminal deficits that many inmates of jails have, this transition back into the community is likely to be fraught with difficulties. That is why any successful **reentry** program must include a recognition of the problems individual inmates may have (e.g., mental illness, physical illness, joblessness, and homelessness) and address them systematically in collaboration with the client and the community (Freudenberg, 2006; McLean et al., 2006). In a study by Freudenberg, Mosely, Labriola, and Murrill (as cited in Freudenberg, 2006) conducted in New York City jails, the researchers asked hundreds of inmates what their top three priority reentry needs were. For adult women, it was housing, substance abuse, and financial; for adult men, it was unemployment, educational, and housing; for adolescent males, it was unemployment, educational, and financial (Freudenberg, 2006, p. 15).

Effective interventions to improve reentry, in the New York study, included everything from referral to counseling to drug treatment to post-release supervision, depending on the needs of the inmate, his or her unique reentry situation, and the services available in the community. Clearly, reentry is a complex process for people with multiple problems, and it requires that jail personnel prioritize the needs they will target and the interventions they will apply and then network with community agencies to provide the package of services most likely to further the goal of a successful reentry (Freudenberg, 2006; McLean et al., 2006). In fact, Bookman and her colleagues (2005) would argue that jail personnel should expect to engage in collaborative arrangements with community agencies (sounds a bit like community jails, doesn't it?), if they hope to succeed in the reentry process.

Summary

- Jails in the United States are faced with any number of seemingly intractable problems. They are often overcrowded, or close to it, and house some of the most debilitated and vulnerable persons in our communities. They house the accused, the guilty, and the sentenced, as well as low-level offenders and the serious and violent ones. As with prisons, their mission is to incapacitate (even the untried), to deter, to punish, and even to rehabilitate. The degree to which they accomplish any of these goals is in large part determined by the political and social climate that the jail is nested in. Since the 1980s, the political climate has favored "harsh justice" meted out by policy makers and the actors in the criminal justice system and has led to the unrelenting business of filling and building prisons and jails across the country (Cullen, 2006; Irwin, 1985, 2005; Whitman, 2003).

- Jails have also served as a dumping ground for those who are marginally criminal and are unable, or unwilling, to access social services. Too often, the needs of such persons go unaddressed in communities, and as a result these unresolved needs either contribute to their incarceration (in the case of substance abuse and mental illness) or make it likely (such as in the case of homelessness) that they will enter and reenter the revolving jailhouse door.

- Sexual violence in jails remains problematic. It is likely true that the rate of violence between inmates and inmates or between staff and inmates has gone down in recent years. However, increased monitoring of this phenomenon is certainly called for and may serve to further reduce violence through the implementation of violence reduction techniques and training for staff. To that end, the implementation of the Prison Rape Elimination Act of 2003, with its reporting requirement for correctional institutions, represents a positive move in that direction.

- Thankfully, there have been some other hopeful developments on the correctional horizon. Jails in a position to do so have expanded their medical and treatment options to address the needs of inmates. Architectural and managerial solutions have been applied to jails in the form of new generation jails and co-equal pay for staff in sheriff's departments, and some few jails have even experimented with community engagement to ensure that the needs of people in communities are not neglected when such folks enter jails or reenter communities.

Key Terms

Co-equal staffing

Community Jails

New generation jail/podular direct supervision jail

Overcrowding

Prison Rape Elimination Act of 2003

Reentry

Discussion Questions

1. Why are jails the "dumping ground" for so many people in our communities? What are the consequences of this social policy?

2. What is the best use for a jail? What factors might make it difficult to operate jails so that they are able to focus on this best use?

3. What do you think are the best practices (most effective) in managing medically challenged or potentially suicidal inmates?

4. How can jail managers best reduce or eliminate sexual violence against inmates in jails? What do you think keeps managers from being successful at eliminating such violence?

5. What factors are likely to compromise the ability of podular direct supervision jails to achieve their promise?

6. Why are jail staff, in most facilities and sheriff's departments, still paid less than those on patrol? What argument can be made for the same, or even higher, pay for jail staff?

7. What are the relative advantages and disadvantages of community jails?

8. How might reentry programs prevent recidivism?

Useful Internet Sites

American Jail Association: www.corrections.com/aja/

Bureau of Justice Statistics (information available on all manner of criminal justice topics): http://bjs.ojp .usdoj.gov/

National Criminal Justice Reference Service: www .ncjrs.gov

National Institute of Corrections: http://nicic.gov

Vera Institute (information available on a number of corrections and other justice-related topics): www .vera.org

CHAPTER 6

Probation and Community Corrections

❖ Introduction: The Origins of Probation

This chapter focuses on **probation**, which is a sentence imposed on convicted offenders that allows them to remain in the community under the supervision of a probation officer instead of being sent to prison. The term *probation* comes from the Latin term *probare*, meaning "to prove." Because probation is a conditional release into the community, the probation period is a time of testing a person's character and his or her ability to meet certain requirements. That is, convicted persons must prove to the court that they are capable of remaining in the community and living up to its legal and moral standards. About 90% of all sentences handed down by the courts in the United States are probation orders (Kramer & Ulmer, 2009).

The practice of imprisoning convicted criminals is a relatively modern and expensive way of dealing with them. Up to two or three hundred years ago, they were dealt with by execution, corporal punishments such as disfigurement or branding, or humiliation in the stocks. All these punishments took place as community spectacles, and even with community participation in the case of individuals sentenced to time in the stocks. Assuming that a convicted person was not executed, he or she remained in the community enduring the shame of having offended it (think of Hester Prynne's punishment in Nathanial Hawthorne's *The Scarlet Letter*). The only kind of offenders typically subjected to this kind of shaming today are sex offenders whose pictures are displayed on the Internet and who are frequently identified to their neighbors through community notification orders.

In this chapter, we discuss how probation has developed and flourished as a correctional sanction. The purpose probation serves as the most common correctional sentence and what that means for the community corrections officer charged with supervising huge caseloads are also explored. Techniques used by such officers to improve upon programming provided to probationers and to reduce their recidivism are reviewed.

❖ Modern Modes of Reprieve

More enlightened eras saw punishments move away from barbaric cruelties and the emergence of the penitentiary where offenders could contemplate the errors of their ways and perhaps redeem themselves while residing there. But as we have seen, penitentiaries were not very nice places, and some kind souls in positions to do so sought ways to spare deserving or redeemable offenders from being consigned to them. This practice had its legal underpinnings in the concept of *judicial reprieve* sometimes practiced in English courts beginning in the 19th century. A **judicial reprieve** was a delay in sentencing following a conviction, a delay that most often would become permanent if the offender demonstrated good behavior. In those days, there were no probation officers charged with supervising reprieved individuals; the nosey and judgmental nature of the small communities typical of such times was more than adequate for that task.

Early American courts also used judicial reprieve whereby a judge would suspend the sentence and the defendant would be released on his or her own recognizance. Today, an "own recognizance" release is the release of an arrested person without payment of bail who promises to appear in court to answer criminal charges. In early America, it was granted to persons already convicted as a form of probation, although offenders received no formal supervision or assistance to help them to mend their ways. The first real probation system in which a reprieved person was supervised and helped was developed in the United States in the 1840s by a Boston cobbler named *John Augustus*. Augustus would appear in court and offer to take carefully selected offenders into his own home where he would do what he could to reform them as an alternative to imprisonment. Probation soon became his full-time vocation and he recruited other civic-minded volunteers to help him. By the time of his death in 1859, Augustus and his volunteers had saved more than 2,000 convicts from imprisonment (Schmalleger, 2001). It should be noted, however, that Augustus only worked with first offenders and excluded the "wholly depraved" (Vanstone, 2004, p. 41), a luxury modern probation officers do not enjoy.

In 1878, the Massachusetts legislature authorized Boston to hire salaried probation officers to do the work of Augustus's volunteers, and a number of states quickly followed suit. This legislation grew out of the need to enforce the conditions of a suspended sentence as well as the need to help offenders to change their lives (Vanstone, 2008). However, the probation idea almost died in 1916 when the United States Supreme Court ruled that judges may not indefinitely suspend a sentence (*Ex parte United States [Killits]*, 1916). In this case, an embezzler was sentenced to 5 years imprisonment, which the judge (federal judge John Killits) suspended, contingent on the embezzler's good behavior. What Killits had done was place an offender on probation without there having been such a system established by law. Probation was such a popular idea with legislators at this time, however, that this ruling led to the passage of the National Probation Act of 1925, allowing federal judges to suspend sentences and place convicted individuals on probation if they found that circumstances warranted it.

❖ Why Do We Need Community Corrections?

Community corrections may be defined as any activity performed by agents of the government to assist offenders to establish or reestablish law-abiding roles in the community while at the same time monitoring their behavior for criminal activity. In theory, monitoring and assisting offenders while in the community protects society from criminal predation without taxpayers having to shoulder the financial cost of incarcerating all its offenders. Even if, as a society, we were willing and able to bear the monetary cost of imprisoning all offenders, incarceration imposes other costs on the community. These costs can and must be borne where seriously violent and chronic criminals are concerned, but to send every felony offender to prison would be counterproductive. Yet the general public is not well-disposed to the idea of probation because "[i]t suffers from a 'soft on crime' image" and is seen as "permissive, uncaring about crime victims, and blindly advocating a rehabilitative ideal while ignoring the reality of violent, predatory criminals" (Petersilia, 1998, p. 30). However, allowing relatively minor offenders to remain in the community under probation supervision to prove that they can live law-abiding lives offers many benefits to them, as well as to their community.

The general public's notion that a probation sentence is "getting away with it" is a notion not shared by many offenders. The probationer receives a prison sentence upon conviction that is suspended during the period of proving that he or she is capable of living a law-abiding life. This sentence hangs over probationers' heads like a guillotine ready to drop if they fail to provide that proof. It may be for this reason that a number of studies have found that "experienced" offenders who have done prison time, probation time, and parole time often prefer prison to the more demanding forms of probation such as day reporting and intensive supervision probation (Crouch, 1993; May, Wood, Mooney, & Minor, 2005). Probation requires offenders to work, submit to treatment schedules, and do lots of other orderly things that many hardened criminals simply do not have the inclination to do. Numerous interviews with active "street criminals" (e.g., burglars, robbers, carjackers) show that such things are treated with disdain (B. Jacobs & Wright, 1999; Mawby, 2001). Serving time in prison is less of a hassle for many of them, and many know they would end up there anyway because they would not live up to probation conditions (May et al., 2005).

Other less criminally involved offenders prefer a probation sentence so that they can retain their jobs and maintain connection to their families and communities. As we saw in Chapter 1, perceptions of the severity of a punishment, and thus its deterrent effect, are a function of the contrast between one's everyday life and life under punishment conditions (the contrast effect). Of the more than 4 million Americans on probation in 2008, a total of 59% of them successfully completed their conditions of supervision and were released from probation (Glaze & Bonczar, 2009). In the event of a failure to live up to the conditions of probation, the prison sentence is then typically imposed. Thus, while many fail the probation period, the majority succeed, so surely providing nonviolent offenders the opportunity to try to redeem themselves while remaining in the community is sensible criminal justice policy. But what are the benefits for the community?

(1) Probation costs between $700 and $1,000 per year as opposed to $20,000 to $30,000 per year (more for women, juveniles, and the elderly) for imprisonment, saving the taxpayer at least $19,000 per year per non-incarcerated felon (Foster, 2006). Many jurisdictions require probationers to pay for their own supervision, which means that the taxpayer pays nothing. However, while economic considerations are vitally important to policy makers, they are not the primary concern of corrections; protecting the community is. Community-based corrections is the solution only for those offenders who do not pose a significant risk to public safety.

(2) Employed probationers stay in their communities and continue to pay taxes; offenders who were unemployed at the time of conviction may obtain training and help in finding a job. This adds further to the tax revenues of the community and, more importantly, allows offenders to keep or obtain the stake in conformity that employment offers. A job also allows them the wherewithal to pay fines and court costs, as well as restitution to victims.

(3) In the case of married offenders, community supervision maintains the integrity of the family, whereas incarceration could lead to its disruption and all the negative consequences such disruption entails.

(4) Probation prevents felons from becoming further embedded in a criminal lifestyle by being exposed to chronic offenders in prison. Almost all prisoners will leave the institution someday, and many will emerge harder, more criminally sophisticated, and more bitter than they were when they entered. Furthermore, they are now ex-cons, a label that is a heavy liability when attempting to reintegrate into free society.

(5) Many more offenders get into trouble because of deficiencies than because of pathologies. Deficits such as the lack of education, a substance abuse problem, faulty thinking patterns, and so forth can be assessed and addressed using the methods that will be discussed in Chapter 14 on treatment. If we can correct these deficits to some extent, then the community benefits, because it is a self-evident truth that whatever helps the offender, protects the community.

We do not wish to appear naïve about this; there are people who are unfit to remain in the community and could not lead a law-abiding life even if given everything they needed to start over. For instance, among criminals in the witness protection program, despite being given new identities, jobs, housing, and basically a new start in life, 21% are arrested under their new identities within 2 years of entry (Albanese & Pursley, 1993, p. 75). The type of criminals in the witness protection program are used to relatively high incomes made in a life filled with excitement and personal power and independence, and they often find it difficult to adjust to an ordinary job with minimal pay.

Although community corrections is sensible policy, and there are more people under such supervision than ever before, there has been a decline in its use from its heyday in the rehabilitation-oriented 1970s and 1980s relative to the use of incarceration. Figure 6.1 shows that the probation population has gone up from just over 1 million in 1980 to over 4 million in 2008, which amounts to about 1 for every 45 adults in the United States (Glaze & Bonczar, 2009). However, the ratio of probationers to prison inmates was 4.02:1 in 1980 and 2.96:1 in 2008. This is more the result of the increased reliance on incarceration than on decreased use of probation as "get tough" sentencing policies have been implemented. In 2008, a total of 76% of probationers were male. Figure 6.2 provides the racial/ethnic breakdown of 4,270,917 probationers in 2008.

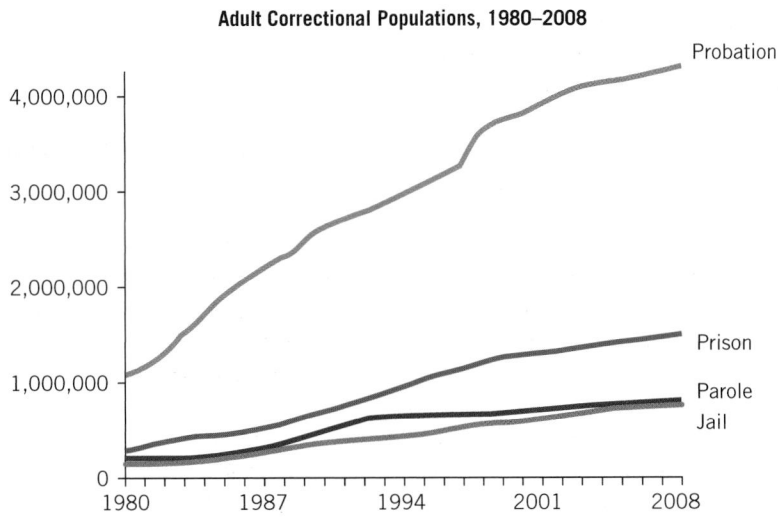

Adult Correctional Populations, 1980–2008

Figure 6.1

Number of Persons Under Correctional Supervision, 1980–2008, by Type of Supervision

Source: Bureau of Justice Statistics (2010).

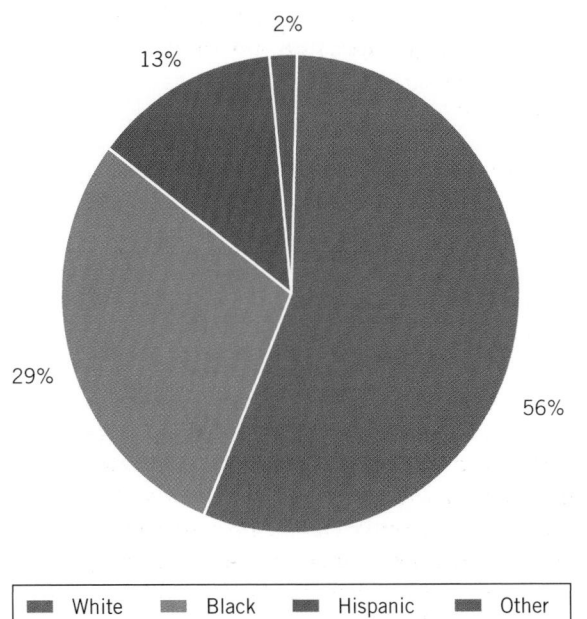

Figure 6.2

Percentage of Individuals on Probation in 2008 by Racial Category

Source: Glaze & Bonczar (2009).

❖ Probation Officer Role and Models of Probation Supervision

Probation and parole officers (the roles are combined in some jurisdictions) have two common roles: to protect the community and to assist their probationers and parolees to become productive and law-abiding citizens. The dual roles mark them as law enforcement officers (their legally defined role in most states) and as social workers. The offenders they supervise may be on probation or parole depending on their legal status. Probation is a judicial function, meaning that the offenders are under the ultimate supervision of the court. Probationers may or may not have served jail time prior to community supervision by a probation officer. Offenders who have served time in prison are placed on parole upon release and are under the ultimate supervision of the executive branch of government, typically the state department of corrections.

Photo 6.1

A probation officer meets with a probationer.

Probation officers are officers of the courts, and in this capacity they are responsible for enforcing court orders, which may require them to monitor programs such as drug and alcohol treatment and develop plans to assist their probationers' transition to a free society. They are also required to make arrests, to perform searches, and to seize evidence of wrongdoing. Officers may have to appear in court occasionally to present evidence of violation of probation orders and to justify their recommendations for either termination of probation and subsequent imprisonment or continued probation with additional conditions. This is part of the officers' law enforcement role. Probation/parole officers work with criminal offenders who may be dangerous and often live in areas that may also be dangerous, which is why 35 states require their officers to carry a firearm (Holcomb, 2008). Nevertheless, it is important that officers spend a lot of time in those communities to learn about their culture, customs, and values, and to learn what resources are available to assist with the rehabilitation of offenders.

As in any other occupation, the effectiveness of probation/parole officers' performance ranges from dismal to outstanding. One of the biggest problems in probation and parole work is gaining the trust of probationers and parolees and developing rapport with them. Most officers are white and middle class, whereas many of their "clients" are minorities, and whereas most probationers and parolees are male, about half of probation officers are female (Walsh & Stohr, 2010). It is very difficult for both officers and probationers to overcome the class, race, and gender divide and for each to understand and appreciate where the other is "coming from." Nevertheless, the job must be done as effectively as possible.

The measures of correctional effectiveness are how well the community is protected from the offenders on an officer's caseload (law enforcement role) and how well their offenders are able to resolve their criminogenic problems and become decent law-abiding citizens (social work role). There is often tension between these supposedly contradictory roles (Skeem & Manchack, 2008). Some officers take on the law enforcement role and embrace working values emphasizing strict compliance with probation conditions and holding offenders strictly accountable. Other officers take on the counselor role, providing offenders with whatever is available in the community to bring about behavior change. The extent to which officers follow these different models depends not only on the personalities and training of the individual officers, but also on the overall model dictated by their agency, which in turn is dictated by the ideology of politicians and whether their correctional philosophy is punitive or rehabilitative.

A third group of officers combine the two roles and follows a "hybrid" approach. Skeem and Manchak (2008) view the law enforcer as authoritarian and the counselor as permissive, but see the hybrid officer as authoritative, the kind of parenting style that psychologists tell us the most effective parents adopt (Grusec & Hastings, 2007). Authoritarian officers are inflexible disciplinarians who require unquestioning compliance with their demands. Such a style often leads to hostility and rebelliousness among those at whom it is directed. Permissive officers set few rules and are reluctant to enforce those that are set. This style often results in the perception of others that the officer is a "pushover," and it practically invites noncompliance and lack of respect. Authoritative officers are hybrids who are firm enforcers but fair, knowing that boundaries must be set and consequences endured for venturing beyond them. They clearly describe those boundaries and the consequences for crossing them (the law enforcement role), but also offer guidance and support to probationers (the social work role) so that they may be better able to stay within those boundaries.

How well do these different styles do with respect to the dual roles of community protection and offender rehabilitation? One study found that in terms of technical violations such as failure to comply with some condition of probation, 43% of probationers supervised by a law enforcement–oriented officer, 5% of probationers supervised by a treatment-oriented officer, and 13% of those supervised by a "hybrid" officer received a technical violation (Paparozzi & Gendreau, 2005). These findings are as expected: Law enforcers do not tolerate any violations, counselors tolerate almost every violation, and hybrids tolerate selectively. New criminal convictions are a better measure of supervision effectiveness than technical violations, however, because while technical violations are largely in the hands of officers, new criminal convictions are out of their hands. Offenders with treatment-oriented officers were convicted of new crimes at twice the rate of those supervised by law-enforcement officers (32% vs 16%), while only 6% of offenders supervised by hybrid officers were convicted of a new crime.

Probation Officer Stress

Supervising criminal offenders is not the easiest or most lucrative job in the world. In common with police officers and correctional officers, probation and parole officers are dealing with difficult human beings on a daily basis, often without the tools and support needed to do the job. Doing a demanding and sometimes dangerous job under less than adequate conditions can, and does, lead to stress (Slate, Wells, & Wesley Johnson, 2003). For instance, one study of officers in four states found that 35% to 55% reported that they had been victims of threatened or actual violence (Finn & Kuck, 2005). Stress is a physical and emotional state of tension that occurs as the body reacts to environmental challenges (stressors). No one can be expected to do a very good job while experiencing stress.

The most important job stressors identified by the officers surveyed by Slate et al. (2003) were poor salaries, poor promotion opportunities, excessive paperwork, lack of resources from the community, large caseloads, and a general frustration with the inadequacies of the criminal justice system. These stressors eventually lead to psychological withdrawal from the job, which means that probationers, and thus the community, are getting shortchanged. High stress levels in the department also lead to frequent absenteeism and high rates of employee turnover. Thus, it is imperative that the issue of probation/parole officer stress be meaningfully addressed.

Slate et al. (2003) emphasize that attempts to address the problem of probation officer stress should not simply involve counseling officers on how to cope with stress, because the problem is organizational (inherent in the probation system), not personal. They suggest that participatory management strategies be instituted so that each person in the department participates in the decision-making process, and thus feels valued and empowered. The researchers found that personnel who did participate in decision making reported fewer stress symptoms and were happier on the job. Participatory management (workplace democracy) leads to a happier and more productive workforce, even if nothing else changes—"contented cows give better milk."

❖ Community Corrections Assessment Tools

Officers have a variety of tools for helping them to do their jobs more efficiently. Among the various assessment instruments, the most widely used one at this time is the **client management classification system (CMC)**. The CMC contains offender risk and needs scales that embody the principle of responsivity discussed in the chapter on offender treatment (Chapter 14), and is also used to determine the level of supervision that offenders receive. The average caseload of a probation officer is around 139 (Finn & Kuck, 2005), which means that there is precious little time to devote to each offender, especially given that there are other duties an officer must perform. This is why a method of organizing officers' time is needed so that they do not find themselves floundering all over the place, and is a major reason why correctional agencies use instruments such as the CMC. Some offenders need more services and higher levels of monitoring than others, and to treat them all as equals in this regard would be both counterproductive and wasteful. The CMC places probationers into the four supervision-level categories, according to Walsh and Stohr (2010), which are outlined below.

Selective Intervention. These are low-risk and low-needs offenders; that is, they are minimally criminally involved and they have a stake in conformity. These folks require little of the officers' time or resources, which means that there are more available for others that

Photo 6.2

A probation officer at work conducting research for a PSI report.

need them. There is evidence that low-risk offenders actually become worse if they are over-supervised and subjected to treatment modalities that they do not need (Lowenkamp & Latessa, 2004). Placing such offenders in the same restrictive programs as high-risk offenders exposes them to bad influences and may disrupt the very factors (family, employment, pro-social activities and contacts) that made them low-risk in the first place. Officers are advised only to intervene in the lives of such people "selectively," that is, only in very special circumstances such as a new arrest.

Environmental Structure. These offenders are on the low end of medium-risk and require regular supervision. Officers work with these individuals to channel them into a number of services such as educational, vocational, and substance abuse programs. These are not necessarily people deeply embedded in a criminal lifestyle but rather people with a number of social deficits that can be corrected. This type of offender probably constitutes the majority of probationers, and perhaps a few parolees.

Casework/Control. These offenders are at the high end of medium-risk or the low end of high-risk. They require intensive casework and their activities should be closely monitored. They require the same services as environmental structure probationers, but they are less likely to benefit from them. They tend to be more entrenched in the criminal lifestyle and more likely to have severe drug and/or alcohol problems.

Limit-Setting. Offenders in this category are firmly embedded in a criminogenic lifestyle and are thus at a high risk for probation failure. They are often supervised by officers with an intensive supervision caseload (see below) and require severe limits to be placed on their activities. Protection of the community through surveillance and strict controls is of primary concern with offenders of this type.

Comparative Perspective: Community Corrections

The probation concept is the child of the Enlightenment of the 18th century and the Christian missionary and temperance movements of the 19th century, and it grew rapidly in the late 19th and early 20th centuries (Vanstone, 2004). Between 1878 (the date of the Massachusetts legislation) and 1920, probation statutes were in place in countries on every continent in the world (Vanstone, 2008). John Augustus can rightfully claim the title of "father of probation" (he coined the term for what he was doing), but in the very same year (1841), a British magistrate in Birmingham, England, named Matthew Davenport-Hill was laying the foundations for probation in Britain. The efforts of Augustus and Davenport-Hill combined to form what is now known as the Anglo-American model of probation (Linder, 2007). Like Augustus, Davenport-Hill was a deeply religious man and an enemy of alcohol. Unlike Augustus, in addition to helping offenders overcome their problems, Davenport-Hill also implemented the supervision of offenders and kept records of their behavior. He used what were known as Police Court Missionaries who were middle-class volunteers animated by strong Christian and temperance values and augmented by appointed police officers. The first full-time professional probation officers in England were appointed in 1907 (Gard, 2007). As in the United States, probation is the most common disposition of a criminal case in Great Britain.

Countries in Asia with communitarian traditions such as China, Korea, and Japan have long relied on informal methods of social control rather than the criminal law to maintain social peace and harmony. The modern Japanese probation system still retains much of the early volunteer tradition of the American and British probation systems. It is an ancient tradition in Japan to share in the control of public behavior and to take an interest in one's neighbors. This interest was first formalized, with the trend toward Westernization, in 1880 with a group known as the Shizuoka Prefecture Discharged Prisoners Aid Association, which ran halfway houses and other programs (Eskridge, 1989). The first paid professional probation officers in Japan were appointed in 1922. There are only about 1,000 professional officers in Japan, as opposed to about 36,000 in the United States (Gardner, 1996). Each officer has up to 300 offenders assigned to him or her, but the officers' primary task is to supervise at least 100 volunteer probation officers (VPOs), of which there are at least 80,000 in Japan. A VPO position is much sought after because a high level of public esteem is afforded VPOs, and thus there is much competition for these (unpaid) positions. Candidates must show excellent character, enthusiasm, and financial security, and because the position requires considerable time, many are retired; others come from the middle and upper classes possessing sufficient time and resources to be able to concentrate on their probation tasks (Eskridge, 1989).

Each VPO is assigned only two or three offenders with the VPO's primary task being to help them to find suitable housing, employment, and community reintegration. The latter task is particularly important because a criminal conviction in Japan carries great shame, and may result in family members shunning the offender. These VPOs are thus often the only source of support and friendship for offenders, who are often quite moved by the fact that someone has taken an interest in their welfare and is willing to help them. According to Gardner (1996), this realization is often a strong motivator for offenders to work toward rehabilitation. It is because of the strong communitarian spirit in Japan and the strong sense of shame generated when one offends it that community involvement through volunteers is the backbone of the Japanese correctional system.

❖ Engaging the Community to Prevent Recidivism

The Comparative Perspective box on community corrections emphasized how probation began as a voluntary community concept in the United State and Britain, and is a concept that still dominates the Japanese system. We in the United States probably cannot revive the old level of community involvement, and cultural differences preclude volunteerism at anywhere near Japanese levels. For instance, to achieve the average 2.5 ratio of probationers to volunteer probation officers that they enjoy in Japan with our approximately 4 million probationers would require about 1.6 million volunteer officers. Nevertheless, we can engage our communities in the process of offender rehabilitation, realizing that whatever helps the offender helps the community.

The criminological literature provides abundant support for the notions that social bonds (Hirschi, 1969) and social capital (Sampson & Laub, 1999) are powerful barriers against criminal offending. *Social bonds* are connections (often emotional in nature) to others and to social institutions that promote prosocial behavior and discourage antisocial behavior. *Social capital* refers to a store of positive relationships in social networks upon which the individual can draw for support. It also means that a person with social capital has acquired an education and other solid credentials that enable him or her to lead a prosocial life. Those who have opened their social capital accounts early in life (bonding to parents, school, and other prosocial networks) may spend much of it freely during adolescence, but nevertheless manage to salvage a sufficiently tidy nest egg by the time they reach adulthood to keep them on the straight and narrow. The idea is that they are not likely to risk losing this nest egg by engaging in criminal activity. Most criminals, on the other hand, lack social bonds, and largely because of this, they lack the stake in conformity provided by a healthy stash of social capital.

If we consider the great majority of felons in terms of deficiency (good things that they *lack*) rather than in terms of personal pathology (bad things they *are),* we are talking about a deficiency in social capital. The community can be seen as a bank in which social capital is stored and from which offenders can apply for a loan. That is, the community is the repository of all of those things from which social capital is derived, such as education, employment, and networks of prosocial individuals in various organizations and clubs (e.g., Alcoholics Anonymous, churches, hobby/interest centers). Time spent in involvement in steady employment and with prosocial others engaged in prosocial activities is time unavailable to spend in idleness in the company of antisocial others planning antisocial activities. The old saying that "the devil finds work for idle hands" may be trite, but it is also very true.

Thus, good case management in community corrections requires community involvement. No community corrections agency is able to deliver the full range of offender needs (mental health, substance abuse, vocational training, welfare, etc.) by itself. Probation and parole officers must not only assess the needs of their charges, but must also be able to locate and network with the social service agencies that address those needs as their primary function. In fact, there are those who maintain that the probation/parole officer's relationships with community service agencies are more important than their relationship with their probationers/parolees. Officers must be skilled at networking with the various agencies if they are to help provide offenders with the services they need. The community corrections worker is a broker who takes input about offenders from offenders themselves, police, courts, friends, neighbors, family, employees, and so on, and refers the problem out to the appropriate agency. This brokerage function can be best achieved with fewer offenders who are intensively supervised on an officer's caseload than with many who are infrequently seen and haphazardly supervised.

❖ Intermediate Sanctions

Intermediate sanctions refers to a number of innovative alternative sentences that may be imposed in place of the traditional prison/probation dichotomy. Such sanctions are considered intermediate because they are seen as more punitive than straight probation but less punitive than prison. They are also a way of easing prison overcrowding and the financial cost of prison while providing the community with higher levels of security from victimization through higher levels of offender surveillance than is possible on regular probation. As we shall see, these supposed benefits are not always realized. We have already seen that many experienced offenders would choose prison over some of the stricter community-based alternatives. Furthermore, since offenders placed in them have recidivism rates not much different from offenders released from prison within the first and subsequent years, the costs of state incarceration are deferred rather than avoided (Marion, 2002). The first alternative we examine is intensive supervision probation.

Intensive Supervision Probation

Intensive supervision probation (ISP) is typically limited to offenders who probably should not be in the community but have been allowed to remain, either in the belief that there is a fighting chance they may be rehabilitated or in an effort to save the costs of incarceration. ISP officers' caseloads are drastically reduced (typically a caseload of 25) to allow officers to more closely supervise, often on a daily basis with frequent drug testing, using the surveillance model. Burrell (2006) describes ISP officers as "aggressive in their surveillance and punitive in their sanctions" (p. 4). Liberal critics of the tactics of ISP officers describe their model of supervision colorfully as one of "pee 'em and see 'em," or "tail 'em, nail 'em, and jail 'em" (Skeem & Manchack, 2008). However, this type of law enforcement surveillance happens to be the kind of supervision recommended for high-risk probationers by the client management classification system discussed earlier.

MacKenzie and Brame (2001) hypothesize in their study of ISP that such supervision coerces offenders into prosocial activities, which in turn leads to a lower probability of them reoffending. Intensive supervision means that probation officers maintain more frequent contact with probationers and intrude into their lives more than is the norm with other probationers. Intensive supervision offenders are supervised at that level because they have the greatest probability of reoffending (they are high-risk) and are the most deficient in social capital (they have high needs). Higher levels of supervision allow officers to coerce offenders into a wide variety of educational and treatment programs and other prosocial activities designed to provide offenders with social capital. MacKenzie and Brame found that ISP supervision did result in offenders being coerced into more prosocial activities and that there was a slight reduction in recidivism. The issue the study left unresolved is whether participating in prosocial activities enabled offenders to acquire skills that provided them with social capital they could put to good use, or if intensive supervision per se accounted for their findings.

The term *coercion* has negative connotations for the more libertarian types among us ("You can lead a horse to water . . ." and all that), but the great majority of people being treated for problems such as substance abuse have very large bootprints impressed on their backsides. Probationers and parolees, almost by definition, will not voluntarily place themselves in the kinds of programs and activities we would like them to be in—they are simply not motivated in that direction. The criminal justice system must provide that motivation via the judicious use of carrots and sticks. Reviews of the U.S. (Farabee, Pendergast, & Anglin, 1998) and UK (Barton, 1999) literature on coerced substance abuse treatment concluded that coerced treatment often has more positive outcomes than voluntary treatment, probably because of the threat of criminal justice sanctions.

Work Release

Work release programs are designed to control offenders in a secure environment while at the same time allowing them to maintain employment. Work release centers are usually situated in or adjacent to a county jail, but they can also be part of the state prison system. Residents of work release centers have typically been given a suspended sentence and placed on probation with a specified time to be served in work release. Work release residents may also be parolees under certain circumstances, such as when there is the need to closely supervise a new parolee, or when a parole violator is given another opportunity to remain in the community rather than being sent back to prison. Surveillance of work release residents is strict; they are allowed out only for the purpose of attending their employment, and are locked in the facility when not working. The advantage of such programs is that they allow offenders to maintain ties with their families and with employers. Such programs also save the taxpayer money because offenders pay the cost of their accommodation with their earnings.

Offenders on work release are generally the least likely of all community-based corrections offenders to be rearrested and imprisoned within 1 year and 5 years of successful completion. However, 64% of offenders successfully released and 71% of those unsuccessfully released had further arrests within 5 years (Marion, 2002). Offenders chosen for work release are typically selected because, although they have committed a crime deemed too serious for regular probation, they are usually employed, although unemployed probationers can be placed in work release contingent on their finding employment within a specified time (Abadinsky, 2009). Being employed is incompatible with a criminal lifestyle (although obviously from the above statistics, not completely), especially if the offender is a probationer rather than a parolee.

Shock Probation/Parole and Boot Camps

Shock probation (mentioned in Chapter 4) was initiated in Ohio in the 1970s and was designed to literally shock offenders into desisting from crime by briefly exposing them to the horrors of prison. It was limited to first offenders who had perhaps been unimpressed with the realities of prison life until given a taste. Under this program, offenders were sentenced to prison and released after (typically) 30 days and placed on probation. In some states, a person may receive shock parole, which typically means that he or she has remained in prison longer than the shock probationer and is released under the authority of the parole commission rather than the courts. Most of the research on this kind of treatment was conducted in the 1970s and 1980s and concluded that shock probationers/parolees had lower recidivism rates than incarcerated offenders not released under shock conditions (Vito, Allen, & Farmer, 1981). This should not be surprising, however, given the fact that those selected for shock probation/parole were either first offenders or repeaters who had not committed very serious crimes.

When we hear of shock incarceration today, it is typically incarceration in a so-called boot camp. **Correctional boot camps** are facilities modeled after military boot camps. Relatively young and nonviolent offenders are the most typical kinds of offenders sent to correctional boot camps for short periods (90–180 days) where they are subjected to military-style discipline and physical and educational programs. Boot camps are most unpopular with offenders. In May et al.'s (2005) analysis of "exchange rates" discussed earlier, offenders who had served time in prison would only be willing to spend an average of 4.65 months in boot camp to avoid 12 months in prison. Interestingly, judges and probation/parole officers also agree that boot camp is more punitive than prison, indicating they were willing to serve only 6.19 and 6.05 months, respectively, to avoid 12 months in prison (Moore, May, & Wood, 2008).

The idea of boot camps for young adult offenders was once a popular idea among the general public, as well as among a considerable number of correctional personnel and criminal justice academics. Boot camps conjured up the movie image of a surly, slouching, and scruffy youth forced into the army who 2 years later proudly marched back into the old neighborhood, crew-cut, sparklingly clean, and properly motivated and disciplined. Yes, the drill sergeant with righteous fire and brimstone would do what the family and social work–tainted juvenile probation officers could never do.

Such magical transformations rarely happen in real life. The army merely provides many such youths with new opportunities to offend, and they spend much of their time either avoiding the MPs or lodged in the brig while awaiting their dishonorable discharges. Botcher and Ezell's (2005) evaluation of offenders sent to correctional boot camp in California revealed the same sorry outcome. Specifically, they found no significant differences between their experimental group (boot campers) and a control group of similar offenders not sent to boot camp in terms of either property or violent crime reoffending. In other words, boot camps have joined the woeful list of correctional programs that appear ineffective.

Victim–Offender Reconciliation Programs (VORPs)

VORPs are programs designed to bring offenders and their victims together in an attempt to reconcile ("make right") the wrongs offenders have caused and are an integral component of the **restorative justice** philosophy. This philosophy differs from models (such as retributive, rehabilitative, etc.) that are offender driven (What do we do with the offender?) in that it considers the offender, the victim, and the community as partners in restoring the situation to its pre-victimization status. Restorative justice has been defined as "every action that is primarily oriented toward justice by repairing the harm that has been caused by the crime," and it "usually means a face-to-face confrontation between victim and perpetrator, where a mutually agreeable restorative solution is proposed and agreed upon" (Champion, 2005, p. 154). Restorative justice is often referred to as a **balanced approach** in that it gives approximately equal weight to community protection, offender accountability, and offender competency.

Many crime victims are seeking fairness, justice, and restitution *as defined by them,* as opposed to revenge and punishment. Central to the VORP process is the bringing together of victim and offender in face-to-face meetings mediated by a person trained in mediation theory and practice. Meetings are voluntary for both offender and victim and are designed to iron out ways in which the offender can make amends for the hurt and damage caused to the victim.

Victims participating in VORPs gain the opportunity to make offenders aware of their feelings of personal violation and loss and to lay out their proposals for how offenders can restore the situation. Offenders are afforded the opportunity to see firsthand the pain they have caused their victims, and perhaps even to express remorse. The mediator assists the parties in developing a contract agreeable to both. The mediator monitors the terms of the contract and may schedule further face-to-face meetings.

VORPs are used most often in the juvenile system but are rarely used for personal violent crimes in either juvenile or adult systems. Where they are used, about 60% of victims invited to participate actually become involved, and a high percentage (mid- to high 90s) results in signed contracts (Coates, 1990). Mark Umbreit (1994) sums up the various satisfactions reported by victims who participated in VORPs:

1. Meeting offenders helped reduce their fear of being revictimized.

2. They appreciated the opportunity to tell offenders how they felt.

3. Being personally involved in the justice process was satisfying to them.

4. They gained insight into the crime and into the offender's situation.

5. They received restitution.

VORPs do not suit all victims, especially those who feel that the wrong done to them cannot so easily be "put right," and who want the offender punished (Olson & Dzur, 2004). In addition, the value of VORPS for the prevention of further offending has yet to be properly assessed.

❖ Legal Issues in Probation and Parole

Both probation and parole are statutory privileges granted by the state in lieu of imprisonment (in the first case) or further imprisonment (in the second case). Because of their conditional privilege status, it was long thought that the state did not have to provide probationers and parolees any procedural due process rights in the granting or revoking of either status. Today, probationers and parolees are granted some due process rights, although like inmates there are restrictions on them that are not applicable to citizens not under correctional supervision.

The first important case in this area was *Mempa v. Rhay* (1967). Mempa was a probationer who committed a burglary, which he admitted, 4 months after he was placed on probation. His probation was revoked without a proper hearing and without the assistance of legal counsel, and he was sent to prison. The issue before the Supreme Court was whether probationers have a right to counsel at a deferred sentencing (probation revocation) hearing. The Court ruled that under the Sixth and Fourteenth Amendments, probationers do have that right because Mempa was being sentenced, and the fact that sentencing took place subsequent to a probation placement does not alter the fact that sentencing is a "critical stage" in a criminal case. The Court further stated that probationers facing revocation should have the opportunity to challenge evidence by cross-examining state witnesses (typically only the probation officer), to present exculpatory witnesses, and to testify him- or herself.

A further advance in granting due process rights to offenders on conditional liberty status came in *Morrissey v. Brewer* (1972). Morrissey was a parolee who was arrested by his parole officer for a number of technical violations and returned to prison without a hearing. Morrissey's petition to the Supreme Court claimed that because he received no hearing prior to revocation, he was denied his rights under the due process clause of the Fourteenth Amendment. The Court agreed that when a liberty interest is involved, certain processes are necessary (A **liberty interest** refers to government-imposed changes in someone's legal status that interferes with his or her constitutionally guaranteed rights to be free of such interference.). The ruling by the Court in *Morrissey* noted that parole revocation does not call for all the rights due an unconvicted defendant, but that there were certain protections under the Fourteenth Amendment to which the person is entitled. These rights were laid out by the Court as follows:

(a) Written notice of the claimed violations of parole.

(b) Disclosure to the parolee of evidence against him.

(c) Opportunity to be heard in person and to present witnesses and documentary evidence.

(d) The right to confront and cross-examine adverse witnesses (unless the hearing officer specifically finds good cause for not allowing confrontation).

(e) A "neutral and detached" hearing body such as a traditional parole board, members of which need not be judicial officers or lawyers.

(f) a written statement by the fact finders as to the evidence relied on and reasons for revoking parole.

While individuals are on probation or parole, they have limited constitutional rights, and their probation/parole officers have broader powers to intrude into their lives than police officers. Because probationers and parolees waive their Fourth Amendment search and seizure rights, probation/parole officers may conduct searches at any time without a warrant and without the probable cause needed by police officers. Evidence seized by probation/parole officers without a warrant can be used in probation or parole revocation hearings, but not as trial evidence in a new case (*Pennsylvania Board of Probation and Parole v. Scott,* 1998). The Court ruled that to exclude evidence from a parole hearing would hamper the State's ability to ensure the parolee's compliance with conditions of release and would yield the parolee free of consequences for noncompliance. This "special needs" (of law enforcement) exception to the Fourth Amendment has been extended to the police under certain circumstances. The Supreme Court has held that if a probation order is written in such a way that provides for submission to a search "by a probation officer or any other law enforcement officer," then the police gain the same rights to conduct searches based on less than probable cause as probation and parole officers (*United States v. Knights,* 2001).

Summary

■ Community-based corrections are used to control the behavior of criminal offenders while keeping them in the community. Conditional release (judicial reprieve) was practiced in ancient times in common law, but probation was not really established until the 20th century. Although often considered too lenient, community corrections benefits the public in many ways. Probation helps offenders by giving them a second chance to demonstrate that they can be law abiding in the community, and what helps offenders automatically helps the communities they live in.

■ Because of many "get tough" policies, and although there are more Americans than ever under community supervision, the use of community corrections is declining relative to the use of incarceration.

■ The client management classification system (CMC) is a tool designed to guide supervision and treatment strategies for offenders placed on probation. As with any tool, however, the CMC is only as good as the skill and conscientiousness with which it is used. The level of stress of the probation officers' job leads many of them to burn out and to put less than adequate care and effort into attending to the many and varied tasks they need to carry out.

■ Intermediate sanctions are considered to be more punitive than regular probation but less punitive than prison, although experienced criminals do not necessarily share that view. Some of these programs, particularly work release, show positive results, although this may be more a function of the kinds of offenders placed in them rather than the programs themselves. Most participants in these programs, however, tend to recidivate at rates not significantly different from parolees.

■ Victim–offender reconciliation programs (VORPs) are a fairly recent addition to community corrections. They consider the victim, the offender, and the community as equal partners in returning the situation to its pre-victimization status. This idea of restorative justice is mostly used with juvenile offenders and minor adult offenders.

Key Terms

Balanced approach

Casework/Control

Client management classification system (CMC)

Community corrections

Correctional boot camps

Environmental Structure

Intensive supervision probation (ISP)

Intermediate sanctions

Judicial reprieve

Liberty interest

Limit-Setting

Probation

Restorative justice

Selective Intervention

Shock probation

Victim–offender reconciliation programs (VORPs)

Work release programs

Discussion Questions

1. Looking at all the pros and cons of community-based corrections, do you think probation is too lenient for felony offenders? If so, what do we do with them?

2. In your opinion, what is the single biggest benefit of probation for the community and its single greatest cost?

3. Studies show that about 95% of probation officers' sentencing recommendations to the courts are followed. Why do you think this is so, and is it a good or bad thing that probation officers control the flow of information to judges?

4. What is the advantage of tools such as CMC over "good old-fashioned" experience to determine how a probationer should be supervised?

5. Boot camps have full and total control of offenders for up to 6 months, so why do you think they are unable to change offenders' attitudes and behaviors?

6. Do you think police officers should be given the same powers of search and seizure as probation and parole officers for the purposes of controlling the activities of probationers and parolees? Why or why not?

Useful Internet Sites

American Probation and Parole Association: www.appa-net.org

Probation and Pretrial Service, United States Courts: www.uscourts.gov

CHAPTER 7

Prisons

❖ Introduction: The State of Prisons

It has become axiomatic to say that correctional programs and institutions are overcrowded, underfunded, and unfocused these days. As the drug war rages on and as mandatory sentencing has its effect, probation and parole caseloads and incarceration rates spiral past any semblance of control. As a consequence, though spending on corrections has steadily, and steeply, climbed over the last several years, it is nearly impossible for most states and localities to meet the needs for programs, staff, and institutions. So they do not. As a consequence, the corrections experience for offenders is shaped by shortages.

But, as has been discussed already in this text, this has always been somewhat true. If it is built, or in the case of probation and parole, offered, they will come—because, as with all corrections sentences, they are forced to. Cases in point, almost immediately after the first American prisons were built, the Walnut Street Jail (1790), the Auburn Prison (1819), the Western Pennsylvania Prison (1826), and the Eastern Pennsylvania Prison (1829) were full, and within a few years, they were expanded or new prisons were under construction.

To say that crowding and corrections have always been linked, of course, is not to dismiss the negative effects of overfilling institutions or to argue that it might not be worse than ever now. Certainly, the United States' combined incarceration rate for jails and prisons has climbed steadily and remains the highest in the Westernized world at 748 per 100,000 U.S. residents as of 2009 (West, 2010a, p. 4). As indicated in Chapter 5, jails on average are at about 90% full, whereas prisons were running at 109% of capacity in 2008 (the latest year for which we have data at the time of writing), not leaving much room for flexibility in classification (Sabol, West, & Cooper, 2009).

In this chapter, we discuss the structure and operation of prisons. The inmate subculture that flourishes in prisons, and the violence and gangs that bedevil them, will also be reviewed. The nature of the correctional experience for individuals incarcerated in prisons is somewhat different from what those in jails or community corrections encounter, and those differences are explored here.

❖ Prison Organizations

Classification

As inmates enter the prison system from the courts, they are usually assessed at a classification or reception facility based on their crime, criminal history, escape risk, behavioral issues (if any), and health and programming needs. Women and children are classified in separate facilities from adult males. This assessment includes the review of materials related to the inmate, by reception center personnel, and tests and observation of the inmate regarding his or her dangerousness and amenability to treatment. After being assessed by prison personnel for a period of weeks or months, inmates are sent to the prison that the personnel believe is the best fit, based first on security needs, followed by space available, and finally the inmate's needs.

Comparative Perspective: Canada's Debate Regarding Social Support Programming Versus More Prisons

In an editorial from the Canadian newspaper *The Globe and Mail* from March, 2011, the writer described the "Pathways to Education" program for youth that is being funded by Canada's national government. Instead of more incarceration for troubled youth, which is a more typical American response to juvenile delinquency and which has been adopted by the Canadian government to some extent, this program, developed by a community health center in Toronto's inner city, provides mentoring and support for these youth and mandatory tutoring for those whose grades fall below a predetermined threshold. Troubled youth also have access to free lunches and bus tickets should they need them. They also get a $1000 a year that they can use toward postsecondary education. According to an assessment done by the Boston Consulting Group, the program cut the drop-out rate of 700 troubled youth from 56% to 12% and increased college attendance to the level of 80% from 20%. Ninety-three percent of those eligible for the program were enrolled and the assessors determined that there was a $600,000 lifetime benefit to society for each of these enrollments. At the time of this editorial the program was available in 11 locations in four provinces, and was slated to be up and running in 20 locations by 2016.

Inmates generally have no control over which prison they are sent to. Once they have done some time, inmates may request that they be moved to a facility that is closer to their family and friends, but such considerations are not a priority for classification and are more an option for adult males, as the facilities available for transfer for adult females and juveniles are much more limited because there are fewer of them.

Prison Types and Levels

Prisons were and are used for long-term and convicted offenders who are to be simultaneously punished (experience retribution), deterred, and reformed (rehabilitated) while being isolated (incapacitated) from the community and, for most, reintegrated back into that community. As the number of prisons has expanded across the United States, their diversity has increased. Rather than just an all-purpose maximum/medium security prison, as was the norm when prisons were first built, there are state and federal prisons with myriad security levels, including super maximum, maximum, medium, and minimum. There are prisons for men, for women, for men and women, for children, and for military personnel. Prisons come in the form of regular confinement facilities, but also prison farms, prison hospitals, boot camps, reception centers, community corrections facilities (sometime known as work release or day reporting facilities), and others (Stephan, 2008).

As indicated in Table 7.1, as of 2005 (the latest Census of State and Federal Correctional Facilities data available at the time of writing), there were 1,821 state and federal prisons in the United States, 1,406 of which were public and 415 were under private contract with either a state or the federal government (Stephan, 2008, p. 2). Most prisons were operated or under contract with the states (1,719) rather than at the federal level (102). Although only about one-fifth of the prisons in the United States are designated as maximum security, because of their size they hold about a third of the inmates incarcerated in this country (Stephan, 2008, pp. 2, 4). In contrast, medium security prisons constitute about one-fourth of prisons, but hold two-fifths of inmates—again, perhaps because of their relatively large size compared to minimum security prisons, which are about a half of all prisons, but hold only about one-fifth of inmates (Stephan, 2008, p. 4). These data do indicate that the popular and academic depictions of maximum security prisons as the norm in America are incorrect.

Table 7.1

State and Federal Correctional Facilities

All	Public	Private	Max	Med	Min
1,821	1,406	415	372	480	969

Source: Stephan (2008).

Supermax Prisons

When states were first building prisons, they tended to be a combination of a maximum and medium security type (think industrial and big house prisons). The exterior of these prisons was very secure, but internally inmates were given some, though restricted, freedom to move about and were often expected to do so for work, dining, and related purposes. **Supermax prisons** developed later, and arguably the first of these was at the federal level with the Alcatraz Prison, which was built in 1934 to hold the most notorious gangsters of its era (featured in In Focus 7.1). Today, supermax prisons at the federal and state level are not all operated exactly the same, though certain characteristics do appear to be common: Inmates are confined to their windowless cells 24 hours a day, except for showers 3 times a week (where they are restrained) and solo exercise time a couple times a week; they eat in their cell, and often it is nutraloaf (a bland, but nutritious food that requires no utensils); if any limited rehabilitation is provided, the treatment personnel stand outside the cell and talk to the inmate within; physical contact is prohibited unless inmates are in restraints (Pizarro & Narag, 2008).

In Focus 7.1

Alcatraz: The United States' First Supermax

Photo 7.1
Alcatraz Prison

As Ward and Kassebaum (2009) note in their book *Alcatraz: The Gangster Years,* when Alcatraz was first opened as a federal prison in 1934, it was created in response to a perceived national crisis. This crisis involved gangsters and outlaws who were terrorizing communities with kidnappings, bank and train robberies, and organized crime. Moreover, there were a number of scandals where these gangsters and outlaws were effectively corrupting state and federal prison officials and staff with their money, contacts, infamy, and through intimidation.

Corrupt and poorly managed, they (federal prisons of the time) were widely perceived as coddling influential felons by permitting special privileges and allowing them to continue involvement in criminal enterprises from behind bars, while flaws in their security systems offered them opportunities for escape. (Ward & Kassebaum, 2009, p. 2)

(Continued)

(Continued)

Al Capone, one of the most notorious gangsters of this era, when incarcerated at the Eastern State Penitentiary in Pennsylvania, was provided by the warden with a single cell, furnished like a home; a cushy library clerk job; unlimited visits by family and friends; and use of the telephone to contact his lawyer, crime partners, and politicians. Though he and his gang were implicated in a number of murders and graft of all sorts, he was able to evade prosecution for most of those crimes, and it was a tax evasion conviction that landed him in federal prison.

Alcatraz, on an island in the San Francisco Bay, was created to end the undue influence on incarceration that the Capones of the world exerted. "Surrounded by cold ocean currents, it was intended to hold the nation's 'public enemies' to an iron regimen, reduce them to mere numbers, cut them off from the outside world, and keep them locked up securely for decades" (Ward & Kassebaum, 2009, p. 2). News reporters were prohibited from interviewing Alcatraz staff or inmates, and staff were told not to talk to reporters or they would be fired. Visits were limited to a few blood relatives, and then only once per month and through a guard-monitored telephone. Inmates were not allowed to talk about their life on the inside during those visits. That life was very controlled and monotonous and consisted of strict adherence to silence at night, work (when earned as a privilege), and not much else. No effort was made to rehabilitate these inmates, as they were supposed to be the most incorrigible troublemakers of all federal prisoners. There was no commissary, so there was no underground economy in goods. Inmates were out of their cells from 6:30 in the morning, if they had a job, until 4:30 in the afternoon. From then on, all they had were small crafts, reading, letter writing, and time for reflection.

In practice, over the 30 years of its existence (it closed in 1963), Alcatraz never lived up completely to the ideal, though Ward and Kassebaum (2009) claim it may have rehabilitated a number of inmates. The maintenance of silence, except at prescribed times, had long since been abandoned. Many inmates had attempted escape, inmates held strikes to protest conditions, and a few inmates and staff were assaulted—including a warden—or killed by inmates in fights and escapes (It is possible that two men did succeed in escaping, but there is some debate about this.). The internal controls were never as harsh or effective as they were supposed to be, either, though more than one inmate of the early years characterized incarceration there as tantamount to being buried alive (Ward & Kassebaum, 2009). Ex-inmates also talked to reporters once released, and the gunfire, fires, and sirens from the various events were noticed and speculated on by San Francisco reporters across the bay. Moreover, as there were not enough notorious and incorrigible inmates in the federal system to fill Alcatraz, about two-thirds of its inmates were less dangerous and influential than had been planned for (Ward & Kassebaum, 2009, p. 459). Yet Alcatraz became the model for other federal and state supermaxes that followed as, for the most part, it had avoided most of the corruption and violence apparent in other prisons and it had, for the most part, effectively controlled the gangsters and outlaws of its time (Ward & Kassebaum, 2009).

Alcatraz closed in 1963, and other federal supermaxes in Marion, Illinois (opened in 1963 to take some of the Alcatraz inmates) and later Florence, Colorado, in 1994, took its place. States began building supermaxes in earnest in the 1980s and 1990s, in reaction to the felt need to control more dangerous and disruptive inmates and to "get tough" with them (Mears, 2008; Olivero & Roberts, 1987; Richards, 2008). As of 2004, a total of 44 states had some form of a supermax (Mears, 2008, p. 43), though definitions of what a supermax is does vary across the states (Naday, Freilich, & Mellow, 2008). As with Alcatraz, these supermax prisons at the federal and state levels are supposed to hold the most dangerous offenders, violent gang members, those who cannot behave well in lower-security prisons, and those who pose an escape risk.

Because of the heightened security requirements, incarceration in a supermax is expensive at about $50,000 per year as compared to $20,000 to $30,000 to incarcerate an adult male in lower-security prisons (Richards, 2008, p. 18), and because of the materials used in their construction, they are at least 2 to 3 times more expensive to build than a "regular" prison (Mears, 2008). The research indicates that wardens believe that the presence of a supermax in a prison system deters violent offenders, increases order and control, and reduces assaults on staff in the other prisons in that system, not just in the supermax itself (Pizarro & Narag, 2008). Several states (e.g., Texas, Colorado, and California) have claimed that violence decreased in their systems once they opened a supermax. Sundt, Castellano, and Briggs (2008) reported that inmate assaults on staff decreased in Illinois once the supermax was opened in that state, though there was no effect for inmate assaults on inmates. Clearly, there is room for more systematic examination of the data, as in this study, of these and other issues as they relate to supermaxes.

For instance, critics and some researchers claim that inmates' mental health is impaired after a stay in a supermax, because of the sensory deprivation, and there is some evidence to support this assertion. There is also evidence that supermaxes are sometimes used to incarcerate those who are merely mentally ill or who have committed more minor infractions (Mears, 2008; O'Keefe, 2008). In addition, the effect of incarcerating less serious or mentally impaired offenders in a supermax, as King, Steiner, and Breach (2008) note, can be a self-fulfilling prophecy of exacerbating inmate mental and behavioral problems through such secure and severe confinement. On the other hand, there are researchers who have found evidence that such a stay had a calming effect on inmates, allowing them to reflect on their wrongs and how they might change their behavior (e.g., see Ward & Kassebaum, 2009). Pizarro and Narag (2008) note, however, that the evidence is weak on both sides of this argument and that more, and more rigorous, research is merited before we know the true effect of supermaxes on inmates or on prison systems.

Maximums, Mediums, and Minimums

State and federal and military laws, traditions, and practices differ on how each type of prison operates, but some generalizations about how prisons with different security levels operate are usually accurate. Those prisons with the greatest internal and external security controls are the super maximums, and next in security are the **maximum security prisons**. Inmates in supermaxes, and less so in regular maximum security prisons, are often locked up all day, save for time for a shower or recreation outside of their cell, and they are ideally in single cells deprived of other sensory stimulation. Visits and contact with the outside are very restricted. The maximum and supermax exterior security consists of some combination of layers of razor wire, walls, lights, cameras, armed guards, and attack dogs on patrol.

As the states that have a supermax usually only have one, maximum security prisons are responsible for holding most of the serious offenders, and those who could not handle themselves in the relatively freer environment of the medium and minimum security prisons. The latter type of inmate might be able to qualify for a medium or minimum security level classification, but instead is in maximum security because the inmate is unable to control his or her behavior.

In many states where the death penalty is legal, their death row is located at a maximum security prison. Death rows are usually wholly separate areas of the prison, sometimes in a different building, and often have their own separate designated staff and procedures. (For a more involved discussion of the death penalty, see Chapter 13.)

Maximum security prisons may have the same exterior security controls as supermaxes, but inside, inmates are not locked down as much, though the treatment and work programming is much more constricted than in the medium security prisons. Maximum security inmates may or may not be double-bunked, depending on the crowding in the institution, and unless under some special classification, they have some access to the yard (a large gathering area for inmates), the cafeteria, and the chapel. Visiting and contact with the outside world are less restricted than in the supermax, and inmates are usually not in some kind of restraint when it occurs.

In **medium security prisons**, the exterior security can be as tight as it is for the supermax and the maximum security prisons, but internally the inmate has many more opportunities to attend school, treatment, and church programming and to work in any number of capacities. There is also greater diversity in rooming options, from dormitories to single cells, with the more preferred single or double cells used as a carrot to entice better behavior. Visiting and contact with the outside world are less restricted. Some medium security inmates may even be allowed to leave the institution for work-related deliveries or on furloughs, though this is much more common in minimum security prisons.

Medium security prisons will hold a mix of people in terms of crime categories, all the way from the convicted murderer doing life, but who programs well, down to the lowly burglar or drug user who is awaiting transfer to a lower-security prison or who is engaged in the substance abuse programming that the prison affords. Medium security prisons are more likely to have a college campus type interior, with several buildings devoted to distinct purposes. There might be a separate cafeteria building, a separate programming and treatment building, a separate gym and recreation building, and separate work and housing buildings. Medium security prisons are heavily engaged in industrial work such as building furniture, making clothing, and printing license plates for the state. In some cases and states, the goods produced in the prison are sold on the open market.

Minimum security prisons have a much more relaxed exterior security; some do not even have a wall or a fence. Inmates are provided with far more programming, either inside the institution or outside in the community. The housing options are often as diverse as in the medium security prisons, and inmates can usually roam the facility much more freely, availing themselves of programming, recreation, the yard, the chapel, and the cafeteria at prescribed times. With the recognition that inmates in a minimum security prison will often be free within a year or two, visiting options are more liberalized to make the transition from prison to the community smoother. Work is promoted, and inmates are often encouraged, or in the case of work release facilities with a minimum security classification, are expected, to work in the community.

Inmates confined to minimum security prisons are usually "short timers," or people who are relatively close to a release date. These could be people who have been classified directly to this prison or work release facility because they received a sentence of a year or two and because they are not expected to be an escape risk or behavioral problem. Whether they can work might also be a consideration in classification, as this is often a central element of these prisons. Other inmates who might do time in a minimum security prison or work release facility are more serious offenders who have moved through, or "down," the other classification levels and are relatively close to their release. Minimum security prisons thus also hold the most serious offenders, including murderers, rapists, and child molesters, along with those convicted of burglary and substance abuse and trafficking offenses. The difference is that in minimum security prisons, all such offenders, no matter their offense, are believed to be a good risk for behavior and in need of preparation for their imminent release.

❖ Attributes of the Prison That Shape the Experience

Total Institutions, Mortification, Importation, Prisonization

Once classified to a given prison, whether maximum or otherwise, the inmate experience is shaped by several factors, including the *operation* of it. One central component of that operation is the "totality" of the organization.

As will also be discussed in Chapter 8 in reference to staff, Erving Goffman (1961) coined the term "total institution" to describe the nature of mental hospitals, but also prisons, in the United States in the 1950s. For one year, he served as a staff member (athletic director's assistant) and did ethnographic research in a federal mental health hospital in Washington, D.C. While avoiding sociable contact with staff, he immersed himself in the inmate world, or as much as he could without being admitted to the hospital, and what he observed allowed him to learn a great deal about that kind of institution and about roles for staff and inmates.

Goffman (1961) defined a **total institution** as "[a] place of residence and work where a large number of like-situated individuals, cut off from the wider society for an appreciable period of time, together lead an enclosed, formally administered round of life" (p. xiii). Another key component of this total institution is the defined social strata, particularly as that includes "inmates" and the "staff" (p. 7). Specifically, there are formal prohibitions against even minor social interactions between these two groups in a total institution, and all of the formal power resides with one group (the staff) over the other group (the inmates).

This definition is directly applicable to prisons, even today, though it more aptly described both prisons and jails of the past. For prison inmates, the institution is where they live, and often work, with people who are like themselves, not only in terms of criminal involvement, but also largely in terms of their social class and other background characteristics. Though there is some ability to visit with others, the mode and manner of this contact with the outside world are quite limited in prisons and are also dependent on the security status of the institution (e.g., whether it is a work release facility or a maximum security prison). The formal rules of prisons also closely control inmate behavior and movement. As already mentioned, another key formal attribute of total institutions governs interactions between staff and inmates. Simply put, staff are to restrict such interactions to business only and are to parcel out information only as absolutely necessary. As Goffman (1961) put it, "Social mobility between the two strata is grossly restricted; social distance is typically great and often formally prescribed" (p. 7).

How do these aspects of total institutions affect the lives of inmates? In the 1950s, Goffman (1961) believed that total institutions had the effect of debilitating their inmates. As he saw it, upon entrance into the institution, the inmate may become *mortified* [known as **mortification**], or suffered from the loss of the many roles he or she occupied in the wider world (see also Sykes, 1958). Instead, only the role of "inmate" is available, a role that is formally powerless and dependent.

In addition, though each person entering a prison *imports* (known as **importation**) aspects of his or her own culture from the outside, to some extent inmates are likely to experience **prisonization,** whereby they adopt the inmate subculture of the institution (Carroll, 1974, 1982; Clemmer, 2001). Couple this mortification, and subsequent role displacement, with the prisonization into the contingent inmate subculture, and you have the potential for the new inmate to experience a life in turmoil while he or she adjusts, and some difficulty when reentering the community.

Pains of Imprisonment

Part and parcel of this inmate world are what Gresham Sykes (1958)—based on his research in a New Jersey maximum security prison—described as the **pains of imprisonment**. Such pains include "the deprivation of liberty, the deprivation of goods and services, the deprivation of heterosexual relationships, the deprivation of autonomy, and the deprivation of security" (pp. 63–83). Inmates in a prison (or jail) are not free to leave, or even to move about the institution without the permission of their keepers (staff). But for Sykes, the worst of the liberty restrictions meant that inmates were cut off, for the most part, from family and friends. They cannot call whomever they like or visit with whom they want, when they wish to do so. As many inmates are functionally illiterate, and poor, they also have difficulty writing letters and affording the postage. This deprivation of contact with family members, particularly their children, is a severe pain that many inmates experience when, as an artifact of their incarceration, they are unable to have regular interactions with their own children or to have any control over their child's environment on the outside (more about this in the chapter on gender, Chapter 10) (Gray, Mays, & Stohr, 1995; Stohr & Mays, 1993).

As to the pain related to goods and services, inmates are required to surrender all of their property upon entrance into the prison system and, in most cases, they cannot have it back until they leave. The property they are allowed to legally possess is very limited and monitored closely by staff. Relatedly, they cannot choose who will cut their hair or where they will get their nails done, nor can they choose their doctor or schedule a visit. As Sykes (1958) noted—and this is perhaps even truer today in many prisons because of court intervention—most inmates' basic needs for food, shelter, space, and health care are met, yet it is the perception of deprivation in this material society that matters, too.

In light of the greater knowledge we have regarding sexual orientation of human populations today, as opposed to 50 years ago, we might amend Sykes's (1958) "deprivation of heterosexual relations" to a more generalized *deprivation of sexual relations*. An inmate's access to significant others in the wider world is limited to visiting where touching is only minimally sanctioned (e.g., a brief kiss or hug at the beginning of the visit). Though much is made of conjugal visits for prison inmates, in reality there are few prisons that allow these, and a miniscule number of inmates even in those prisons are granted access to such visits. A very few prisons do allow conjugal visits between inmates and their gay or lesbian partners. Though, as with the free population, there are likely to be 3% to 5% of prison inmates who are gay or lesbian (see a brief discussion related to this topic later in this chapter), sexual intimacy between same-sex inmates is against the rules, though it does occur, and illegal as well between same and opposite sex staff and inmates.

Autonomy for the inmate is also severely restricted in the rule-bound prison world. When, how, where, and with whom they live, eat, work, and play are all determined by the rules of the institution. Inmates can make few choices regarding their lives while imprisoned, and all of those choices are shaped by their imprisonment.

Photo 7.2

Visiting hours at the Youngstown Prison, Ohio. An inmate's access to significant others is limited to visiting where touching is only minimally sanctioned (e.g., a brief kiss or hug at the beginning of the visit).

Due to their imprisonment, inmates are thrown together with others, some of whom are aggressive and violent or become so in a prison environment, perhaps particularly in the maximum security environment that Sykes was studying. Because of the circumstances surrounding incarceration in a supermax, but also to a lesser degree in medium and minimum level prisons, inmates are deprived of their security, a basic human need as defined by Maslow (1943/2001). Quoting an inmate in his study, Sykes (1958) repeated that "the worst thing about prison is you have to live with other prisoners" (p. 77), meaning that even if one is prone to violence or manipulation (termed an "outlaw" inmate by Sykes [p. 77]), and not all inmates are, it is unnerving for even an outlaw inmate to have to live with others who are also so inclined. This lack of security, according to Sykes, can lead to anxiety on the part of inmates and the belief that at some point they, whether an outlaw among outlaws or not, are likely to be forced to fight to defend themselves or to submit to the abuse of others.

Sykes (1958) argued that these pains, though not physically brutalizing, have the cumulative effect of destroying the psyche of the inmate. In order to avoid this destruction, inmates in prisons may be motivated to engage in deviance while incarcerated as a means of alleviating their pain. So bullying other inmates, involvement in gangs, buying items through the underground economy, and homosexual acts might all be motivated in fact by the need for some autonomy, liberty, security, goods and services, and sexual gratification (Johnson, 2002). Extrapolating from this point, the extent to which female inmates form pseudo-familial relationships may be a means of alleviating the pain experienced due to the separation from children and other close family members (Owen, 1998).

One final note regarding the pains of imprisonment: Sykes (1958) did not believe that all inmates experienced or perceived these pains in the same way. He acknowledged that the way in which one experiences these pains does vary some by individual and by background, as well as by the prison one is incarcerated in. However, he argued that, at least among the inmates he studied, there was a consensus that "life in a maximum security prison is depriving or frustrating in the extreme" (p. 63).

Photo 7.3

Inmates in the yard at Mule Creek State Penitentiary in California.

❖ The Prison Subculture

Prison subculture, or a subset of culture with its own norms, values, beliefs, traditions, and even language, tends to solidify when people are isolated from the larger culture and when members have regular and intense contact with each other for an extended period of time. In other words, it would appear that the "total institution" nature of prisons provides the perfect environment for an inmate subculture to form. Accordingly, the degree to which a correctional environment fits the definition of a "total institution" will determine the extent to which a client subculture exists. It is also possible that the shared experiences of deprivation, as detailed by Sykes (1958), can further solidify a subculture for inmates.

Thus, research on inmate subcultures has tended to focus on prison inmates and specifically on medium or maximum security prison inmates. This is not to say, of course, that those in a jail or a minimum security prison do not have distinguishable "norms, values, beliefs, and language" that sets them apart from the wider community, but it is much less likely. By definition, the longer inmates are in an institution, associating with others like them, and the more "total" the institution is in its restrictions on liberty and contact with "outsiders," the more subjected inmates are to the pains of imprisonment, and the more likely they are to become "prisonized" in that they adopt the inmate subculture.

Indicators of such a subculture, as identified by prison researchers, include prescribed values and defined roles for inmates (Clemmer, 2001; Owen, 1998). For instance, Clemmer in 1940 (reprinted in 2001) broadly defined criminal subcultural values as including "the notion that criminals should not betray each other to the police, should be reliable, wily but trustworthy, coolheaded, etc." (p. 7). Also emphasized among these values are ultra masculinity and displays of toughness, and solidarity among inmates and against staff (Lutze & Murphy, 1999; Sykes, 1958). Accordingly, the types of roles (inmate typology) that Sykes and Messinger (1960) and Sykes described for adult male prison inmates in their own *argot*, or language, are detailed in In Focus 7.2. Though these researchers identified these roles for inmates in prisons over 50 years ago, current researchers still see them in prisons of today. Of course, any given inmate might be expected to change roles from time to time during his or her incarceration, or to engage in more than one role simultaneously.

In Focus 7.2

Inmate Roles Identified by Sykes (1958) and Sykes and Messinger (1960)

- **The right guy:** the inmate who fully supports and embraces the inmate code.
- **The rat or squealer:** the inmate who "snitches" on others to staff. Usually such inmates are despised by other inmates as they violate the prohibition in the inmate code regarding this sort of behavior. Because they are so disliked, rats are vulnerable to attacks by other inmates.
- **The center man:** the inmate who agrees with the prison rules and procedures either because he is trying to curry favor from staff or because he believes that is the correct way to behave.
- **The tough:** the inmate who is aggressive, has anger issues, and is touchy, and so is willing to fight at will.
- **The hipster:** the inmate who aspires to be a tough, but who is really all talk and little action. He chooses his victims selectively to be sure he can in fact conquer them, whereas the tough will fight both the weak and the strong.
- **The gorilla:** the inmate who may be as aggressive as the tough, but uses that aggression, or just the threat of it, to gain something from other inmates.
- **The merchant or peddler:** the inmate who engages in the underground black market to supply illicit goods and services to other inmates for material advantage.
- **The weakling:** the inmate who is vulnerable to exploitation by others and cannot stand up to them. Someone who submits to the coercion, or threatened coercion, of the gorilla.
- **The fish:** the inmate who has just arrived and is not yet adjusted to the ways of the prison.
- **The wolf:** the inmate who aggressively and sexually pursues other inmates. He is believed to play a "masculine" role. There is no emotion or connection in the sexual act for the wolf; rather, he pursues men in the prison and rapes them.
- **The "fag":** the inmate who plays the passive, though not unwilling, role in the sexual relationship with another inmate. He is believed to play a "feminine" role. (Note: The authors of this text acknowledge that the term "*fag*" is derogatory, and the assumption that male and female roles are tied to aggression and passivity is dated and limiting, but this term and these assumptions are the ones used by Sykes [1958] and Sykes and Messinger [1960], who were repeating the argot used by inmates of the late 1950s. As Sykes explains, inmate argot was often meant to be derogatory and inflammatory as it marked and labeled the roles of others.)
- **The punk:** the unwilling inmate who is coerced or bribed into the passive sexual role vis-à-vis other inmates. This inmate does not adopt the "feminine" role.
- **The ball buster:** the inmate who continually struggles against the system and staff, despite the futility of it, often to the point of foolishness.
- **The innocent:** the inmate who repeatedly claims his innocence of the crime he is incarcerated for.
- **The square john:** the inmate who does not become prisonized, but identifies with the free world values of staff and the outside community.

These roles are played out in the prison in a criminal subculture, which becomes a "convict subculture" for Clemmer (2001) when such inmates seek power and information so that they might get the goods and services they desire to alleviate those pains of imprisonment. Owen (1998) noted that some women engage in a version of this subculture and these roles, although it might be tempered by the relationships they had and the goods and services they needed (more about the different roles women inmates adopt in Chapter 10). Notably, both Clemmer and Owen, however, found that a significant portion of inmates in the male and female prisons they studied were not at all interested in being involved in the convict subculture or the "mix" of behavior that can lead to trouble in prisons. Such inmates, in the argot identified by Sykes (1958) and Sykes and Messinger (1960), were "square johns." These inmates either chose to not connect to the inmate subculture, or they held on to more traditional and legitimate values from the larger culture.

Moreover and relatedly, more recent research confirms that inmates are not as solidly aligned against staff as the early works would indicate (Hemmens & Marquart, 2000; J. B. Jacobs, 1977; Johnson, 2002; Jurik, 1985a; Jurik & Halemba, 1984; Lombardo, 1982; Owen, 1998). Many inmates identify with free world values, as much or more than "inmate values" and inmate subculture.

Other recent researchers and writers on prisons (e.g., see Conover, 2001; Johnson, 2002; Rideau, 2010) find that staff and inmates engage in more personal and informal relationships with each other than is formally acknowledged, a reality that Sykes (1958) noted as well. The diversification of staff by race, ethnicity, and gender has changed the old dynamic between staff and inmates, making staff less dissimilar to inmates and the inmate world less "masculinized" than it previously was (for a more involved discussion of these matters, see Chapters 10 and 11), perhaps continuing to break down some of the more formal barriers between staff and inmates in a total institution.

❖ Gangs and the Prison Subculture

Gangs, or groups of people with similar interests who socialize together and who may engage in deviant or criminal activities, are a common phenomenon in jails and prisons. According to the U.S. Department of Justice (2010) website, gangs in prisons and jails are by definition engaged in criminal activities and are connected through members and criminal involvement with communities. Prison gangs have a hierarchical organizational structure and a set, and often strict, code of conduct for members. As reported by the United States Department of Justice,

> Prison gangs vary in both organization and composition, from highly structured gangs such as the Aryan Brotherhood and Nuestra Familia to gangs with a less formalized structure such as the Mexican Mafia (La Eme). Prison gangs generally have fewer members than street gangs and are structured along racial or ethnic lines. Nationally, prison gangs pose a threat because of their role in the transportation and distribution of narcotics. Prison gangs are also an important link between drug-trafficking organizations (DTOs), street gangs, and OMGs [outlaw motorcycle gangs], often brokering the transfer of drugs from DTOs to gangs in many regions. Prison gangs typically are more powerful within state correctional facilities rather than within the federal penal system. (p. 1)

Correctional scholars and practitioners believe gangs are so ubiquitous in corrections because they meet the needs of inmates for security, goods and services, power, and companionship. They lessen the pains of imprisonment by providing protection in numbers and the potential to respond with force to any threats an inmate might face. They are conduits for the supply of illicit goods like tobacco, drugs, alcohol, and sex in prisons and jails. They also provide some substitution for the diminished relationships that inmates have with those family and friends on the outside.

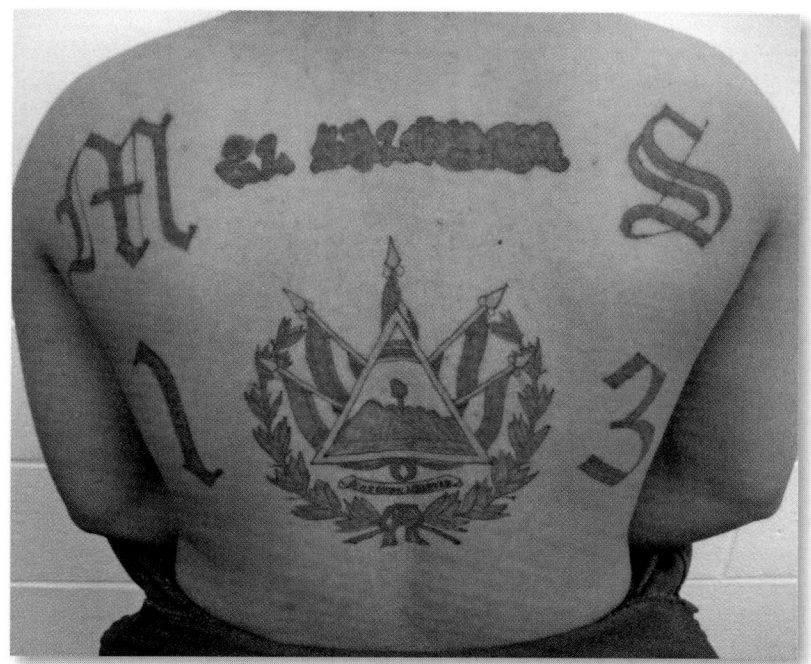

Photo 7.4

A California prison inmate displays his tattoos.

The history of gangs in prisons is a long one. Sykes (1958) noted that the first investigation of the New Jersey State Prison, in 1830, found what they called a "Stauch-Gang" firmly entrenched there and engaged in terrorizing both inmates and staff, while also planning escapes (p. 92). Ward and Kassebaum (2009) also point out the importation and exportation of gang-related criminal activity between state and federal prisons and the streets in the 1920s. J. B. Jacobs (1977), in his history of the Stateville Prison in Illinois, observed that prison gangs have existed in that state for decades, as imports from the streets of Chicago, though he thought their ferocity and strength increased in the late 1960s and early 1970s.

The prison gangs of today are almost too numerous for correctional authorities to keep track of, but do tend to have in common a criminal focus. According to the Florida Department of Corrections (FDOC) (2010), most prison gangs these days recruit their membership based on ethnicity or race. Both the federal government and the FDOC report that gangs are much stronger in male prisons, and gangs will conspire with others, even rival gangs, so as to provide protection and increase their criminal reach (e.g., the Aryan Brotherhood might sometimes work with members of the Black Guerrilla Family—despite the racial hatred of their members for each other—if it will increase their drug sales). The FDOC website identifies the six major prison gangs in America as follows:

1. Neta (Puerto Rican-American/Hispanic)

2. Aryan Brotherhood (white)

3. Black Guerrilla Family (black)

4. Mexican Mafia (Mexican-American/Hispanic)

5. La Nuestra Familia (Mexican-American/Hispanic)

6. Texas Syndicate (Mexican-American/Hispanic)

Because of their underground engagement in prison crime and the rivalries that develop between gangs, even those with members of the same ethnic backgrounds (e.g., the Mexican Mafia and La Nuestra Familia) are sworn enemies. Moreover, both the Mexican Mafia and La Nuestra Familia are rivals of the Texas Syndicate. Because of these rivalries, jails and prisons constantly have to consider gang membership in classification decisions. Whether it is fights over turf, protection of members, or some other issue, the presence of gangs and gang activities leads to disruption and even murder in prisons. For instance, according to the FDOC (2010) the Aryan Brotherhood (AB) disruptions of prisons include the following:

- The main activities of the AB are centered on drug trafficking, extortion, pressure rackets, and internal discipline.
- Prison activities include introduction of contraband, distribution of drugs, and getting past facility rules and regulations.
- Traditionally, targets have been non-gang inmates and internal discipline.
- From 1975 to 1985, members committed 40 homicides in California prisons and local jails, as well as 13 homicides in the community.
- From 1978 to 1992, AB members, suspects, and associates in the federal system were involved in 26 homicides, 3 of which involved staff victims.

Because of the threat prison gangs present for the security of the institution and the safety of staff and inmates, managers try to control or suppress gang involvement in their facilities. The first step in this process is the identification of gang members and their leaders. Once identified, correctional staff will try to separate members and leaders from each other. However, given the crowding of most prisons and prison systems, it is almost impossible to always employ the separation tactic as a means of control and suppression. Therefore, what they are often left with is the monitoring of gang activity and, as much as possible, punishing or neutralizing gang members and reducing their impact in a given system.

❖ Violence

Why Prisons Are Violent

Violence is endemic to prisons. There is violence in prisons because incarcerated people are there unwillingly; forced to do things they normally would not do, with people they may not like; and most important of all, some of them are inclined to be violent. According to the *Sourcebook of Criminal Justice Statistics* (2008), in 2006 about 50% of the state prison population were incarcerated for violent offenses. As was mentioned earlier, prisons as a whole are running at 109% of capacity. Maximum and medium security prisons tend to hold more inmates convicted of violent offenses or who have problems with following the rules in prisons, and as a result they are more likely to experience violent outbursts. Add to this mix the presence of gangs and their willingness to use force to achieve their criminal ends, and the possibility of violence in prisons rises.

The Amount of Violence

However, with the exception of deaths due to violence, it is difficult to determine the exact amount of violence in prisons. Correctional institutions tend to underreport its incidence, and there is variation across facilities about what constitutes violence (Byrne & Hummer, 2008). Relatedly, inmates are reluctant to report the violence they experience or witness to staff. With these caveats regarding official statistics in mind, Stephan and Karberg (2003) found, using data collected from correctional institutions around the United States, that the number of assaults on staff and inmates increased in state and federal prisons from 1995 to 2000 and that the size of the increase was greatest for private institutions. They do note, however, that the rate of assault on staff (or the number of assaults per staff person) decreased slightly during this time period. We also know, based on the BJS data presented in Figure 7.1, that both the suicide and homicide rates in jails and prisons have decreased from 1980 to 2003.

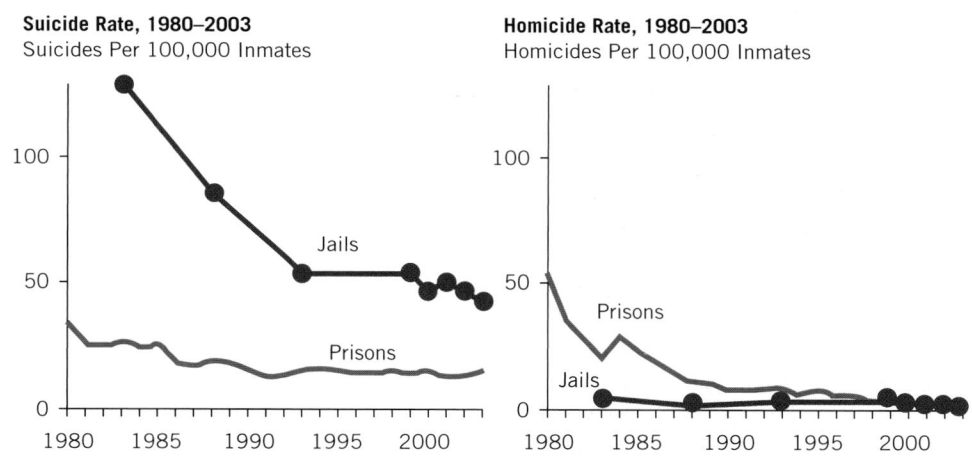

Suicide Rate, 1980–2003
Suicides Per 100,000 Inmates

Homicide Rate, 1980–2003
Homicides Per 100,000 Inmates

Figure 7.1

Source: Bureau of Justice Statistics (2010).

Sexual Assaults

As was discussed in Chapter 5, the amount of sexual violence in prisons and jails, based on inmate surveys and official statistics, is becoming increasingly clear. The data indicate that 4.4% of prison inmates experienced sexual assault in a given year, which was perpetrated by either other inmates or staff (Beck & Harrison, 2010, p. 1). Whether the amount of such assaults is trending up or down and why it might be more prevalent in some facilities over others are matters yet to be determined by more longitudinal research. Certainly, though horrific for the inmate experiencing it, and keeping in mind the likely underestimation of victimization, a 4.4% sexual victimization does not fit the "myth" of sexual victimization that many people believe regarding prisons, a myth that predicts that everyone, particularly young males, who goes to prison will be raped (Fleisher & Krienert, 2009).

It might well be that this myth has more than a grain of truth in it, in that prison rapes are underreported or that in the past it was much more common. The percentage of those raped or sexually assaulted may have declined, the way the rate of prison assaults appears to have done (though not the number) over the years as corrections has become less "total" in its isolation and operation, and as the courts have recognized some inmates' rights and training of staff has increased. As Byrne, Hummer, and Stowell (2008) note, there are many factors (e.g., staffing levels and crowding) whose effect on violence is not yet established in the research.

Rioting

Rioting is another form of violence. It is group violence. Rioting presents a direct threat to the security of the institution and the inmates in it and is often met with reciprocal force by the staff and administration of the prison. Prison riots have existed as long as there have been prisons, in fact before there were prisons. Recall the Newgate "prison" of Connecticut (see Chapter 2), where inmates were confined to a copper mine for much of their incarceration. Inmates at Newgate repeatedly rioted throughout the history of its operation. In fact, virtually every maximum and medium security prison, with any longevity, has experienced some form of rioting by inmates. Rioting, and violence in general, is engaged in by inmates to achieve some end like better food or housing or power, or inmates might riot out of anger or frustration. When violence is used to achieve some end, it is known as *instrumental violence*, but when it is just an angry outburst, then it is known as *expressive violence*. Of course, inmates engaged in violence or a riot could be involved for *both* instrumental and expressive reasons. An inmate who wants to protest the overcrowding of her institution may riot to let the world know about the conditions of confinement (instrumental violence), but she might also be angry about the effect such crowding has on housing and the ability to sleep, and become violent as a means of expressing it. When enough inmates engage in this violence together, it is called a riot.

The two most notorious instances of inmate rioting to date occurred in the **Attica** (New York) **Prison Riot** of 1971 and the **New Mexico Prison Riot** of 1980. At Attica, the riot began with a spontaneous act of violence by one inmate against an officer when the officer tried to break up a fight. The violence spread when other inmates became involved the next day to avenge the punishment of the two fighting inmates (Public Broadcasting Service [PBS], 2000). The riot also spread because inmates were frustrated and angry about the overcrowded conditions and lack of programming for inmates, among other problems with the conditions of confinement; even showering and toilet paper were rationed. There were charges of racism by the mostly African American inmates regarding their treatment by the mostly white staff at Attica as well (Useem & Kimball, 1989). Add to this the student protests against the Vietnam War and the civil rights movement that had roiled the country outside of the Attica prison walls in the late 1960s/early 1970s, and it was clear why there was tension within them.

Being that the prison staff were unprepared to respond to a riot, the inmates easily took over the Attica prison, burning some buildings and eventually congregating in one yard with their 40 hostages (PBS, 2000). In the negotiations between the inmate leaders and the administration, the inmates asked for better food, health care, and the ability to practice their religion. A number of observers, composed of politicians and media members, tried to intervene in the negotiations, but to no avail. Some inmates killed three other inmates and one hostage, which also impaired the ability to negotiate. Moreover, Governor Rockefeller, who was considering a run for the White House at the time, did not want to appear soft on crime by being too soft on the rioting inmates.

In the end, the inmates and administrators could not come to an agreement (the inmates wanted amnesty for the rioters) and eventually, on Governor Rockefeller's orders, the prison was stormed by the state police and by correctional staff. Tear gas was dropped from police helicopters into the occupied yard, and the inmates and their hostages were indiscriminately fired upon with shotguns by the staff and police. As a consequence, 10 hostages and 29 inmates were dead or dying when the prison was secured, and another 80 inmates had gunshot wounds (Useem & Kimball, 1989). It was the bloodiest riot in American history. Inmates, even injured ones, were then beaten and humiliated (forced to stand naked in the yard for hours), and medical care was delayed or denied.

The state indicted 60 inmates for a number of crimes including sodomy and murder arising out of the riot, but only eight inmates were convicted (Gonnerman, 2001, p. 1). Years of legal wrangling eventually led to the $8 million award by the state of New York to the inmates who were beaten or tortured after the riot (PBS, 2000). In 2005, the state paid out another $12 million to the survivors and families of employees killed in the aftermath of the riot (Kirshon, 2010).

In 1980, the New Mexico Prison also exploded in a riot over the conditions of confinement, which were deplorable, and crowding, which was at epidemic levels. Despite repeated warnings that a riot was going to occur, the administration and staff failed to adequately prepare. When a staff member slipped up in a security measure when locking down a dormitory for the night, he was grabbed, along with his keys, and inmates quickly advanced through the prison, taking control of several cellblocks, including the pharmacy and shops. Drugs and weapons were readily available as a result, and brutal inmate-on-inmate violence ensued. Some of this violence was particularly focused on inmate snitches and child molesters who were housed in a separate cellblock. Rioting inmates broke into this cellblock and gruesome and vicious assaults and murders of these inmates were committed. The state eventually retook the prison without the resulting bloodshed that happened at Attica. However, over the course of 3 days, 33 inmates were killed by other inmates. Numerous other inmates, along with staff hostages, were beaten or raped, and millions of dollars in damage was done (Useem, 1985; Useem & Kimball, 1989). In the aftermath of this riot, New Mexico was sued several times. The state did build a number of medium and minimum security prisons following the riot, which eased the overcrowding at the main facility.

Mature Coping

Prison violence, whether by individuals or groups of rioters, occurs in part because some inmates are not capable of interacting, or do not know how to interact, with others without violence. In his research on corrections, Johnson (2002) noticed that despite the mortification, prisonization, and pains experienced to different degrees by incarcerated individuals, some were able to adjust prosocially, even to grow, in a prison setting. Though the exception rather than the rule, he noted that some inmates developed another means of adjusting. This alternative means of handling incarceration, or supervision in the case of probationers and parolees, is **mature coping**. As identified and defined by Johnson,

> Mature coping means, in essence, dealing with life's problems like a responsive and responsible human being, one who seeks autonomy without violating the rights of others, security without resorting to deception or violence, and relatedness to others as the finest and fullest expression of human identity. (p. 83)

As indicated by this definition, the offender needs to learn how to be an adult with some autonomy in an environment where formally the individual has little power (although the informal reality may be different) and his or her status is almost subhuman by wider community standards. Moreover, offenders must accomplish this feat without doing violence to others—though Johnson (2002) allows that violence in self-defense may be necessary—and they need to exercise consideration of others in their environment.

Johnson (2002) notes that mature coping is relatively rare among the inmate population for a number of reasons. He argues that inmates are typically immature in their social relations to begin with, which, of course, is one of the reasons they are in prison in the first place. Because of impoverishment, poor or absent or abusive parenting, mental illness, schools that fail them or that they fail, offenders enter the criminal justice system with a number of social, psychological, and economic deficits. They are often not used to voluntarily taking responsibility for their actions as one would expect of "mature" individuals, nor are they typically expected to "[e]mpathize with and assist others in need," especially in a prison or jail environment (p. 93).

Secondly, Johnson (2002) argues that for inmates to maturely cope, it is helpful if they are incarcerated in what he terms a *decent prison*. Such a facility does not necessarily have more programming, staffing, or amenities than the norm, though he thinks it might be helpful if it did; rather, such institutions or programs would be relatively free of violence and would include some opportunities so that inmates might find a *niche* (defined below) to be involved in. In order for inmates to find this niche, however, decent prisons need to include some opportunities for inmates to act autonomously.

Being secure from violence, like autonomy, is basic to human development. In fact, according to Maslow (1998), if the security need is not fulfilled, it will preoccupy offenders and motivate them to engage in behaviors (e.g., bullying or gang activity) that they normally might avoid if they were not feeling continually threatened (Johnson, 2002). Then, assuming that the offender perceives that he is relatively safe, there need to be prosocial activities, including work, school, athletic, church, treatment, or art programs, that provide some sort of means for positive self-value reinforcement. Such places are termed *niches* by Johnson, and the opportunities they afford provide redress for the mortification and pains that offenders, particularly those who are incarcerated, experience.

In Focus 7.3

The Story of Wilbert Rideau

Wilbert Rideau, an African American, grew up poor to laborer parents in Lawtell, Louisiana, in the 1940s and 1950s. The family moved around the segregated state from small town to small town looking for work, eventually settling in Lake Charles, Louisiana. His father drank, womanized, and abused his mother and later the children; he abandoned the family when Wilbert was a teenager, and his family went on welfare.

Lake Charles of the 1950s was segregated and restricted the opportunities of African Americans to advance in any profession. After he was denied advancement at his low-paying job at a sewing shop, and as a means of leaving town for better opportunities, Wilbert Rideau attempted to rob a nearby bank. The robbery went horribly wrong. He kidnapped three bank employees, killing one and shooting another, in the botched bank robbery attempt. Once caught, almost immediately after committing the crimes, he was nearly lynched instead of being tried, his confession was coerced, the evidence presented was false, and he had inadequate counsel. But he admitted that he had committed the murder, had shot another woman, and had kidnapped three people.

In 1961, at the age of 19, Rideau was convicted and received a death sentence. He then spent 10 years on death row at the Angola Prison in Louisiana. He was eventually released to the general population of the prison when the *Furman v. Georgia*

(1972) Supreme Court case invalidated death sentences around the country. He served the rest of his sentence, another 33 years, at the infamous Angola prison, and during most of it was the editor of the *Angolite,* an inmate newspaper.

In his book *In the Place of Justice: A Story of Punishment and Deliverance* (2010), Rideau discusses how as an inmate he grew—you might say he maturely coped—by finding a niche at the *Angolite* and doing work that was worthwhile. He won critical acclaim for the uncensored writing about prison life that he was allowed to present to the world by one forward-thinking warden. Once this pattern of an uncensored prison paper had been established, other subsequent wardens were reluctant to shut him down, though they tried. Because his writing became known in the outside world, and Rideau was even allowed by several wardens to leave the prison to talk to community groups, he had a form of power that allowed him access to wardens and directors of prisons. According to Rideau, he used this access to help some inmates, to avert violence against inmates, and to steer wardens in the direction of treatment and programming. He also had plenty of time to regret his crimes and to reflect on not only the racism he had confronted in his life, but also his own racial stereotypes. He observed the workings of the inmate code in prison and learned to walk a fine line between upholding it and staying within the rules.

Rideau was released from prison in 2005 after a court found that his original trial had been mishandled. He had spent a total of 44 years in jails and prisons since his conviction, when the norm for his crime was less than 10. He lives with his wife whom he met 20 years before on one of his speaking gigs outside of prison. She had spent many years of her life working to free him.

❖ Special Populations

The Elderly and the Physically and Mentally Ill

As mentioned in Chapter 5, the number of the elderly in jails and prisons is increasing at an exponential rate. As America ages and mandatory sentences and other such laws

Photo 7.5

The number of ill people incarcerated in America's prisons and jails has grown in tandem with the number of elderly inmates.

lengthen sentences, correctional populations are graying. There are a number of collateral consequences that derive from this fact, most of them unintended:

1. The cost of incarceration increases to accommodate the extra medical care needed to maintain older people;

2. Elderly inmates are less able to work in, or for, the prison, making them a further economic drain on the system;

3. Elderly inmates may require housing that is separate from younger inmates who may prey on them;

4. Elderly inmates, particularly those who have spent much or all of their adult lives in prisons, are less likely to have a supportive family or friends waiting for them on the outside, which makes the development of a parole or reintegration plan even more challenging for them.

As elderly inmates necessarily present such a drain on state and federal correctional budgets, it might make sense for states to rethink the sentencing laws and correctional practices that led to the graying of prison populations nationally (more about this rethinking in the last chapter of the book). To not do so is to support the continued exponential growth in correctional budgets at the expense of all other budget priorities.

The number of ill people incarcerated in America's prisons and jails has grown in tandem with the number of elderly inmates. At this juncture, more than 33% of inmates in jail, 44% in state prison, and 39% in federal prison report an illness more serious than a cold or the flu (Maruschak, 2008, p. 1; see also 2006). According to a 2004 survey of state and federal prison inmates by the Bureau of Justice Statistics, the two most prevalent medical problems for prison inmates were arthritis and hypertension (Maruschak, 2008). Women and elderly inmates in prisons, as in jails, report more medical problems than do other inmates.

The extent of medical care provided for such inmates depends on the jurisdiction, with some larger counties and some states, and the Bureau of Prisons at the federal level, providing better care than other jurisdictions. According to that 2004 study, about 70% of the state and 76% of the federal prison inmates with medical problems reported seeing a medical professional at the prison about their illness, and more than 80% reported receiving a medical exam since their admission (Maruschak, 2008, p. 1). However, even in those jurisdictions that can afford to, and do, provide decent medical care, it is often minimal. Dentistry typically consists of pulling teeth rather than crowning or even filling them. Not much preventive medical care is provided, and the common response to complaints is the provision of medication.

Most larger jails have a section devoted to their inmates with medical complaints. Larger prisons or prison systems often have buildings or whole institutions devoted to inmates with medical maladies. The staffing of such sections, buildings, or institutions again varies by jurisdiction and the ability and willingness to pay the high cost for qualified staff (Vaughn & Carroll, 1998; Vaughn & Smith, 1999). Working in a jail or prison medical facility has not usually been the first choice of medical personnel, and so it is not surprising that it might be hard to recruit and keep the best personnel.

The number of mentally ill inmates has also grown in America's prisons, though not to the same extent as it has in the jails (see Chapter 5). As jails became dumping grounds for the mentally ill after mental health hospitals closed in the 1970s, some of these inmates with chronic mental illnesses have found themselves in a prison environment (Slate & Johnson, 2008).

The **deinstitutionalization of the mentally ill** in the United States came about as a result of the civil rights movement and the related effort to increase the rights of powerless people (Slate & Johnson, 2008). Too many people were civilly committed to mental health institutions for years without any legal recourse or protection, it was thought. In addition, the pharmaceutical company Smith, Kline, & French (now GlaxoSmithKline) pushed its drug, Thorazine, as a potential "cure" for mental illness with state legislators who were eager to save money by closing mental health institutions (Slate & Johnson, 2008). As legal restrictions on civil commitment of the mentally ill spread across the country, and as state legislators believed the claims (which turned out to be unfounded) of the pharmaceutical company, states and counties closed their mental health hospitals or reduced their capacities significantly. Congress passed the Community Mental Health Act in 1963, which ended much of the federal support for mental health hospitals. Instead, Congress was to fund less restrictive institutional alternatives such as halfway houses, and outpatient facilities were either underfunded or shunted by community members who did not want such facilities in their neighborhoods (Slate & Johnson, 2008). Thus, an unintended consequence of this deinstitutionalization movement was that there were few public services available in communities to assist the mentally ill and their families. Jails, and then prisons, became the de facto mental health patient dumping ground.

Unfortunately, and as with those who have major medical problems, most prisons and jails are ill-equipped and -staffed to handle mentally ill inmates. There are difficulties in diagnosis, management of people who do not understand how to behave in a prison, programming and developing appropriate prison employment, and in devising a reentry plan (Slate & Johnson, 2008). Any treatment programming available has long waiting lists. Sometimes staff need to be concerned that the mentally ill inmates require protection from predation, and to protect other inmates they will need to keep an eye on the violent outbursts of mentally ill patients as they might injure others.

Needless to say, the cost of providing medical and mental health care to an aging and ill correctional population is cost-prohibitive. However, as these inmates are unable to access such services in communities, due to their incarceration, such costs must be borne.

Gay, Lesbian, Bisexual, and Transgender Inmates

The true number of gay, lesbian, bisexual, and transgender (LGBT) inmates in corrections is not known. According to Gary Gates (personal communication, March 14, 2011), a demographer with the Williams Institute at the University of California Los Angeles School of Law, estimates, based on surveys that query people about gender identity, suggest that about 3% to 5% of the free community are gay, lesbian, or bisexual. He notes that the number of transgender people

Photo 7.6

Transgender jail inmates, exercising separately from the main jail inmate population.

in the community is also unknown, but is likely 1% or less. Based on these estimates, we might reasonably expect that similar percentages for each group are represented in prisons. In addition, as heterosexual relations are formally restricted in prisons, there is some percentage of male and female inmates who engage in homosexual relations while incarcerated, though they may never have done so when they were free. Lieb and his colleagues (2011) found that the number of men who have ever had sex with men in the free community ranges around 6.4%. We might expect, therefore, that the number of men who engage in sexual relations with other men while in prison is likely around 6% or higher.

As with gays and lesbians, the exact number of transgender people in communities is unknown, being that they often keep their feelings on this matter hidden knowing they might be marginalized (Tewksbury & Potter, 2005). In prisons, the number of transgender inmates is also difficult to know, although large male prisons will often house many (Sexton, Jenness, Sumner, 2010). Jenness and her colleagues (2007, as cited in Sexton et al., 2010) found in their study of transgender female inmates in California prisons that they are much more likely than other inmates (59% as compared to 4.4% for non-transgender inmates) to report being sexually assaulted while in prison.

Protecting sexual orientation and gender identity minorities in prisons presents a challenge for administrators, particularly in male prisons. Male bravado and posturing to show strength and ward off attacks is common in male prisons, which makes males who are unable or unwilling to put on such fronts targets for abuse or predation. If one is a transgender inmate in a male prison, or a person who was once biologically male, but through dress, hormonal treatment, and/or surgery is now a female, the challenges mount. Not only must administrators be concerned about the safety of the inmate, but they must also figure out a way to accommodate her needs for privacy, medical treatment, and housing. In response, some few prisons are considering recognizing the likelihood of assault in classification decisions for transgender inmates and accommodating their requests for housing that fits their gender (Sexton et al., 2010).

In women's prisons, as with racial and ethnic minorities, there appears to be more acceptance of both lesbian and transgender inmates; part of this greater acceptance may have to do with the sense that women inmates have direct experience with marginalization and therefore are more understanding of those who vary from the norm in sexual orientation or gender identity. Or it might have to do with the lesser need of women in prisons to defend themselves physically from predation of other inmates; the amount of violent and sexual attacks in women's prisons, as far as this can be determined, is much lower than it is in men's facilities.

Summary

- Prisons come in various shapes and security levels.
- To varying degrees, inmates experience mortification and pain related to their status, and as humans they will behave in either pro- or antisocial ways to ameliorate that pain.
- Inmates adopt certain roles and engage in certain behaviors because they are prisonized and adopt the subculture, or because they import aspects of the culture from the outside community into the prison.

- "Total institutions" exist to different degrees, depending on the security level and operation of prisons.
- Gangs and violence are one way that inmates "adjust" to their environment and have their needs met and their pain alleviated.
- Mature coping is one way that correctional clients can fruitfully "adjust" and perhaps reform in that environment.
- Special population inmates present unique challenges for administrators interested in meeting their needs and keeping them safe in the correctional environment.

Key Terms

Attica Prison Riot

Deinstitutionalization
of the mentally ill

Gangs

Importation

Mature coping

Maximum security prisons

Medium security prisons

Minimum security prisons

Mortification

New Mexico Prison Riot

Pains of imprisonment

Prison subculture

Prisonization

Prisons

Supermax prisons

Total institution

Discussion Questions

1. Which prison, from our history of prison chapters, most reminds you of super maximum security prisons? What were the problems with this historical prison? Based on what we know of that prison from our history, what problems do you foresee arising with the supermax prisons of today?

2. Define what a total institution is and how it might vary by type of correctional arrangement (e.g., probation, parole, jail, prison) and inmate status.

3. Inmate subcultures are thought to be related to the concepts of prisonization, importation, and the pains

of imprisonment. Discuss how and why this might be so.

4. What are the attributes of gangs that make them appealing to inmates in prisons? How might that appeal be reduced by prison managers?

5. How might correctional clients configure their environment to ensure their own reform? How might we, as citizens, assist them in that endeavor?

6. What might be the most effective strategies for managing special populations in prisons?

Useful Internet Sites

American Correctional Association: www.aca.org

American Friends Service Committee (a Quaker organization interested in correctional reform): www.afsc.org

Bureau of Justice Statistics (information available on all manner of criminal justice topics): http://bjs.ojp.usdoj.gov/

Federal Bureau of Prisons: www.bop.gov

The Sentencing Project: www.sentencingproject.org

The Williams Institute—UCLA School of Law: www.law.ucla.edu/williamsinstitute

Vera Institute (information available on a number of corrections-related topics): www.vera.org

CHAPTER 8

Parole and Prisoner Reentry

❖ Introduction: What Is Parole?

The term *parole* comes from the French phrase *parole d' honneur,* which literally means "word of honor." In times when a person's word really meant something, parole was used by European armies to release captured enemy soldiers on condition (their word of honor) that they would take no further part in the present hostilities (Seiter, 2005). The practice also existed for a brief time in the American Civil War, but it was soon realized that neither side was honoring it because paroled soldiers were back in the fight just weeks after release (www.civilwarhome.com/prisonsandprisoners.htm). Modern **parole** refers to the release of convicted criminals from prison under the supervision of a parole officer before the completion of their full sentences on their promise of good behavior. Parole is different from probation in two basic ways. First, parole is an administrative function practiced by a parole board, which is part of the executive branch of government, while probation is a judicial function. Second, parolees have spent time in prison before being released into the community, whereas probationers typically have not. In many states, both parolees and probationers are supervised by state probation and parole officers or agents, while in others they are supervised by separate probation or parole agencies.

❖ Brief History of Parole

The philosophical foundation of parole, as applied to convicted criminals, was laid by the superintendent of the Norfolk Island penal colony off the coast of Australia in the 1830s. As mentioned in other chapters, the superintendent was an ex-British naval officer and geography professor named Alexander Maconochie, who had experienced imprisonment himself as a captive of the French during the Napoleonic Wars after his ship was wrecked (Morris, 2002). The horrendous conditions in the colony offended the compassionate and, like Augustus and Davenport-Hill, deeply religious Maconochie, who was a firm believer in the primacy of human dignity. As prison superintendent, he operated on three basic principles: (1) Cruel and vindictive punishment debases both the criminal and the society that allows it, making both worse; (2) the purpose of punishment should only be the reformation (today we would say rehabilitation) of the convict; and (3) criminal sentences should not be seen in terms of time to be served, but rather in terms of tasks to be performed. To implement his programs, Maconochie required indefinite prison terms rather than fixed determinate terms so that convicts would have an incentive to work toward release. As we have seen, the type of sentence Maconochie advocated is known today as an indeterminate sentence.

Photo 8.1

The Norfolk (Australia) penal colony's superintendent in the 1830s, Alexander Maconochie, laid the philosophical foundation of parole, as applied to convicted criminals.

Maconochie was a realist, not a bleeding-heart idealist. The philosophy behind Maconochie's correctional ideas and ideals is best stated by the man himself:

I am no sentimentalist. I most fully subscribe to the right claimed by society to make examples of those who break its laws, that others may feel constrained to respect and obey them. Punishment may avenge, and restraint may, to a certain limited extent, prevent crime; but neither separately, nor together, will they teach virtue. That is the province of moral training alone. (quoted in Morris, 2002, p. xviii)

Maconochie's point is that retribution may deter some offenders and incapacitation is a temporary hold on a criminal career, but the real goal of corrections should be to correct; or in Maconochie's words, "teach virtue." With respect to the third principle (tasks performed rather than time served), Maconochie devised a *mark system* involving credits earned for the speedy and efficient performance of these tasks, as well as for overall good behavior. When a convict had accumulated enough credits, he could apply for a *ticket of leave* (TOL), which was a document granting him freedom to work and live outside the prison before the expiration of his full sentence. TOL convicts were free to work, acquire property, and marry, but they had to appear before a magistrate when required and church attendance was mandatory. Maconochie's system appeared to have worked very well. It has supposedly been determined that only 20 out of 900 of Maconochie's TOL convicts were convicted of new felonies (many more crimes were felonies in the 19th century than today), a recidivism rate of 2.2% that modern penologists are scarcely able to comprehend, and many dismiss as too good to be true (R. Hughes, 1987). Nevertheless, and perhaps predictably, when Maconochie returned home to England and tried to institute his reforms there, he was accused of coddling criminals and relieved of his duties (Petersilia, 2001).

Nevertheless, the TOL system was adapted to the differing conditions in Britain under Sir Walter Crofton, director of Irish prisons, who devised the so-called **Irish system** in the early 1850s. This system involved four stages, beginning with a 9-month period of solitary confinement, the first 3 months with reduced rations and no work. This period of enforced idleness was presumed to make even the laziest of men yearn for some kind of activity. The solitary period was followed by a period in which convicts could earn marks through labor and good behavior to enable their transfer to an open prerelease prison when enough marks had been accumulated, and finally a TOL. TOL convicts were supervised in the community by either police officers or civilian volunteers (forerunners of the modern parole officer) who paid visits to their homes and attempted to secure employment for them (Foster, 2006). Of the 557 men released on TOL under the Irish system in the 1850s, only 17 (3.05%) were revoked for new offenses (Seiter, 2005); again, if true, this is an extraordinary level of success.

Elements from both Maconochie's and Crofton's system were brought into practice in the United States in the 1870s by Zebulon Brockway, superintendent of the Elmira Reformatory in New York. Brockway's system required indeterminate sentencing so that "good time" earned through good conduct and labor could be used to reduce inmates' sentences (Roth, 2006). However, there were no provisions for the supervision of offenders who obtained early release until 1930 when the U.S. Congress established the United States Board of Parole. Eventually, parole came to be seen not as a humanistic method of dealing with "reformed" individuals, but rather as a way of maintaining order in prisons by holding out the prospect of early release if convicts behaved well. It has also become a way of trying to reintegrate offenders back into the community by offering programs to prepare them for life outside the walls and as a partial solution to the problem of prison overcrowding. Because of these functions, parole became an essential and valued part of the American correctional system.

❖ The Modern Parole System

In 2008, there were 828,169 state and federal parolees in the United States, 88% of whom were males. Figure 8.1 provides a racial/ethnic breakdown of parolees in 2008 (Glaze &

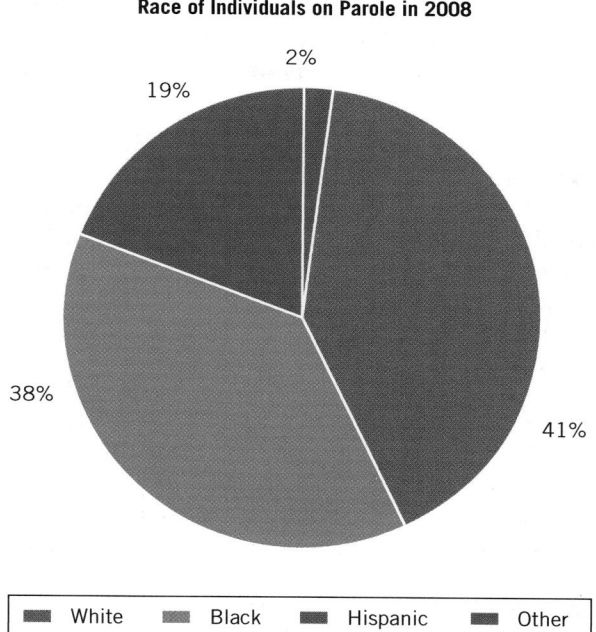

Race of Individuals on Parole in 2008

2%
19%
38%
41%

White   Black   Hispanic   Other

Figure 8.1

Percentage of Individuals on Parole in 2008 by Racial Category

Source: Glaze & Bonczar (2009).

Bonczar, 2009). Drug offenses were the most serious for 37% of these parolees, followed by violent (26%), property (23%), and other (14%) offenses.

The skyrocketing crime of the 1970s, 1980s, and early 1990s, much of it committed by offenders on probation or parole, led to the "tough on crime" approaches to punishment discussed in other chapters. The heinous kidnapping, rape, and murder of 13-year-old Polly Klaas by parolee Richard Davis and of 7-year-old Megan Kanka by parolee Jesse Timmendequas led to calls for the abolition of parole from a fearful public and their representatives. With public outrage at a fever pitch, the federal government and a number of states abolished parole, substituting the return of the fixed determinate sentence that Maconochie so disliked. This "no parole" system essentially means that prisoners are released after the completion of their sentences without supervision or reporting requirements. This is known as **unconditional release**. Inmates who are either required to "max out" their time or choose to rather than be placed on parole, have less incentive to enter rehabilitation programs or to abide by prison rules. A number of states had already made the switch to mandatory sentences for their own reasons prior to these hideous crimes committed by Davis and Timmendequas.

Abolishing parole sounds very tough and goes over quite well politically, but the reality is something different. Prisoners are still released early for reasons of overcrowding and budgetary concerns, but there is much less rational control today over who is released than there was in the past. It is discretionary parole that has really been reduced or abolished in some states in favor of mandatory parole. **Discretionary parole** is parole granted at the discretion of a parole board for selected inmates who have earned it. Prisoners earn discretionary parole by avoiding disciplinary infractions and engaging in programs that prepare them for reentry into the community. Discretionary parole also allows parole board members to assess the probability of a given offender's risk to society based on the crime for which he or she was incarcerated, his or her criminal history, prison behavior, and psychological assessments.

Mandatory parole, on the other hand, is automatic parole for almost all inmates in states that have a system of determinate (i.e., fixed) sentencing (Petersilia, 2001). This system is used by the federal government and about half of the states. Mandatory parole still has provisions for earning "good time" credits. Tragically, both Davis and Timmendequas were granted mandatory parole determined by mathematical norms generated by a computer, based solely on time served, that is, without any kind of consideration of the risk these individuals posed to society. Had their cases gone before a parole board, where board members can peruse parole applicants' criminal history and target violent and dangerous criminals for longer incarceration, odds are that neither man would have been released (Petersilia, 2001). Davis, for instance, had been released after serving only one-half of his 16-year sentence because of his supposed "good behavior" while institutionalized for a previous kidnapping, robbery, and assault of a woman (Skolnick, 1993).

As we see from the graph in Figure 8.2, discretionary parole releases have been dropping significantly since 1980, while mandatory parole releases have gone up significantly. Both discretionary and mandatory release parolees are supervised post-release, but that 17% to 18% of inmates who are released at the expiration of their sentences (unconditional release) are not supervised (Hughes, Wilson, & Beck, 2001). The graph in Figure 8.3 supports those who favor a return to discretionary parole and the elimination of mandatory parole. Note that in 1999, about 52% of discretionary parolees successfully completed parole, while only about 32% of mandatory parolees did. This discrepancy was still very much in evidence in 2008 (Paparozzi & Guy, 2009).

Figure 8.2

Percentages of Offenders Released From Prison by Various Methods

Source: Hughes et al. (2001).

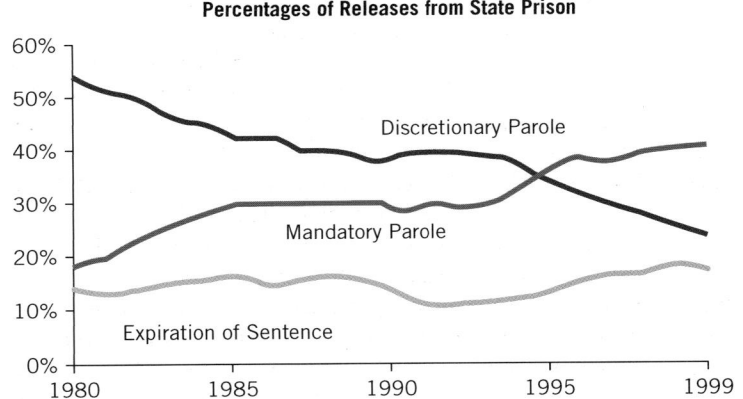

Percentages of Releases from State Prison

Figure 8.3

Percentages of Parolees Successfully Completing Parole by Release Type

Source: Hughes et al. (2001).

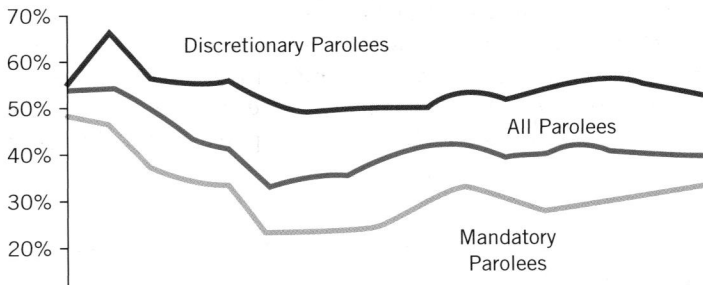

Percentages of Parole Discharges Successfully Completing Supervision

❖ Parolee Recidivism

The last nationwide study of parolee recidivism looked at almost 300,000 parolees from 15 states released from prison in 1994 (Langan & Levin, 2002). Within 3 years of release, 67.5% were rearrested for a new offense, with the entire sample accumulating an astounding 744,000 new offenses in those 3 years. Property offenders (over 70%) had the highest rate of recidivism, while murderers (40.7%) and sex offenders (41.4%) had the lowest. A. Solomon, Kachnowski, and Bhati (2005) analyzed a subset of 38,624 parolees for whom the type of release was known. There were surprisingly few differences between the parolees other than the fact that discretionary release parolees had a lower average number of prior arrests (7.5) than mandatory (9.5) or unconditionally released parolees (9.6). Among those rearrested after release, the unconditionally released were rearrested an average of 9.9 months after release, mandatory parolees an average of 10.4 months, and discretionary parolees an average of 11.5 months. The average time served, as one would expect, was also noticeably different, with the unconditionally released serving an average of 32 months; mandatory parolees, 18.5 months; and discretionary parolees, 21.3 months. This suggests that supervision is particularly vital at the early stages of the parole process when parolees are still struggling to find their way back into their communities and are thus most susceptible to returning to their criminal ways.

Figure 8.4 provides an example of a mandatory parole release form showing the "word of honor" promises that parolees are supposed to abide by in Alaska. The listed conditions are fairly standard from state to state and are indistinguishable from the conditions set for probationers. These are general conditions with which all parolees must comply, but often there are also additional conditions set for individual offenders. These conditions may be things such as requiring alcoholics or drug addicts to enroll in and attend certain programs, or requiring sex offenders to stay away from all contact with children or to refrain from Internet pornography. These additional conditions will likely be set by the parole board or by each individual parole officer after reviewing each parolee's case. It is then the responsibility of the parole officer to make sure that the parolee follows the conditions that have been mandated.

Parole boards are only required in states that employ a discretionary system of granting parole. A **parole board** is a panel of people presumably qualified to make judgments about the suitability of a prisoner to be released from prison after having served some specified time of his or her sentence. This is termed an inmate's *parole eligibility date,* which is the earliest possible time that he or she can be released from prison. Board members are appointed by the governor for a fixed (renewable) term in most states. Some states mandate that a board appointee be well-versed in criminology and corrections, while other states have no such requirement. Although many board members serve on a part-time basis and collect a minimal salary and per diem expenses, chairpersons and vice-chairpersons are full-time salaried individuals in almost all states (Abadinsky, 2009).

In making their decision whether to grant or deny parole, board members assess a variety of information about the inmate and interview him or her at a parole hearing to discuss the assessed information and to gain some face-to-face insight about the inmate. Among the information considered by the parole board to help them make their decision are the following:

- The nature of the offense for which the inmate is currently incarcerated
- The criminal history of the inmate
- Indications by word and/or deed that the inmate is repentant
- The inmate's mental health via psychological and psychiatric reports

Figure 8.4

Example of a
Mandatory Parole
Form

Source: Department
of Corrections, Form
902.05D, *Rev. 07/03.*

State of Alaska   Department of Corrections

Order of Mandatory Parole

Parolee: _____ DOB: _____ OTIS#_____ Released:_____ Supv. Expires: _____

The following terms and conditions are effective on the release date shown on the Certificate of Good Time Award (AS 33.20.030) for all prisoners released pursuant to AS 33.16.010 or AS 33.20.040. I understand I am required by law to abide by the conditions imposed, whether or not I sign these conditions. The Parole Board may have me returned to custody at any time when it determines a condition of parole has been violated.

Conditions of Mandatory Parole

1. Report Upon Release: I will report in person no later than the next working day after my release to the P.O. located at: _____ and receive further reporting instructions. I will reside at: _____.

2. Maintain Employment/Training Treatment: I will make a diligent effort to maintain steady employment and support my legal dependents. I will not voluntarily change or terminate my employment without receiving permission from my Parole Officer (P.O.) to do so. If discharged or if employment is terminated (temporarily or permanently) for any reason, I will notify my P.O. the next working day. If I am involved in an education, training, or treatment program, I will continue active participation in the program unless I receive permission from my P.O. to quit. If I am released, removed, or terminated from the program for any reason, I will notify my P.O. the next working day.

3. Report Monthly: I will report to my P.O. at least monthly in the manner prescribed by my P.O. I will follow any other reporting instructions established by my P.O.

4. Obey Laws/Orders: I will obey all state, federal, and local laws, ordinances, and court orders.

5. Permission Before Changing Residence: I will obtain permission from my P.O. before changing my residence. Remaining away from my approved residence for 24 hours or more constitutes a change in residence for the purpose of this condition.

6. Travel Permit Before Travel Outside Alaska: I will obtain the prior written permission of my P.O. in the form of an interstate travel agreement before leaving the State of Alaska. Failure to abide by the conditions of the travel agreement is a violation of my order of parole.

7. No Firearms/Weapons: I will not own, possess, have in my custody, handle, purchase, or transport any firearm, ammunition, or explosives. I may not carry any deadly weapon on my person except a pocket knife with a 3" or shorter blade. Carrying any other weapon on my person such as a hunting knife, axe, club, etc. is a violation of my order of parole. I will contact the Alaska Board of Parole if I have any questions about the use of firearms, ammunition, or weapons.

8. No Drugs: I will not use, possess, handle, purchase, give, or administer any narcotic, hallucinogenic (including marijuana/THC), stimulant, depressant, amphetamine, barbiturate, or prescription drug not specifically prescribed by a licensed medical person.

9. Report Police Contact: I will report to my P.O., not later than the next working day, any contact with a law enforcement officer.

10. Do Not Work as an Informant: I will not enter into any agreement or other arrangement with any law enforcement agency which will put me in the position of violating any law or any condition of my parole. I understand that the Department of Corrections and Parole Board policy prohibits me from working as an informant.

- The original presentence investigation report
- Reports of institutional conduct, including disciplinary reports and participation in religious and rehabilitative programs
- The inmate's parole plan, that is, where does the inmate plan to reside, does the inmate have guaranteed or possible employment waiting, does the inmate enjoy a good support system in the community
- Statements made by others supporting (family members, counselors, ministers) or opposing (the victim, prosecutor, other criminal justice officials) the inmate's request for parole

Photo 8.2

Charles Manson reads a statement at his parole hearing in 1985. As of this writing, Manson has been denied parole 11 times and remains incarcerated.

Of course, parole boards are composed of human beings who have their own reasons for granting or denying parole. Although a majority vote wins the day, there are always members (especially chairpersons) who wield more influence than others. They can never be certain who will or will not fail on parole, but they know that on average about two-thirds will fail. They are thus more likely to err on the side of caution (not releasing someone who would have made it successfully) than to endanger the public and embarrass themselves by releasing someone who then proceeds to commit heinous and highly publicized crimes such as those committed by Davis and Timmendequas.

❖ What Goes In Must Come Out: Prisoner Reentry Into the Community

Numerous commentators of all political persuasions consume an incredible amount of space writing about America's imprisonment binge, but rarely outside of the academic literature do we find much concern about the natural corollary of the binge: What goes in must come out. Travis (2005) introduces us to the idea of prison reentry thusly: "Reentry is the process of leaving prison and return to society. Reentry is not a form of supervision, like parole. Reentry is not a goal, like rehabilitation or reintegration. Reentry is not an option. Reentry reflects the iron law of imprisonment: they all come back" (p. xxi). "They all come back," refers to prisoners returning to the communities from which they came.

Understanding the process of prisoner reentry and reintegration into the community is a very pressing issue in corrections. In 2008, a total of 503,189 adult offenders entered

American prisons and 483,200 left them (Glaze & Bonczar, 2009). Except for those few who leave prison in a pine box or who make their own clandestine arrangements to abscond, all prisoners are eventually released back into the community. Unfortunately, among this huge number, 1 in 5 will leave with no post-release supervision, rendering "parole more a legal status than a systematic process of reintegrating returning prisoners" (Travis, 2000, p. 1). Pushing inmates out the prison door with $50, a bus ticket to the nearest town, and a fond farewell is a strategy almost guaranteeing that the majority of them will return. With the exception of offenders who max out, prisoners will be released under the supervision of a parole officer charged with monitoring offenders' behavior and helping them to readjust to the free world.

Because parolees have been in prison, and are thus on average more strongly immersed in a criminal lifestyle, we should expect them to be more difficult to supervise than probationers, and they are, and to have lower success rates, and they do. While Glaze and Palla (2005) indicate a success rate of 60% for probationers in 2004, the same figure for successful completion of parole was only 46% (although note the difference between the success rates of discretionary vs. mandatory release parolees above).

The longer people remain in prison, the more difficult it is for them to readjust to the outside world. Inmates spend a considerable amount of time in prison living by a code that defines as "right" almost everything that is "wrong" on the outside. Adherence to that code brings them acceptance by fellow inmates as "good cons." Over time, this code becomes etched into an inmate's self-concept as the prison experience becomes his or her comfort zone. When inmates return to the streets, they do not fit in, they feel out of their comfort zone, and their much sought-after reputations as good cons become liabilities rather than advantages. As prison movie buffs are aware, these readjustment problems were dramatically presented in the suicide of Brooks in *The Shawshank Redemption* and in the final crazy hurrahs of Harry and Archie in *Tough Guys*.

Brooks, Harry, and Archie were all old men who had served very long periods of incarceration and who had thoroughly assimilated the prison subculture by the time of their release into an alien and unaccepting world. Thus, one recommendation might be to reduce the length of prison sentences so that those unfortunate enough to be in them do not have time to become "prisonized." Such a recommendation gains support from statistics showing that the shorter the time spent in prison, the greater the chance of success on parole (Travis & Lawrence, 2002), but if we are looking for something causal in those statistics, we are surely sniffing around the wrong tree. Shorter sentences typically go to those committing the least serious crimes and who have the shortest arrest sheets; such people are already less likely to commit further crimes than those who commit the more serious crimes and have long rap sheets. It is not for nothing that former U. S. Attorney General Janet Reno called prisoner reentry "one of the most pressing problems we face as a nation" (as quoted in Petersilia, 2001, p. 370).

The Impact of Imprisonment and Reentry on Communities

Because crime is highly concentrated in certain neighborhoods, a disproportionate number of prison inmates come from and return to those same neighborhoods. There are those who assert that although high incarceration rates may reduce crime in the short run, the strategy provides only a temporary reprieve and that it will eventually lead to higher crime rates by weakening families and communities and reduce the supervision of children (DeFina & Arvanites, 2002). According to this body of literature, the loss of individuals concentrated in certain communities reduces community organization and cohesion, disrupts families economically and socially, and adds many other problems that will eventually lead to more crime than would have occurred if offenders had been allowed to remain in the community. Clear, Rose, and Ryder's (2001) interviews of residents of high-incarceration neighborhoods are in this tradition. They hypothesize that when public control (incarceration) occurs at high levels, private control (informal control) functions at low levels and ultimately results in more crime. Clear et al. do not deny that neighborhood residents are better off and safer when the bad apples are pulled from the shelf and shipped off to prison. This implies the opposite of their hypothesis; that is, when private control functions at low levels, public control occurs at high levels.

There can definitely be many negative impacts on the community, especially a financial impact on families, when working fathers and mothers are removed from it. However, a Bureau of Justice Statistics report (Mumola, 2000) showed that 48% of imprisoned parents were never married and 28% of those who were ever married were divorced or separated. In addition, very few men lived with their children prior to imprisonment. Moreover, the majority of parents had been convicted of violent or drug crimes and 85% had drug problems, which makes it difficult to see how the presence of antisocial fathers in the community somehow contributes to private control rather than detracts from it (Rodney & Mupier, 1999). For instance, a longitudinal study of 1,116 British families showed that the presence of a criminal father in the household predicted antisocial behavior of his children, and that the harmful effects increased the longer he spent with the family (Moffitt, 2005). Another large-scale study found that when an antisocial father resides with the mother and offspring, there is more risk to children than when the father does not because

> children experience a double whammy of risk for antisocial behavior. They are at genetic risk because antisocial behavior is highly heritable [greatly influenced by genetic factors]. In addition, the same parents who transmit genes also provide the child's environment. (Jaffee, Moffitt, Caspi, & Taylor, 2003, p. 120)

What Makes for a Successful Reentry?

Reentry is a process of reintegrating offenders back into their communities regardless of whether or not they were integrated into it in a prosocial way before they entered prison. Part of that process is preparing offenders to reenter by providing them with various programs that target the risks they pose to the community (e.g., anger management classes) and their needs (e.g., educational and vocational programs). Unfortunately, one of the prices we have paid for the prison expansion in recent years is "a decline in preparation for the return to the community. There is less treatment, fewer skills, less exposure to the world of work, and less focused attention on planning for a smooth transition to the outside world" (Travis & Petersilia, 2001, p. 300). Yet there is always an abundance of suggestions about what we should be doing to prepare inmates to reenter free society. If these programs were to be actually implemented, and if they did what they are supposed to, parolees might have a fighting chance of remaining in the community. The U.S. Department of Justice (A. Solomon, Dedel Johnson, Travis, & McBride, 2004) lists three programmatic phases believed necessary for successful reentry:

Phase 1—Protect and Prepare: The use of institution-based programs designed to prepare offenders to reenter society. These services include education, treatment for mental health and substance abuse issues, job training, mentoring, and a complete diagnostic and risk assessment.

Phase 2—Control and Restore: The use of community-based transition programs to work with offenders prior to, and immediately following, their release from jail or prison. Services provided in this phase include education, monitoring, mentoring, life skills training, assessment, job skills development, and mental health and substance abuse treatment.

Phase 3—Sustain and Support: This phase uses community-based, long-term support programs designed to connect offenders no longer under the supervision of the justice system with a network of social services agencies and community-based organizations to provide ongoing services and mentoring relationships.

The above programming strategy is an ideal rather than a reality most of the time, but many offenders released from incarceration do have access to some parts of each of these phases, and some manage to successfully complete parole. To be successful in phase 1, offenders have to be willing to apply themselves to the training and programming offered to them. They have to make a conscious decision to set long-term goals and convince themselves that a criminal career is not for them. Obviously, these are decisions that are typically only made by individuals not fully immersed in the criminal lifestyle. In phase 2, they have to implement those decisions in the real world where it counts, and not be sidetracked by either the frustration generated by the monitoring they are subjected to from their parole officers or by the criminal opportunities that may present themselves. Phase 3 may be the most difficult of all for some offenders who rely on authoritative figures to give their lives direction. This is why it is so important for community corrections officers to plug their clients into agencies outside the criminal justice system that are able to address parolee needs prior to termination of supervision.

Perhaps the most important tool for the successful reintegration of offenders who want to go straight is employment. Unfortunately, the typical offender is not prepared for much other than a low-skill manufacturing job, the kind of job that the United States has been losing in truly staggering numbers due to technological advances and companies moving operations overseas. Job prospects are thus fairly limited unless offenders can improve themselves educationally. Amy Solomon and her colleagues (2004) report that 53% of Hispanic inmates, 44% of black inmates, and 27% of white inmates have not completed high school or obtained a GED, as opposed to 18% of the general population. Add that to their general lack of preparedness, and it becomes understandable why employers are reluctant to hire ex-cons (ex–prison inmates). Even taking into consideration lack of preparedness and employer reluctance to hire offenders, economists find that incarceration reduces employment opportunities by about 40%, wages by about 15%, and wage growth by about 33% (Western, 2003). These facts make finding employment to support oneself, let alone one's children, particularly in times of recession, truly challenging for ex-mates reentering their communities.

❖ Determining Parole "Success"

Although one would not think so, defining parole success is difficult because there is no agreed-upon standard by which we can judge success. Does it mean (1) a completed crime-free/technical violation–free period of parole, or does it mean (2) that the offender was released from parole without being returned to prison due to his or her behavior while on parole? It obviously means vastly different things in different states because parole "success" rates ranged from 19% in Utah to 83% in Massachusetts in 1999 (Travis & Lawrence, 2002). Does this mean that Utah criminals are over 4 times more resistant to taking the straight and narrow road than Massachusetts criminals, or that Massachusetts has much better programming and professional parole officers? Of course not; it most likely means that conservative Utah follows our first definition of success while liberal Massachusetts follows our second definition. Given the national average rate of 42% that year, it is plain that many parolees are forgiven various technical violations, or perhaps even a petty arrest or two, in most states. Thus, "success" has as much or more to do with the behavior of the parole authorities in different jurisdictions as it does with the behavior of parolees. When we speak of "success," then, we are generally speaking in middle-of-the-road terms in which certain parolee misbehaviors are forgiven occasionally in the interest of maintaining him or her on a trajectory that is at least somewhat positive.

An additional problem facing our efforts to reintegrate parolees into their communities is knowing what works best, why, and for whom. After a review of 32 studies that examined the process of prisoner reentry, Richard Seiter and Karen Kadela (2003) identified programs that work, that do not work, and that are promising in helping prisoners in the long process of successfully reentering the community. In their research, they looked at transitional community programs such as halfway houses and work release programs as well as programs that initiated treatment for inmate deficits (drug dependency, low education, poor life skills, etc.) while they were in prison and continued in the community after release. Programs that worked best were concrete programs that provided offenders with skills to compete in the workforce and intensive drug programs. Programs that were located in the community such as halfway houses were more effective than prison-based programs. We have explored or will explore many treatment programs in other chapters, so we will limit ourselves to discussing halfway houses and electronic monitoring in this one.

❖ Halfway Houses

As the name implies, **halfway houses** are transitional places of residence for offenders that are, in terms of strictness of supervision, "halfway" between the constant supervision of prison and the much looser supervision in the community. As with probation and parole, early halfway houses were organized and run by private religious and charitable organizations designed to assist released prisoners to make the transition back into free society. The earliest such home was set up in New York City in 1845 by Quaker abolitionist Isaac T. Hopper (Conly, 1998). Also as with probation and parole, federal and state governments adopted the principles of halfway houses as a good idea. Many states use halfway houses as transition points between prison and full release into the community, and the U.S. federal prison system releases about 80% of its inmates into halfway houses ("Director Addresses Changes," 2006).

In addition to being a tradition between prison and the community, such places (also referred to as community residential centers) may serve as an intermediate sanction for offenders not sent to prison, but needing greater supervision than straight probation or parole. The rationale behind halfway houses is that individuals with multiple problems such as substance abuse, lack of education, and a poor employment record may have a better chance to positively tackle these problems and to comply with court orders if they are placed in residential centers where they will be strictly monitored while at the same time being provided with support services to address some of the problems that got them there. Halfway houses may be operated by corrections personnel, but they are also likely to be operated by faith-based organizations such as the Salvation Army and Volunteers of America. Nevertheless, residents are still under the control of probation/ parole authorities and may be removed and sent to prison if they violate the conditions of their probation or parole.

In times of rising costs and prison overcrowding, cost-conscious legislators tend to view community-based alternatives to prisons like diet-conscious beer drinkers—"prison lite." Community-based residential programs supposedly provide public safety at a fraction of the cost while allowing offenders to remain in the community and at work earning their own keep, and, best of all, they are assumed to reduce recidivism. Halfway houses are also a valuable resource in that they provide offenders released from prison, who would otherwise be homeless, with an address for employment purposes.

Nancy Marion (2002) questions some of these assumptions and generally paints a disappointing picture for those who believe that keeping offenders out of prison aids in rehabilitation. She sees programs such as halfway houses admitting individuals who would not have gone to prison anyway, which means such programs increase rather than decrease correctional budgets. This phenomenon of bringing in more people to the criminal justice system when the intent is to decrease such contact is known as **net widening**. Many residents are "unsuccessfully released" from halfway houses because they used alcohol or drugs while there, and even of those successfully released in Marion's multiyear study, between 10.8% and 50.6% (depending on the year of release) were later imprisoned. However, Lowenkamp and Latessa's (2002) more comprehensive examination of 38 such facilities in Ohio, while finding that all were not effective, showed that most were. They found that community-based programs were of no use for low-risk offenders (indicating that they did not need them in the first place), but that the majority of them were effective in substantially reducing recidivism among medium- and high-risk offenders. Another review of the relative success of halfway houses concluded that, "[r]elative to individuals discharged into the community without supportive living environments, those men and women who found residence in halfway houses had better substance abuse, criminal justice, and employment outcomes" (Polcin, 2009, p. 11).

Halfway houses should not be viewed as another way of coddling criminals. The "exchange rate" (i.e., how much time in an alternative sanction an offender is willing to serve to avoid 12 months in prison) for halfway house placement in the May et al. (2005) study discussed in Chapter 6 was an average of 12.77 months for offenders who had served time in prison (14.42 months for all offenders). "Experienced" offenders therefore see it as almost as punitive as prison, despite the relative freedom that halfway residency affords offenders to reintegrate themselves into the community. Much of this has to do with the level of responsibility expected of residents of halfway houses. Living in a halfway house, offenders are expected to take more responsibility for their lives than prison inmates. Halfway house residents are expected to be in programming and working or looking for work, are subjected to frequent and random testing for drug/alcohol intake, and some halfway houses augment all this with electronic monitoring. None of these expectations is "suffered" by prison inmates (Shilton, 2003).

❖ House Arrest, Electronic Monitoring, and Global Positioning Systems

House arrest is a program used by probation and parole agencies that requires offenders to remain in their homes at all times except for approved periods, such as travel to work or school, and occasionally for other approved destinations. As a system of social control, house arrest is typically used primarily as an initial phase of intensive probation or parole supervision, but can also be used as an alternative to pretrial detention or a jail sentence. As is the case with so many other criminal justice practices, house arrest was designed primarily to reduce financial costs to the state by reducing institutional confinement.

House arrest did not initially gain widespread acceptance in the criminal justice community because there was no way of ensuring offender compliance with the order, short of having officers constantly monitoring the residence. It was also viewed by the public at large as being soft on crime—"doing time in the comfort of one's home." However, house arrest gained in popularity with the advent of **electronic monitoring (EM)**. EM is a system by which offenders under house arrest can be monitored for compliance using computerized technology. In modern EM systems, an electronic device worn around the offender's ankle sends a continuous signal to a receiver attached to the offender's house phone. If the offender moves beyond 500 feet of his or her house or apartment, the transmitter records it and relays the information to a centralized computer. A probation/parole officer is then dispatched to the offender's home to investigate whether the offender has absconded or removed or tampered with the device. As of 2004, almost 13,000 offenders were under house arrest, with 90% of them being electronically monitored (Bohm & Haley, 2007).

An even more sophisticated method of tracking offenders is that of a **global positioning system (GPS)**. GPS requires offenders to wear a removable tracking unit that constantly communicates with a non-removable ankle cuff. If communication is lost, the loss is noted by a Department of Defense satellite, which records the time and location of the loss in its database. This information is then forwarded to criminal justice authorities so that they can take action to determine why communication was lost. Unlike EM systems, the GPS can be used for surveillance as well as detention purposes. For instance, it can let authorities know if a sex offender goes within a certain distance of a schoolyard, or if a violent offender is approaching his or her victim's place of residence or work (Black & Smith, 2003). As of 2007, a total of 28 states had legislation calling for some form of electronic monitoring of sex offenders (Payne, DeMichele, & Button, 2008).

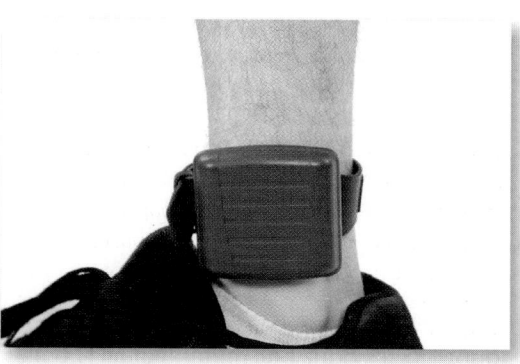

Photo 8.3

An electronic monitoring device, attached to an offender's ankle, sends a continuous signal to a receiver which helps to ensure compliance without probation/ parole officers having to constantly monitor the offender.

Payne and Gainey (2004) indicate that detractors of electronic monitoring tend to criticize it as intruding too much into the realm of privacy, and even as barbaric. Of course it is intrusive; that is the point! But it is far less intrusive than prison, and Payne and Gainey state that offenders released from jails or prisons and placed in EM programs are generally positive about the experience (not that they enjoyed it, but that it was better than the jail or prison alternative). Their findings mirror those from a larger sample of offenders on EM programs in New Zealand (Gibbs

& King, 2003). Many see it as jail or prison time simply served in a less restrictive and less violent environment (so much for the charge that it is barbaric), although the average experienced offender would exchange 11.35 months on EM for 12 months in prison, and although offenders overall would exchange 13.95 months on EM for 12 months in prison (Moore et al., 2008). It would seem from this and similar studies that inveterate offenders tend to prefer prison over virtually any other correctional sentence, other than straight probation.

Although the authors of these studies appear positive about the alleged rehabilitative promise of allowing offenders to serve time at home and thus maintaining their links to family, and although successful completion rates are high, recidivism rates, which are the litmus tests for any corrections program aimed at rehabilitation, were not any better than for probationers/parolees not on EM programs matched for offender risk in several Canadian provinces (Bonta, Wallace-Capretta, & Rooney, 2000). This may be viewed positively, however, as a function of the greater ability to detect noncompliance with release conditions among those under EM supervision.

An additional problem with EM is that because its low cost relative to incarceration is alluring to politicians, it may be (and is) used without sufficient care regarding who should be eligible for it. While offenders can be monitored and more readily arrested if they commit a crime while on EM, EM does not prevent them from committing further crimes. Several high-profile cases including rapes and murders have been committed by offenders who succeeded in removing their electronic bracelets (Reid, 2006). When cases such as these are reported, the public (which by and large would rather see iron balls and chains attached to offenders rather than plastic bracelets) responds with charges of leniency. This is unfortunate because EM does appear to have a significant impact on prison overcrowding and on reducing correctional costs. Of course, EM can only be considered to reduce correctional costs if it is used as a substitute for incarceration, not as an addition to normal probation and parole, in which case it is an added cost.

Comparative Perspective: Electronic Monitoring in Europe

Because several European nations have found themselves burdened by overcrowded prisons over the past two decades, they have turned to electronic monitoring to ease the burden. The table produced below shows that there is widespread variation in the use of EM, ranging from a mere 0.4 per 100,000 of the population in Austria to a high of 33.3 per 100,000 in England and Wales. These rate differences reflect a number of factors, including the penal policies of the country involved, the strength of the financial and prison overcrowding burden, and the demands and fears of the public. The large differences in daily costs (the costs were converted from euros to dollars by the authors using Internet currency converters) also reflect national differences in the economy, the type of monitoring used, officer caseload sizes, and how much EM is supplemented by other surveillance and treatment methods such as probation/parole officer and social worker home visits. Some countries use the private sector to install and monitor EM equipment and others do not, while some use a combination of private and public sector workers. The sophistication of the monitoring equipment also varies from simple telephone voice recognition methods to GPS systems.

The completion rates are fairly consistent across countries, and they are refreshingly high. This is possibly a function of the lower-risk offenders that are typically placed on EM in Europe. It is generally required that for an offender to be placed on EM, he or she must have a suitable residence, a functional phone line, and be working (Havercamp, Mayer, & Levy, 2004). There is also a wide range of times to be served on EM, ranging from about 3 months in England and Wales for probationers to 13 to 23 months for parolees in France (Wennerberg & Pinto, 2009).

Country	Population	EM Rate per 100,000	Completion Rate	Daily Costs
Austria	8,350,000	0.4	N/A	N/A
Belgium	10,584,000	6.3	90%	$52
Denmark	5,490,000	2.6	N/A	N/A
England & Wales	54,670,000	33.2	90%	$50
France	62,100,000	5.5	94%	N/A
Netherlands	16,440,000	5.8	93%	N/A
Norway	4,740,000	2.9	N/A	N/A
Portugal	10,640,000	4.8	N/A	$20
Scotland	5,190,000	15.4	N/A	NA
Spain (Catalonia only)	7,200,000	1.1	85%	$10
Spain (Madrid only)	5,840,000	30.9	N/A	N/A
Sweden	9,170,000	12.1	94%	$98

Source: Havercamp, R., M. Meyer, R. Levey (2004). Electronic monitoring in Europe. *European Journal of Crime, Criminal Law & Criminal Justice,* 12: 36-45. Copyright © Brill.

One large-scale study of parolees released under home detention curfew (using electronic monitoring to enforce it) conducted in Britain provided very positive results (Dodgson et al., 2001). Six months after release, only 9.3% (118 out of 1,269) of the EM parolees had been reconvicted of a new crime, compared with 40.4% (558 out of 1,381) of the prisoners who were unconditionally released. Of course, all of this difference cannot be attributed to the EM program, since the groups were not matched for criminal history and those on the program were already considered to have a lower risk of reoffending. Nevertheless, the EM groups were positive about the program (it got them early release from prison), and the net financial saving to the prison service over 12 months was estimated to be £36.7 million, or about $56.25 million. The consensus in the European literature reviewed seems to be that electric monitoring "works" and that it is here to stay.

❖ Concluding Remarks on Reentry and Recidivism

We have learned that reentry into the community, whether on supervised parole or not, is an extremely difficult process. Everything appears to be working against offenders' successful reintegration, not the least of these factors being the offenders themselves. Some criminals are committed to their lifestyles and simply cannot or will not lead a law-abiding life. Numerous lines of evidence show that work and other normal responsibilities that go with the straight life do not mesh well with their "every night is Saturday night" lifestyles (B. Jacobs & Wright, 1999; Mawby, 2001; Rengert & Wasilchick, 2001). Nevertheless, correctional work is premised on the assumption that people can change, and not all criminals, by any means, are resistant to change. When we find the best reentry programs (programs that have repeatedly been shown empirically to significantly reduce recidivism) and implement them, what kind of success can we expect? Joan Petersilia (2004), the preeminent reentry researcher today, sums up the combined Canadian and American reentry "what works" literature and states that

> they took place mostly in the community (as opposed to institutional settings), were intensive (at least six months long), focused on high-risk individuals (with risk level determined by classification instruments rather than clinical judgments), used cognitive-behavioral treatment techniques, and matched therapist and program to the specific learning styles and characteristics of individual offenders. As the individual changed his or her thinking patterns, he or she would be provided with vocational training and other job-enhancing opportunities. Positive reinforcers would outweigh negative reinforcers in all program components. Every program begun in jail or prison would have an intensive and mandatory aftercare component. (pp. 7–8)

Pertersilia suggests that if we could design programs that combined all these things, we might be able to reduce recidivism by about 30%. So, with all the best methods currently available—caring and knowledgeable counselors, legislators willing to provide a budget sufficient to meet the needs of all the identified programs—the best we can hope for is a 30% reduction in recidivism. Even this 30% figure is a "best guess" premised on Petersilia's faith in the efficacy of rehabilitative programs. As much as we would all love to find a way to turn offenders into respectable citizens, Petersilia's estimate reminds us that human beings are not lumps of clay to be molded to someone else's specifications. Although correctional workers might regret that they cannot mold their charges' minds as they might wish to, the fact that they cannot do so without their owner's consent is a vindication of human freedom and dignity.

Summary

- Parole is the legal status of a person who has been released from prison prior to completing his or her full term. The concept can be traced in the 19th century to Maconochie's "ticket of leave" system in Australia and Walter Crofton's Irish system. Parts of these systems were brought to the United States in the 1870s by Zebulon Brockway, superintendent of the Elmira Reformatory in New York. Brockway's system required indeterminate sentencing so that "good time" earned through good conduct and labor could be used to reduce inmates' sentences.

- In the modern United States, we have two systems of parole—discretionary and mandatory. Discretionary parole is granted by a parole board based on its perceptions of the inmate's readiness to be released; mandatory parole is based simply on a mathematical formula of time served. Discretionary parolees are significantly more likely to successfully complete parole than mandatory parolees.

- The reentry of prisoners into the community is a very difficult process. The ex-con stigma makes getting employment problematic, and the period of absence makes it tough to reestablish relationships. Successful reentry depends on several factors, not the least of which are the policies of the parole authorities, as the huge gap between the Massachusetts and Utah "success" rates indicates. Nevertheless, providing parolees with concrete help such as job skills and drug rehabilitation programs can go a long way in helping them to remain crime-free. This effort may be particularly fruitful if it is made in some form of community-based residential program.

- Electronic monitoring and global positioning systems technology is increasingly used in corrections. It helps by increasing the level of offender monitoring and apprehension, but it cannot altogether prevent additional crimes committed by offenders while on the program, which is why candidates for this type of supervision must be chosen carefully. In sum, few programs can be said to work for most offenders if we define "working" unrealistically. Human nature is complicated, often ornery, and resistant to change. Even "ideal" programs such as those defined by Petersilia could only be expected (according to her) to reduce recidivism by about 30%.

Key Terms

Discretionary parole	House arrest	Parole
Electronic monitoring (EM)	Irish system	Parole board
Global positioning system (GPS)	Mandatory parole	Unconditional release
Halfway houses	Net widening	

Discussion Questions

1. Compare the recidivism rates claimed for the TOL parolees with those of modern American recidivism rates of parolees. What do you think may account for the huge differences?

2. Explore and discuss why it is that we still continue to utilize mandatory parole in the face of evidence that discretionary parole is a safer bet?

3. What do you think may be the single most difficult problem to overcome by a parolee who has just been released after serving 5 years, in staying out of trouble?

4. Would it be a good thing to have a number of community-based residential facilities located in high-crime communities so that some of the problems noted by Clear, Rose, and Ryder might be avoided?

5. Given that expert opinion says that a 30% reduction in the recidivism rate is about the best we can accomplish, do you think that trying to rehabilitate criminals is a waste of time and that the money would be better spent on keeping them locked up? Why or why not?

Useful Internet Sites

American Correctional Association: www.aca .org

American Probation and Parole Association: www .appa-net.org

CHAPTER 9

The Corrections Experience for Staff

❖ Introduction

Work in corrections has changed a great deal from the *shire reeves* (the Old English name for sheriffs) who ran the jails in the Middle Ages and the guards who staffed the first Pennsylvania and Auburn prisons. For one thing, security staff in corrections are no longer called "guards," because it is thought to reflect a more primitive role and are instead formally referred to as correctional officers—perhaps a greater reflection of the move to professional status for these staff.

Yet the public does not generally see corrections work as a profession. Other than probation and parole work (which not coincidentally usually requires more education and pays more), most college students do not identify work in jails or prisons as their career goal either. Also, despite a century of effort by some determined correctional administrators, corrections organizations (such as the American Correctional Association, the American Jail Association, and the American Probation and Parole Association) and some academicians, many correctional jobs are not structured like a profession. A **profession** is typified by five things: (1) prior educational attainment involving college, (2) formal training on the job or just prior to the start of the job, (3) pay and benefits that are commensurate with the work, (4) the ability to exercise discretion, and (5) work that is guided by a code of ethics. Yet most jobs in corrections still do not adequately meet the first three of these criteria for professional status, and though most correctional workplaces either have their own code of ethics, or are nominally guided by that of a corrections organization (e.g., see the American Correctional Association's Code of Ethics in In Focus 9.2), there is no enforcement of this code as there is for medical doctors and lawyers.

In this chapter, we will explore the nature of correctional work as it has evolved and as it is shaped by professionalism, the nature of the work, clients and inmates, the structure of its organizations, and other environmental forces. We will review the factors that lead to stress and turnover for officers and how those might be mitigated by organizational changes. Though correctional work is often not the first choice of a college graduate, it does have its own appeal, and we will discuss why that might be so.

❖ The State of the Work in Correctional Institutions and Programs

Growth in Staff and Clients/Inmates

In addition to exploding inmate and offender populations, the number of employees in corrections, albeit often undereducated, undertrained, and underpaid for their work, has grown astronomically in the last 20 years. From 1982 to 2006, there was almost a 600% increase in direct expenditures for all criminal justice agencies (i.e., police, courts, and corrections) (see Figure 9.1). During that same time period, just the expenditures for corrections increased by 660%! Not surprisingly, employment in corrections more than doubled from 300,000 to 765,466 during this time period, with the majority of those employees being correctional officers (66%) working in state and local, public and private, correctional facilities and programs (Hughes, 2006; Perry, 2008; Stephan, 2008, p. 4).

Figure 9.1

Direct Expenditure by Criminal Justice Function, 1982-2006

Source: Bureau of Justice Statistics (2008).

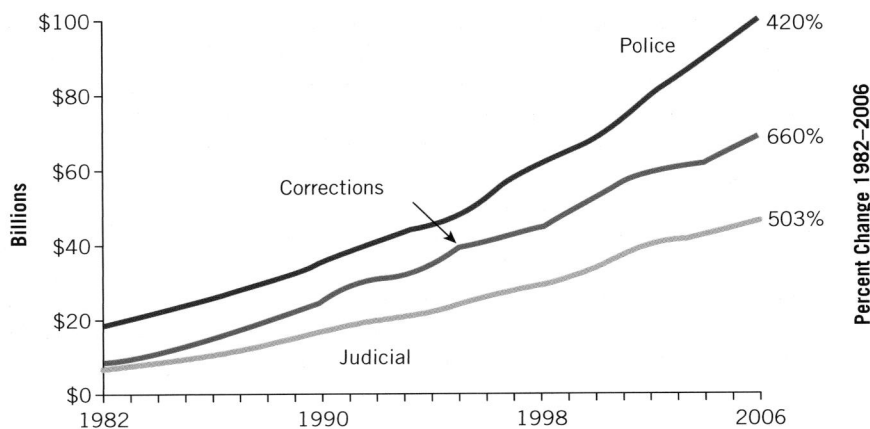

Of these employees in 2005 (the latest date for which we have figures), the majority, particularly those who work directly with inmates as correctional staff, were males by a 3-to-1 ratio (Stephan, 2008, p. 2). In federal facilities, the gender differences were largest with 87% of correctional officers being male and only 13% being female. The smallest gender employment difference was in the 400-plus private facilities where only 52% were male and 48% were female. In the more numerous state correctional facilities, there were 74% male correctional officers compared to 26% female (Stephan, 2008, p. 2). Notably, these gender differences by level and type of facility very closely align with pay differences. Among prisons, federal correctional officers, who are much more likely to be male, are the best paid, and private facility correctional officers, who are almost as likely to be female, are paid the least (see Table 9.1 regarding private vs. public pay for correctional work).

Even with this growth in staff numbers overall, whether male or female, the proportional growth in inmates has been much larger. As a result, the ratio of inmates to staff in prisons has grown from 2000 to 2005 (the latest date for which we have figures at the time of writing) with the most inmates to correctional staff in federal prisons (10.3) and private prisons (6.9 vs.

Table 9.1

Full-Time Local, State, Federal, and Private Correctional and Law Enforcement Worker Median Annual Pay (in U.S. dollars)

	All	Correctional Officers & Jailers	Probation and Parole Officers	Police and Sheriff
Local	44,530	37,315	50,109	55,058
State	43,426	36,525	46,714	57,175
Federal*	74,403	53,459	Not available	Not available
Private	35,006	21,216	Not available	45,594

Source: U.S. Department of Labor (2010). Data taken from Tables 6 and 7.

*Source for the federal data: Bureau of Labor Statistics (2009).

5.0 in public) in 2005, as compared to state prisons (4.9) (Stephan, 2008, p. 5). Likewise, jails continue to hold in the neighborhood of 90% of their capacity (Minton, 2010, p. 1), and the caseloads of parole officers at the state levels (data on state or federal probation officer caseloads and parole officer caseloads at the federal level were not available) ranges around 38 per officer (Bonczar, 2008, p. 1). Simply put, staff working in corrections are stretched very thin despite the growth in their numbers.

Given these numbers of inmates and clients, one can appreciate the organizational problems that develop for correctional managers seeking to hire, train, and retain the best employees to do this difficult work. Perhaps in part because of this recognized need to get and keep the best employees, the management of correctional institutions and programs has shifted over the years as efforts to professionalize, democratize (allow more say in the work by those doing it), and standardize work in corrections have had some success.

Pay

In general, compensation that is commensurate with job requirements and skills is a clear indication of the value given to a particular profession. It is true that it would be difficult to regard some correctional institutions or programs or their staff as "professionals" because they do not meet the educational, training, or pay requirements of a "profession." However, there are a number of correctional institutions and programs that have made much progress in this area, though the path to professionalization and the creation of a work environment that is conducive to employee growth and welfare are not always achieved.

So, if we were to imagine a continuum of correctional professionalism across the field, we could probably generalize that community corrections officers (probation and parole officers) and their work are more professionalized—defined as having more education, training, and pay—than are prison and jail correctional officers (see Table 9.1; we will discuss ethics separately at the end of the chapter). Next along that continuum would come correctional officers who work in public prisons and then those who work in public jails. Of course, these generalizations do not hold true for every institution or every locality. For instance, in some larger city jails, officers might be paid and trained as well as or better than prison officers or even adult probation and parole officers. Typically, jail officers in large counties are paid more than those in small counties and often are paid less than, but not always, state- and federal-level correctional officers. In fact, the recent data presented in Table 9.1 indicate that the median pay for probation and parole officers (and correctional officers) at the local level is more than that at the state level. We would attribute this higher pay at the city/county level to the consolidation of local services such as jails into larger counties in the last few decades; large cities and counties are now able to pay as much as or more than their respective states.

Generally speaking, probation and parole officers at the federal level are paid the most, followed by county/city level officers, and then state-level community corrections officers (see Table 9.1). Usually, federal-level community corrections officers make more than prison officers at their level of government (e.g., at the federal or state level) (Stohr & Collins, 2009; see also Table 9.1). These data do indicate that attempts to professionalize corrections, as this pertains to pay, need more effort in some areas than in others.

Note that police officers at both the state and local levels are paid almost $20,000 more per year than correctional officers and $5,000 to $10,000 more than probation and parole officers in any given state (see Table 9.1). Therefore, it is not surprising that both local and state police officers, along with probation and parole officers, are more likely to require at least some college and possibly a college degree. Students of criminal justice tend to recognize this difference in that when asked what they want to do when they graduate, they are more likely to identify police or community corrections (probation and parole) jobs as desirable over work in a prison or jail.

Notably, the push to privatize correctional and law enforcement work since the 1980s has not resulted in better pay for staff. In fact, as is illustrated in Table 9.1, privately employed correctional and law enforcement officers' median pay is at least $15,000 less for the former and $10,000 less for the latter, than for comparable jobs in the public sector. If better pay is correlated with a more professional workforce, then we can conclude that the privatization efforts are not moving us in that direction.

❖ Why the Need to Require More Education and Training Exists

First of all, and unfortunately, most correctional institutions and programs do not have prior educational requirements that would elevate them to the level of a "profession." Though it is true that many probation and parole officers must have a college degree or at least some college to qualify for the job, most jails, prisons, and even juvenile institutions, even those with a greater emphasis on rehabilitative programming, do not require such a qualification from applicants.

Stanford Prison Experiment

Photo 9.1

Stanford Prison Experiment "inmate"

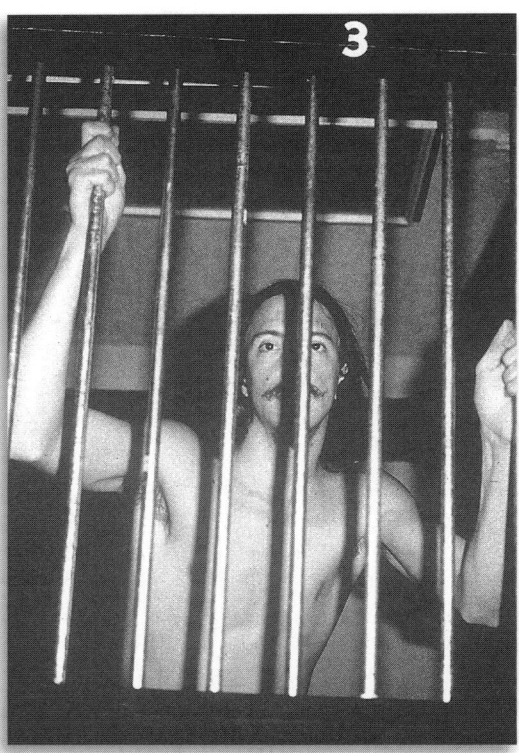

Yet, the oft-cited **Stanford Prison Experiment** provides a powerful argument for the value of formal education and training for correctional staff. In this 1971 experiment, volunteer students, with no training as officers and only their own expectations and beliefs to guide them, were divided into officers and inmates in a makeshift "prison" (Haney, Banks, & Zimbardo, 1981). The "officers" were outfitted in uniforms, including reflective sunglasses, and given nightsticks. The "inmates" were given sack-like attire. Neither "officers" nor "inmates" were told of any rules or policies to guide or restrict their behavior. Predictably, a few of these "officers" or "guards" engaged in verbal and psychological abuse of the "inmates." In the end, about a third of the "officers" engaged in the abuse, and others stood by while it was going on. The experiment was stopped after a few days and is often referenced as an example of how correctional work, and the subcultures that develop as part of the job, can foster corrupt behavior by officers.

The problem with the experiment, however, was that the "officers" were never given any education or training in corrections work. They were directed to exercise their discretion in controlling the inmates, but it was a discretion that was not necessarily anchored to any history or knowledge of "best practices" in corrections. Rather, the "choices" made by the officers in the absence of any education and training were likely shaped by the movies and popular press depictions of corrections that tended to reinforce the stereotypes of the institutions/programs and work. Not knowing how best to "get people to do what they otherwise wouldn't" (Dahl's [1961] definition of **power**), the "guards" used what knowledge of corrections they had, even if it was all wrong.

Abu Ghraib

The *Abu Ghraib scandal* of 2004, in which prisoners were tortured by mostly untrained "correctional officers" in the American-operated Abu Ghraib military prison in Iraq, tends to reinforce the lessons of the Stanford Prison Experiment, even contrived as those circumstances were. At Abu Ghraib, some correctional officers made inmates sleep naked, crawl on the floor, and pose in pyramids naked (while staff took pictures). A number of officers also deprived inmates of food and basic necessities and engaged in physical torture. The U.S. Army's investigation of the abuses at Abu Ghraib found that officers engaged in the following:

> Breaking chemical lights and pouring the phosphoric liquid on detainees; pouring cold water on naked detainees; beating detainees with a broom handle and a chair; threatening male detainees with rape; allowing a military police guard to stitch the wound of a detainee who was injured after being slammed against the wall in his cell; sodomizing a detainee with a chemical light and perhaps a broom stick, and using military working dogs to frighten and intimidate detainees with threats of attack, and in one instance actually biting a detainee. (Hersh, 2004, p. 1)

Though the Army wanted to blame the abuses that occurred at Abu Ghraib on untrained and rogue staffers at the prison (and six of these were prosecuted), the blame for the abuses extended up the chain of command to the Army Reserve brigadier general in charge of all Iraqi prisons, though she had had no previous experience or training in running them (she was relieved of command), and perhaps as high as Defense Secretary Donald Rumsfeld.

> As the international furor grew, senior military officers, and President Bush, insisted that the actions of a few did not reflect the conduct of the military as a whole. Taguba's report [Major General Antonio M. Taguba investigated what happened at Abu Ghraib], however, amounts to an unsparing study of collective wrongdoing and the failure of Army leadership at the highest levels. The picture he draws of Abu Ghraib is one in which Army regulations and the Geneva conventions were routinely violated, and in which much of the day-to-day management of the prisoners was abdicated to Army military-intelligence units and civilian contract employees. Interrogating prisoners and getting intelligence, including by intimidation and torture, was the priority. (Hersh, 2004, p. 4)

Simply put, the lesson from these incidents, and countless corrections scandals over the years that have involved the systematic abuse of inmates by staff, is that some people will not act professionally, or even decently, especially when they have no education or training in that profession. Now, the training and education will not necessarily prevent all such abuses (at least two of the officers involved in Abu Ghraib had prior experience as correctional officers in Virginia, and presumably some training in that state), but they will at least provide the officers with the knowledge and skills to do the job the way it should be done.

Photo 9.2

Ropes in the gallows at Abu Ghraib Prison in Iraq, where Iraqi citizens were tortured by Saddam Hussein's secret police and by American military personnel.

Correctional work often does not resemble other professions because the formal training provided for many new hires, including the number of hours required and the quality of that training, does not approach the level of other professions, which may schedule months of training (e.g., police departments with an average of 749 hours' training for new recruits [Hickman & Reeves, 2006]) or extensive internships of months' or years' duration (e.g., teachers, social workers, doctors).

The typical correctional job has lesser requirements for formal training or structured experience. For instance, in a *Corrections Compendium* (2003) survey, the researchers found that 31 of the reporting United States agencies required at least 200 hours of preservice training for those destined to work in a correctional institution. Likewise, in a quick survey of 150 directors and staff trainers, with responses received from 13 states/agencies in April 2004, the Juvenile Justice Trainers Association found that about 140 to 180 hours of preservice, academy-like training are required for most new hires in juvenile facilities (personal correspondence with B. Collins, May 2004).

Moreover, in some professions, the requisite college or professional degree is geared toward the work itself (e.g., computer programming, law school, or a master's in social work). Yet, when a college degree is *required* for a job in corrections, it is rarely specifically a criminal justice degree, but is typically one or more of the social science degrees, which may include no classes on corrections or the criminal justice system at all.

Of course, both of these deficits in formal education and knowledge base and in training leave correctional workers less suited to perform their job in anything approaching a historical or present-day, research-based context. When they have not studied corrections or been provided with sufficient training, they may not understand the reason some practices are undertaken or why others are abandoned. They do not have the requisite tools to suggest changes or the background in research to know whether something "works" or not. Their ability to behave, and develop, as a professional is limited. So when they use their "discretion" (defined here as *the ability to make choices and to act or not act on them*), they could be making ill-informed choices that are not based on knowledge or experience and are overly influenced by their personal ideology, politics, or the media (Merlo & Benekos, 2000).

Correctional Work Is Little Understood

It is possible that students and the general public may not view correctional work as a desirable career choice because they do not understand it. The truth is that few people outside of correctional work (or academe) probably know how institutions and community supervision actually operate, nor do they hold the roles of staff working in those agencies in very high regard. Students are acculturated by a media preoccupied with violence that tends to depict correctional institutions as dark, corrupted places peopled by abusive, or at a minimum, cynical and distant, "guards" (Conover, 2001; Johnson, 2002; O'Sullivan, 2006). In the movies and television specials, prisons are almost always maximum security, old, and noisy; jails are crowded and huge monstrosities; and juvenile facilities are depressing and havens for child predators. Perhaps as discouraging, community corrections, which is arguably—based on our criteria here—the most professionalized sector of corrections, is rarely depicted in the mass media at all.

Unfortunately, the mass media are not alone in misleading the public and students of criminal justice and criminology about corrections and correctional work. Academics have also tended to focus much of their attention on only the biggest, and the "baddest," of correctional institutions and programs and the labor of their staff. Maximum security institutions, and to a lesser extent metropolitan city jails, have been showcased, though they are not the norm for most corrections in this country (e.g., see Conover, 2001; Hassine, 1996; J. B. Jacobs, 1977; Johnson, 2002; Morris, 2002; Sykes, 1958). Research on these institutions tends to focus on the negative, or what the institutions or staff are doing wrong, rather than on what is working well. Of course, it is understandable that the "negative" shines through when these particular institutions, and their type, are the center of attention: Given the makeup of their inmate population and the likelihood that they are overcrowded and understaffed, there is much that is amiss in such places.

Predictably, work in probation and parole, much like the study of jail staff, receives short shrift by academics who tend, like the media, to be preoccupied with what is "sexy," violent, and controversial. Given these depictions by the media and academics, it is hard to discern the truth about correctional institutions and programs and work in them, because it is clear that the work is underappreciated, little understood, and hampered by misguided perceptions of it.

❖ Organizational-Level Factors That Affect the Correctional Workplace

Bureaucracies

It is often forgotten that people created organizations and that they can change them if need be. Organizations can be reconfigured to suit workers so that they are more satisfied with, less stressed by, and more challenged by, their work.

A **bureaucracy** is a kind of organizational structure most used in corrections. Most correctional institutions and programs are shaped like a bureaucracy. Bureaucracies were created to increase the efficiency of workers and the uniformity of their work. A *bureaucracy*, as defined by the German theorist/philosopher Max Weber (1946), can be distilled down to these three elements of its structure: hierarchy, specialization, and rule of law. The hierarchy refers to the pyramidal shape of the organization, with one person, the leader, at the top and an expanding number of people below him or her at each level of the organization. Most of the formal power of a bureaucratic organization resides at the uppermost portions of the bureaucracy, and most of the formal communications come from the top-level administrators and travel down to the lower-level administrators and then the workers. Specialization occurs in a bureaucracy when the sectors of the pyramid, below the leader, are segmented to concentrate on different aspects of the work, particularly as that work becomes more complicated and requires specific skills and abilities. The rule of law element of a bureaucracy refers to the formal rules, procedures, and laws that govern agency operations. Such regulations, whether imposed by an outside governmental agency such as a legislature, a court, or the executive branch, or whether internally created for the organization, are formal and usually prescribe how the organization works and how the employee should react when faced with given circumstances.

Working in a bureaucracy, as most of us do even outside of corrections, has the effect of making our work routine and predictable. Community corrections officers, working with adults in larger bureaucratic organizations, will specialize in work with either parolees or probationers, with more serious offenders or first-time felons. Their work will be governed by both legal requirements and court decrees, and rules and procedures created by their organization. They will report to, and be accountable to, supervisors, who will report to other supervisors, who will report to managers, all the way up to the director or secretary of corrections in the state who is usually a political appointee of the governor.

Closed Institutions

In addition to being bureaucratic organizations, corrections agencies, particularly prisons and jails and less so community corrections, are thought to be partially *closed* institutions in that they are separated from their outside environments, and unaffected by those environments, to some extent. This closed nature of corrections is much reduced from the past when inmate visits and outside contacts were more restricted; when staff behavior, even off the job, was controlled; and when what went on inside the institution was kept secret from the general public, let alone the courts and the media. Of course, however, even in the early prisons and jails, this "closed" characteristic was never absolutely descriptive of those institutions, as inmates would have visitor contact (albeit usually much less than these days) and they would get out eventually. Also, most staff did go out and about in the community, even if in some cases they slept at the institution. Moreover, outside factors like the changing laws and funding and technology, and even the weather, affected the operation of such facilities (Stohr & Collins, 2009).

But the point here is that the extent to which an organization is open or closed affects the work of employees. If open, the employees are free to discuss their work with members of the outside world, and they are free to gather information, research, and assistance that might improve their work. The more closed the organization is to the outside world, the less employees can seek out these materials and resources.

Total Institutions

A related aspect of correctional organizations, and again this applies more to institutions than community corrections, is the fact that they are *total institutions*. A **total institution**, as defined and described by Goffman (1961) and mentioned in Chapter 7, is a place where

> First, all aspects of life are conducted in the same place and under the same single authority. Second, each phase of the members' daily activity is carried on in the immediate company of a large batch of others, all of whom are treated alike and required to do the same thing together. Third, all phases of the day's activities are tightly scheduled, with one activity leading to a prearranged time into the next, the whole sequence of activities being imposed from above by a system of explicit formal rulings and a body of officials. Finally, the various enforced activities are brought together into a single rational plan purportedly designed to fulfill the official aims of the institution. (p. 6)

In a total institution, the formal and distinct roles of staff and inmates are sacrosanct, and each must stay separate from the other. Staff are invested with formal power, and inmates must not be given any, or allowed to exert any, of their own. Of course, in reality even the most locked-down maximum security prison is not a total institution, nor is it a model of a bureaucracy or completely closed off from the community. Correctional institutions fit these descriptors by degrees, and there are no absolutes. But, by necessity and practice, that maximum security prison is more likely to fit these descriptors than is a probation agency in the community.

❖ Individual-Level Factors That Affect the Correctional Workplace

Race/Ethnicity and Gender

Clearly, people should not be hired who are unqualified for a job, but if you have two people who are roughly comparable in terms of skills and abilities, should the employer look for other personal features, such as race or gender or military background or age, to determine who would be the best hire for the organization? Whether people will admit it or not, such matters do enter the calculus of who gets hired and promoted in organizations (The topics of gender and race will be explored in greater detail in Chapters 10 and 11.). Left to their own devices, and without the pressure that courts brought, the hiring of minorities and women did not occur in most criminal justice agencies to any extent until after the Civil Rights Act (CRA) of 1964 was passed, and modified in 1972 to include gender (Stohr & Collins, 2009). Before the passage of this law and its modification, most employees in criminal justice agencies were very homogeneous in terms of race/ethnicity and gender, and today, about 40 years later, this is still true in some organizations, but to a much lesser extent or not true at all in others.

Correctional organizations, like other criminal justice agencies, did not hire women (except to work with other women and then for less pay) or minorities until they were forced to by the CRA and sometimes lawsuits. One of the authors of this book worked as a correctional officer for an adult male prison in Washington State in 1983, and she was only the second woman hired there; the other had been hired only a month before. The warden told her he had fought central office for years about hiring any minorities or women. When promoted to the counselor position the next year, at that same prison, she was the first woman in the history of that institution to work as a counselor. At the time she worked there, only one Hispanic officer was employed, and the warden did not want him there. The warden told the author that he would never hire a black man or woman for a correctional position. Fast-forward 30 years, and women and minority group members have integrated the correctional workplace at all levels and in every position. There are women correctional officers in maximum security prisons and as juvenile probation officers. There are minority group members who serve as wardens and directors of corrections and who serve on the line in medium security prisons. Is corrections fully integrated? The answer to the question is no, not fully, but the law and courts have made it possible for all qualified applicants to work in correctional organizations so that they better reflect the composition of their communities.

As the correctional workplace has diversified, so has the importance of race as it might affect workers' perceptions of each other and their labor (Camp, Steiger, Wright, Saylor, & Gilman, 1997). Different racial and ethnic group employees perceive that their job opportunities are shaped by their race or ethnicity and the race/ethnicity of others in their workplace. Managers, therefore, need to be cognizant of these perceptions and sensitive to them as they hire, train, and promote people.

As much as race, the change in gender composition of the employee workplace has had an effect on correctional worker attitudes, perceptions, and behavior. We will discuss the effect of gender much more in Chapter 10, but suffice it to say here that some research has found that female staff are as capable as males and bring a different supervisory style to the work from that of men (Jurik & Halemba, 1984). When one of the authors first started in corrections, she had several inmates and officers who approached her and remarked on how the language had become less harsh by staff and inmates since she and another woman had integrated the correctional officer workforce. She and the other female officer had done nothing to change the language; rather, those around them thought the language should change because of their presence.

As regards sexual harassment in the correctional workplace, the research indicates that women workers are more likely to be harassed by male coworkers or supervisors than are male workers to be similarly harassed by females (Stohr, Mays, Beck, & Kelley, 1998). However, when inmates are harassed by staff, the harassment is not merely restricted to female victims and male offenders, nor is it limited to just staff (Marquart et al., 2001). Recent research indicates that female officers are much more involved, particularly in the more minor and "consensual" versions of sexual harassment, of male inmates, than are their male counterparts. In an example of a serious "boundary violation" in this regard, a female McNeil Island Prison (Washington State) nurse, hired in 2005, was fired in 2006 for allegedly having an intimate sexual relationship with a violent sex offender. Yet she continued to call him and have phone sex with him and visit him after her firing, and that was not even the worst of her behavior vis-à-vis this inmate. "In 2007 and 2008, she reportedly smuggled in 50 pornographic movies to the inmate and delivered crack cocaine to him 11 times" (Glenn, 2010, p. A4).

Age

Correctional agencies will also consider the age of the applicant when making a hiring decision. None will hire below age 18, and many will require that the applicant be at least 21. Though correctional agencies do not typically have an upper age limit for hiring, as police departments legally do, rarely will police departments hire someone in their 30s, unless they have prior experience, and almost never will they make a first hire of an officer who is in his or her 40s. Most of the time, the initial hires in corrections are people in their 20s and 30s. Depending on the job specified and the correctional clients worked with, correctional work can be a physically taxing and stressful job, which is why agencies will tend to target youngish workers. When the job has fewer physical requirements (e.g., counseling and treatment programming work or work as a probation or parole officer), the agency is much more likely to hire a worker in his or her 40s and older. What correctional, and police, agencies may fail to realize, however, is that more mature workers are able to bring a level of human experience and wisdom that might compensate for what they lack in physical agility in the management of inmates and clients.

Prior Military Service

Another personal characteristic, beyond race/ethnicity and gender, of workers that tends to shape the correctional environment is prior military service. Such experience is usually considered favorably by correctional agencies seeking to hire. Some agencies will state such a preference explicitly, and others will provide extra points for military service when applications are assessed. Whether such service better prepares workers to handle jobs in corrections is not a settled matter, although the militaristic accoutrements of correctional work (i.e., the uniform, the military titles, the command structure, etc.) would certainly make ex-military officers feel at home in some correctional workplaces.

Photo 9.3

Inmates receiving instructions from a correctional officer

❖ Correctional Roles

The Role Defined

Another aspect of the work that affects staff is the correctional role. The **role** of those working in a correctional workplace, or any other, is determined by what that person does on the job every day. The role of staff in corrections is determined by their job description, their assigned duties, and the type of organization and clientele they work with. Therefore, correctional officers working in a living unit in a medium security prison have a different role from that of probation officers working with lower-level offenders in the community. The first role involves constant supervision and interaction with inmates who live a very restricted lifestyle in a secure institution. The second role, that of probation officer, also involves supervision and interaction (though less intense, as they are likely to see their clients less than daily), but also formally includes assistance to those they supervise. Both the prison and the probation officers usually work with convicted felons, but the prison inmates are more likely to be repeat felons and are usually convicted of more serious offenses. Though the first role involves the maintenance of both safety and security, for the prison officer that usually means for other staff and other inmates, whereas for the probation officer it usually means he or she is concerned about community members the officer's clients interact with. Both are likely involved in facilitating programming, but the prison officer is most often watching inmates engaged in it or escorting them to it—though in some prisons, officers help deliver it—whereas the probation officer might recommend it or even run treatment groups. The prison officer might make recommendations regarding a person's placement in housing or work assignments in the secure facility, whereas a probation officer will monitor a probationer's engagement in these or counsel him or her about how to find them.

Both roles involve paperwork and loads of it, and both involve interactions with supervisors and accountability for what they do. For prison officers, they must adhere to the formal and informal rules that shape prison work, as those are provided by state law, policies and procedures, their administrators, and the subcultural values of coworkers and clients (Lipsky, 1980). For probation officers, they also must work within these formal and informal strictures, but in addition they are often in constant contact with court actors, prosecutors, police officers, and jail workers and these, plus the clients on their caseload and their families, can shape the role they play on the job.

We clearly have overgeneralized in our comparison here of how a correctional officer role in a medium security prison and that of a probation officer role in the community might differ, and be similar. In those states and institutions where treatment is emphasized, some correctional officers are very engaged in providing treatment, not just supervising an inmate's involvement in it, for instance. Moreover, the role of some probation officers, particularly those who have an intensive supervision caseload with more serious offenders, may involve as much direct supervision as correctional staff provide in a prison. The point is that a role for staff is determined by many things, and it is defined by what people actually do in their work.

Hack Versus Human Service

The public may be less inclined to support the increased professionalization of correctional staff when they think their role is limited to the use of brute force, a role termed as "guard" in old-style Big House prisons, and defined as a **"hack"** by current scholars (Farkas, 1999, 2001; Johnson, 2002). If a correctional officer is viewed by the public as a hack, or as a violent, cynical, and alienated keeper of inmates in a no-hope warehouse prison, then there would appear to be little need to encourage education, or provide the training and pay that would elevate such officers to "professional status." Yet there is reason to believe that most officers in prisons, jails, juvenile facilities, and community corrections actually regularly engage in **human service** work, which serves as an alternative, more developed, and more positive role for correctional workers (Lombardo, 2001; Johnson, 2002).

Johnson (2002) defines human service correctional officers as those who provide "goods and services," serve as "advocates" for inmates when appropriate, assist them with their "adjustment" to prison, and use "helping networks" of staff to facilitate that adjustment (pp. 242–259). Such goods and services might involve food and clothing or medication, and advocacy might include helping inmates find jobs or apply for different housing or roommates, while adjustment assistance could include counseling them about how to handle difficult people or situations. Clearly, these kinds of activities are not necessarily in the job description of most officers who work in institutions, but they are often very much a part of what they do in reality and require that the officer be skilled and knowledgeable. When the public does not know about, and the correctional organization does not recognize, the alternate "human service" work role performed by correctional or juvenile justice officers, and probation and parole officers, then, again, there is no perceived need to provide the training and pay that would be commensurate with that more developed professional role.

❖ The Subculture and Socialization

The staff subculture in corrections, much like the roles, varies by facility and by type of organization. A *subculture* might be defined as the norms, values, beliefs, history, traditions, and language held and practiced by a group of people. In corrections, those aspects of a subculture are shaped by what kind of facility or organization you have, what kind of clientele you are dealing with, and how isolated the group of people are from the rest of the community. The more isolated and the more exclusive the interaction of the group, the more likely it is that the subculture's norms, values, and beliefs are distinct from the larger community.

Historically, staff literally lived on the prison grounds with the inmates, in prisons that were not open to the public or the media (Ward & Kassebaum, 2009). In such institutions, a distinct subculture was more likely to form than it is in today's prisons, jails, or community corrections entities, where staff come and go with shifts and where visitors, lawyers, and the media have much more access than they did in years past. However, though correctional organizations of today are less likely to have as strong a staff subculture as they did in years past, this is not to say that they do not have a subculture.

Subcultural Values

Subcultural values in the correctional setting are likely to have an effect on what staff do. Even today, correctional organizations and the people who work in them are somewhat "cut off" from the larger society by the nature of what they do and the need to keep some matters private for legal and security reasons. Moreover, staff in corrections have very intense experiences together, involving violence and strong emotions, experiences that are likely to bind staff together in an "us versus them" stance toward their clients and the larger society. It is for this reason that the role of staff in corrections is likely, for better or worse, to be influenced by subcultural values of the group (e.g., see In Focus 9.1, Subcultural Values of Probation and Parole Officers).

As indicated by the subcultural values delineated in In Focus 9.1, some subcultural values are "positive" in that they facilitate the ability of officers to do their work well (e.g., aiding your coworker and doing your own work). However, other subcultural values sound exactly like those expressed by inmates (e.g., "never rat" and the "us vs. them" mentality) and serve to isolate the work and the workers, making them more likely to either participate in corrupt

In Focus 9.1

Subcultural Values of Probation and Parole Officers

In ethics training exercises in 1994 and 1995, probation and parole managers in a western state identified the subcultural values of the community corrections officers they supervised (Stohr & Collins, 2009, p. 63):

1. Always aid your coworker.

2. Never rat on coworkers.

3. Always cover for a coworker in front of clients.

4. Always support the coworker over the client in a disagreement.

5. Always support the decision of a coworker regarding a client.

6. Don't be sympathetic toward clients. Instead be cynical about them (to be otherwise is to be naïve).

7. Probation/parole officers are the "us" and everyone else is the "them," including administration, the media, and the rest of the community.

8. Help your coworkers by completing your own work and by assisting them if they need it.

9. Since you aren't paid much or appreciated by the public or the administration, don't be a rate buster (i.e., don't do more than the minimal amount of work).

10. Handle your own work and don't allow interference.

activities or to turn their head when they witness them, and to reinforce negative attitudes toward clients and the work.

❖ Staff Interactions With Inmates

The Defects of Total Power

Many of those subcultural values are shaped by the power relationships between staff and inmates. In a classic work by Gresham Sykes (1958), he describes the relationship between staff and inmates in a maximum security prison. Sykes notes that staff at this prison, or any prison, need inmates to comply with orders as it would be difficult, if not impossible, to force inmates to do "what they otherwise wouldn't" (the definition of power mentioned earlier; Dahl, 1961). Use of force to get inmates to comply with basic commands or orders would be inefficient, impossible, and counterproductive. For instance, if an inmate, or several inmates, refused to make their bed, it would be difficult for an officer to call in the emergency response team every time this happened as it would mean that these team members would be pulled from their own duties. It would also make the officer in question, the one who could not get the inmate to do something as simple as make his bed, appear like she could not handle the supervision of inmates. Moreover, force is sometimes impossible to use, at least daily, as the inmates outnumber staff in living units by sometimes as much as 50 or 100 to 1. There are ways to lock down living units and use gas and other measures to suppress disruptions, but if this had to be done on a regular basis, the prison would be in a constant uproar. And if force were used regularly for such trivial matters as making a bed, it is likely that inmates would align themselves more in opposition to staff than they already are, thus making the use of force counterproductive and perhaps requiring the greater use of it.

It is for these reasons that Sykes (1958) found that staff *needed* inmate compliance in prisons as much as inmates needed staff assistance. However, gaining compliance from inmates is not always easy. Staff, according to Sykes, have relatively little that they can give to inmates to motivate them. They can sometimes get them better work or housing assignments, but the amount of these rewards is limited and their power to garner them for inmates may be likewise so. Moreover, staff do not have the ability to reward inmate compliance with what inmates want most—their freedom. Therefore, according to Sykes, given these realities, staff and inmates tend to engage in a "corrupted" relationship whereby inmates comply with staff orders as long as staff overlook some violations, usually minor ones, by inmates. Such a relationship is most likely to develop between staff and those inmates who have the most power to control, and gain compliance from, other inmates.

Of course, Sykes was only describing the dynamics of the staff–inmate relationship in a New Jersey maximum security prison in the 1950s, and much has changed since then, even in the most secure prisons. However, the informal side of the relationship between staff and inmates/clients in corrections is still there, and the need for an exchange relationship between the two is still part of that dynamic. In another classic work on public service workers generally, Michael Lipsky (1980) also recognized the fact that formally staff control inmates, but informally, in prisons and schools and social welfare departments, clients, even ostensibly powerless ones, also exercise some power over staff. By not complying with orders, or failing to review information, or complaining about the service they receive, clients are able—even inmate clients—to force staff to adjust their behavior.

The Correctional Role When Supervising Children

An example of the variability of the correctional role would be the form it takes when officers and counselors supervise children. Inderbitzin (2006) found, in a 15-month ethnographic study of staff members supervising a cottage of boys in a juvenile training facility (the juvenile version of a prison), that they "serve as their adolescent inmates' guardians, keepers, counselors, and role models" (p. 431). Noting that the work with these serious and sometimes violent juveniles could be "frustrating, dangerous, often amusing, and occasionally rewarding," she noted that officers and counselors needed to be "flexible" and "energetic" in order to best cope with a volatile mix of duties that each day would bring (p. 439). The tactics used by staff to gain the boys' cooperation and to supervise them effectively ranged from the more punitive disciplinary actions all the way to reasoning with or cajoling them and using humor. Some staff members tended to emphasize certain tactics over others and some staff members tended to vary their own behavior as the situation required. Inderbitzin concluded that the staff who were most successful in their work, in that they were able to effectively supervise their charges, were those who took on a "people worker" (p. 442) role or the human service role that was described earlier in this chapter (Farkas, 2000). Such staff members were also the most flexible, kind, creative, and respectful in their work with the adolescents.

❖ Other Issues for Staff: Stress, Burnout, Turnover

Work in corrections can be taxing and troublesome. Most correctional work in institutions is shift work and when first started, involves late nights, sometimes all-night shifts, and weekend and holiday work. Even parole and probation officers are often called upon to visit clients in the evenings or on weekends. Needless to say, late-night shifts and weekend work play havoc with family life and children's school schedules, making family obligations challenging to meet.

It does not help matters that the people that correctional staff supervise are often angry and upset or immature (Johnson, 2002). They are usually unhappy about being supervised and sometimes unpleasant to those correctional staff engaged in supervision.

Relatedly, people who are incarcerated or supervised in the community have often led impoverished and tragic lives, riddled with alcoholism, drug abuse, child abuse and neglect, unemployment, early deaths of loved ones, and homelessness. Many have inflicted serious harm on others, including their victims, but also family and community members, and the guilt and regret they shoulder only exacerbate the negativity that surrounds them. Collectively, these tragic lives weigh heavily on the people experiencing them and can create a negative environment for themselves and for those who work with them (Johnson, 2002).

Correctional staff immersed in this environment and required to get unwilling people to do things they are not inclined to do, often experience stress and the related consequences of it: burnout and turnover (Lambert, Hogan, & Tucker, 2009; Slate & Vogel, 1997; Stohr, Lovrich, & Wilson, 1994; Tewksbury & Collins, 2006). In a meta-analysis of studies on stress in prisons, Dowden and Tellier (2004) found that those who adopted a more human service work attitude experienced less stress on the job than those who had a more punitive (i.e., hack) and custody orientation. In a study of five jails by Stohr, Lovrich, Menke, and Zupan (1994), the authors found that those jails that invested in training and pay and who allowed staff to have a voice in how to do their work—all factors related to professionalism that we discussed earlier—were more likely to have greater job commitment by staff and less stress and turnover from them. Relatedly, in a study of a large county correctional system (nine facilities), Paoline, Lambert, and Hogan (2006) found that adherence to policies and standards (of the American Correctional Association, which emphasize professionalism, safety, and humanity in supervision), and positive and noncompetitive work relations with coworkers, decreased the stress of workers.

Collective Bargaining

Correctional staff moved to unionize as a means of gaining power vis-à-vis administrators. As we know from our history of corrections, the wardens of earlier prisons (and this is still true to some extent today in some correctional institutions) often acted like dictators over their fiefdom, the prison (Bergner, 1998; Rideau, 2010). And it was not only inmates who were their subjects, but also staff, who were relatively powerless to voice their concerns or to earn a decent and livable wage. By unionizing, correctional staff working at the state and local levels were able to gain some collective power to bargain with administrators to improve their own work conditions. A "union worker" is defined by the U.S. Department of Labor (2010) as any employee in a union occupation when all of the following conditions are met: a labor organization is recognized as the bargaining agent for all workers in the occupation; wage and salary rates are determined through collective bargaining or negotiations; and settlement terms, which must include earnings provisions and may include benefit provisions, are embodied in a signed, mutually binding collective bargaining agreement. A nonunion worker is an employee in an occupation not meeting the conditions for union coverage. (p. 1)

A number of concerns regarding unionization have been raised, however, and these include the belief that unionization restricts the ability of administrators to fire incompetent people, that union contracts are too restrictive regarding the work that people can do, and that in some cases unions have worked to increase incarceration in states as a means of increasing job opportunities. Although some of these concerns may have merit, it is also true that those states with unionized correctional staff pay those staff more and provide more benefits for them (see Table 9.2), which is likely to increase the ability of those states to attract and keep better and more professional-level workers. If it is harder for administrators to fire incompetent unionized workers, it is also harder for administrators to fire people based on their politics or simply because they do not like them. Such contrarian voices in the workforce (those who disagree with administrative practices) often provide an important check on the power of administrators, and unions give such people protection from the wrath of those administrators who might want to retaliate against "contrarians."

Table 9.2

Mean Hourly Wages of Union and Nonunion State and Local Correctional Workers and Law Enforcement

	Union	Non Union
Probation and Parole	$27.69	$21.00
Correctional Officers & Jailers	23.36	14.74
Police	30.42	20.97

Source: U.S. Department of Labor (2010). Data taken from Table 13.

Though the union for correctional officers in California did lobby to increase the number of institutions and jobs in corrections in that state in the 1990s, it is possible that they were just representing the sentiments of their membership. Correctional staff tend to be somewhat conservative regarding crime issues and so would tend to support, like the general public of the 1980s, 1990s, and 2000s, the creation of more correctional institutions.

Abuse of Power

There are plenty of instances, both historically and currently, of correctional staff abusing their power over inmates and clients. We will discuss such instances in some depth in several chapters in this book. It is important to mention here, however, that the abuse of power is more likely to occur in environments where staff behavior is not supervised closely enough by administrators or those outside the work environment, where inmates have little or no ability to contact the outside, where staff are not sufficiently trained, where there is a higher concentration of young and inexperienced staff, and where there is a higher concentration of disruptive inmates (Antonio, Young, & Wingeard, 2009; Rideau, 2010; Stohr & Collins, 2009).

Use of Force

The actual use of force, or being prepared to use such force, is part of correctional work. Correctional administrators interested in channeling its use in appropriate and legal ways ensure that staff are trained on when to escalate or de-escalate its use based on the situation (Hemmens & Atherton, 1999). In such training, the officer is taught to pay attention to cues that will tell him or her when to increase the force level and when not to do so.

Having said this, the use of force in any correctional environment depends on the type of institution or agency, the clients/inmates served, and the way they are managed. Generally speaking, there is less call for force in probation and parole work with adults or juveniles, though when engaged in arrests it is not uncommon for community corrections officers to have to use force. Minimum security prisons for adults or juveniles generally have less cause for use of force; inmates in such institutions are classified as less prone to violence and/or are on their way out of the system and so are more likely to rein in any violent inclinations. As one proceeds up the correctional security ladder, from minimum to medium and then to maximum, the need to use force increases given the types of inmates incarcerated and the need to maintain stricter controls. It should also be mentioned, however, that in some prisons, under some wardens, the use of force is resorted to more frequently than under other wardens of similar prisons (Rideau, 2010). As Rideau observes after his 44-year incarceration in the Angola Prison in Louisiana and observing the management style of many wardens (see description of Rideau's experience in Chapter 7), those wardens who allow more openness between staff and inmates and the outside, and who have more transparency about their management decisions, are less likely to experience the need for the use of force in their facility. The more ready resort to force may be spurred by the type of inmates held (e.g., more violent) or the conditions of the facility (e.g., more crowded), but it could also just be a management tactic adopted by some wardens, as Rideau and others have argued (e.g., see Reisig, 1998).

❖ Ethics

Ethics involves right or wrong behavior on the job. Knowing what is ethical and following an ethics code is perhaps more important in corrections than in any other area of criminal justice. This is so because in corrections, as opposed to the police and courts, most of the behavior of staff in relation to inmates and clients is hidden from the general public and little known outside the confines of an institution or agency. Moreover, the clients of corrections either convicted or not of a crime (more than half of those in jails are not convicted) are accorded few rights or protections by the law. The combination of these two factors— interactions that are kept from public view, and where one party to the interaction has all of the official power over the other—make correctional work ripe for abuse.

Not surprisingly then, there have been numerous instances of documented ethical abuses by staff that include sexual assaults—the beatings and rapes of clients on probation and parole, as well as of inmates in jails and prisons in the United States (Amnesty International, 2004; Bard, 1997; Schofield, 1997; Serrano, 2006). As a means of preventing such abuses, correctional agencies will often adopt an ethical code, such as those promulgated by the American Correctional Association (ACA; see In Focus 9.2) or the American Jail Association, or others, and train people on what is right and wrong behavior in corrections.

In Focus 9.2

ACA Code of Ethics

Preamble

The American Correctional Association expects of its members unfailing honesty, respect for the dignity and individuality of human beings, and a commitment to professional and compassionate service. To this end, we subscribe to the following principles.

1. Members shall respect and protect the civil and legal rights of all individuals.

2. Members shall treat every professional situation with concern for the welfare of the individuals involved and with no intent to personal gain.

3. Members shall maintain relationships with colleagues to promote mutual respect within the profession and improve the quality of service.

4. Members shall make public criticism of their colleagues or their agencies only when warranted, verifiable, and constructive.

5. Members shall respect the importance of all disciplines within the criminal justice system and work to improve cooperation with each segment.

6. Members shall honor the public's right to information and share information with the public to the extent permitted by law subject to individuals' right to privacy.

7. Members shall respect and protect the right of the public to be safeguarded from criminal activity.

8. Members shall refrain from using their positions to secure personal privileges or advantages.

9. Members shall refrain from allowing personal interest to impair objectivity in the performance of duty while acting in an official capacity.

10. Members shall refrain from entering into any formal or informal activity or agreement which presents a conflict of interest or is inconsistent with the conscientious performance of duties.

11. Members shall refrain from accepting any gift, service, or favor that is or appears to be improper or implies an obligation inconsistent with the free and objective exercise of professional duties.

12. Members shall clearly differentiate between personal views/statements and views/statements/positions made on behalf of the agency or Association.

(Continued)

(Continued)

13. Members shall report to appropriate authorities any corrupt or unethical behaviors in which there is sufficient evidence to justify review.

14. Members shall refrain from discriminating against any individual because of race, gender, creed, national origin, religious affiliation, age, disability, or any other type of prohibited discrimination.

15. Members shall preserve the integrity of private information; they shall refrain from seeking information on individuals beyond that which is necessary to implement responsibilities and perform their duties; members shall refrain from revealing nonpublic information unless expressly authorized to do so.

16. Members shall make all appointments, promotions, and dismissals in accordance with established civil service rules, applicable contract agreements, and individual merit, rather than furtherance of personal interests.

17. Members shall respect, promote, and contribute to a work place that is safe, healthy, and free of harassment in any form.

Source: Reprinted from the ACA website (www.aca.org), with permission from the American Correctional Association.

The importance of training correctional workers initially, and throughout their careers, on what constitutes ethical behavior cannot be overemphasized. Just as important as the training, however, is ensuring that ethical people work in the organization. To do this, managers might consider using selection instruments and practices geared toward weeding out those who have little understanding of what is right and wrong and promoting only those people who exhibit that understanding in their daily work.

❖ Perceived Benefits of Correctional Work

There are many reasons why people who work in corrections might find it gratifying. For one, people are attracted to correctional work because it has been a booming business for a number of decades, creating thousands of jobs in the process. Correctional work also provides a steady, if not lucrative, paycheck (U.S. Census Bureau, 2010a).

People who want to make a difference in other people's lives, to end the tragic circumstances that tend to produce a cycle of street-level criminals, might choose to work with juveniles in corrections or in treatment programming with adults or juveniles on community supervision or in institutions. When they see someone change for the better, and know they might have contributed to that change, this is when correctional work can be personally fulfilling.

It is also possible that those who are attracted to correctional work are curious about the human psyche and how it operates. Working with people who have engaged in seriously deviant behavior enhances the understanding about why people behave the way they do. Some might also find it rewarding to be engaged in keeping these seriously deviant and violent offenders off the streets, and in that sense to be part of the justice-dispensing machinery that is corrections.

Comparative Perspective: Staff in Adult Swedish Prisons and Probation Services

According to the Swedish Ministry of Justice (2004), or *Regeringskansliet*, there are about 7,500 staff who work in probation and prison services in Sweden, most of them in prisons. Twenty-five percent of the prison officers are women and 71% of the probation officers are women (p. 4). During the first year of employment, the prison officers attend a 16-week "special education," or what we might regard as a training academy, and probation officers attend a 7-week "special education" (p. 4). As per Swedish law, these officers are charged with treating probationers and inmates with respect, and their work is to center around helping offenders to adjust or readjust to society, while keeping the need to protect the community paramount. Staff in prisons help inmates to program through work, study, or treatment. A relatively new role for prison officers is that of contact person for an inmate where "he or she is responsible for planning the implementation of the sentence for one or more prisoners" (p. 4).

Summary

- The correctional experience for staff is fraught with challenges and much promise. It involves a diverse role that encompasses work with juvenile and adult inmates in institutions as well as with offenders in the community. It is not as narrow a role as is commonly perceived by the public, and it includes many opportunities to effectuate a just incarceration or community supervision experience for inmates and clients.

- Though many have worked long and hard to "professionalize" correctional work, there is every indication that most jobs in this area do not meet standard professional criteria.

- When correctional workers do not receive the requisite professional training and education they need to do their job in an appropriate manner, both clients and the community are likely to suffer.

- Many who labor in corrections undertake the "human service" role as it is presented by provision of goods and services, advocating for inmates/offenders/clients, and assisting in their adjustment (Johnson, 2002).

- The research presented here would indicate that there are organizational factors that can be manipulated to improve this experience for staff so that they, and their organization, can realize their promise. Namely, better pay and benefits and training can be used to foster the greater development of professional attributes such as education and the consequent reduction in role ambiguity, turnover, and stress, and increase in job satisfaction. This research also indicates that the organization can do much to reduce the problems associated with the greater diversification of its staff. It can be open in its promotion practices so that false impressions regarding unfair advantage are not perpetuated and have a demoralizing effect on the workforce. It is also within the correctional organization's power to prevent most sexual harassment, whether practiced by staff or inmates. In short, the promise of a positive correctional experience for staff is achievable, and as that perception seeps into the public consciousness, a "correctional officer" is much more likely to be perceived as a professional.

Key Terms

Bureaucracy	Human service	Total institution
Ethics	Profession	
Hack	Stanford Prison Experiment	

Discussion Questions

1. Explain why work in corrections is often not regarded as "professional" in comparison to other commonly referenced professions.

2. Note which jobs in corrections are most sought after and why. Discuss how all correctional work might become more appealing to educated workers.

3. Review the events surrounding the Stanford Prison Experiment. If you were going to conduct such an experiment now, how would you go about it? What questions would you like to address with the experiment?

4. Explain and discuss the "hack" versus the "human service" role for staff. Which role do you think the public typically ascribes to correctional staff? Which role do you think is most commonly undertaken by staff?

5. Explain how the correctional organization can provide the right environment to reduce stress, turnover, and harassment of its staff.

6. Answer the question, "Why is ethical behavior such a challenge in correctional work?"

7. Consider the benefits and drawbacks of working in corrections. Do you plan on working in corrections? Why or why not?

Useful Internet Sites

American Correctional Association: www.aca.org/

American Jail Association: www.corrections.com/aja/

American Probation and Parole Association: www.appa-net.org

Bureau of Labor Statistics (2009). *Occupational Employment Statistics Survey Program.*

Office of Juvenile Justice and Delinquency Prevention. Accessed at www.ojjdp.gov

U.S. Department of Labor. (2010). *The National Compensation Survey: Occupational Earnings in the United States, 2009.* www.bls.gov/ncs/ocs/sp/nctb1349.pdf

Vera Institute (information available on a number of corrections and other justice-related topics): www.vera.org

Women and Corrections

❖ Introduction

As far back as anyone can remember, there have been fewer women and girls incarcerated or under correctional supervision than there have been men or boys. Of course, there have been exceptions for some crimes (e.g., prostitution for adult and juvenile females or status offending for girls), but correctional populations have always included more males. Though the percentage of women and girls in those populations has increased in recent years, it is still true that the vast majority of those under correctional supervision in the United States are male.

What this numerical "minority" status for girls and women has meant is that institutions and programming are, and have been, typically geared toward boys and men. As discussed in Chapter 2, the first prisons were built for men, though sometimes a section of them was set aside for women. The United States' very history of institution building illustrates this fact. What Young (1994) found from her research on the construction of juvenile facilities in the southern portion of the United States following the Civil War, was that males, and particularly white males, were much more likely to have juvenile prisons constructed specifically for them than were females. Women and girls accused (in the case of jails) or sentenced (in the case of jails or prisons) were much more likely to "do their time" in male facilities; initially as part of those facilities (e.g., in bridewells and other poorhouses) and later in separate sections of jails and prisons or completely distinct facilities and houses of refuge (Baunach, 1992; Belknap, 2001; Chesney-Lind & Shelden, 1998; Kerle, 2003; Pollock, 2002b; Rafter, 1985).

Part of what shaped the treatment of women and girls in the past was the numerical fact that they constituted fewer offenders and inmates than men and boys. As those numbers have increased, however, and as feminist beliefs regarding the value of women and girls have changed attitudes, concerns about how females are treated as clients, what their needs are, and about female staff and their rights, have reshaped correctional practice. In this chapter, we will explore all of these issues, but will begin with a brief history of the female correctional experience.

❖ History and Growth

Women in Historic Prisons

As mentioned in Chapter 2, the first American correctional facility—though it is not generally acknowledged as the first American prison—to hold felons only was the Newgate New York prison in Greenwich Village, New York City. Opened in 1797, it had a separate wing for women inmates, where they were housed in a group setting. As Rafter (2009) notes, putting the women together, though they had no matron, helped provide protection from "lascivious turnkeys" or guards (p. 51). All indications are that the women at Newgate washed and sewed for the prison (while the male inmates were engaged in production of goods for sale), and were situated close to other inmates.

When the Newgate Prison closed and the inmates were scheduled for transfer to the congregate, but silent and strict disciplinarian prisons of Auburn and Sing Sing, neither prison wanted to take the women, stating that they were difficult to manage (Rafter, 2009). While the matter was debated, the female inmates from New York City were held at the city's Bellevue Penitentiary where the conditions in terms of food, housing, supervision, and classification were poor. Moreover, the "silent" requirement so popular at the men's prisons could not be enforced because of the congregate housing and the lack of a female matron. When a cholera epidemic hit the prison, 8 women died and 11 escaped (Rafter, 2009).

The New York women outside of New York City were sent to the new Auburn Prison in 1825 (Rafter, 2009). However, their treatment there was also subpar as they were housed in a cramped, unventilated attic above the kitchen, without a matron until one was hired in 1832. Because of the congregate nature of their living and working conditions (they were engaged in sewing), it was once again difficult to enforce the "silence" requirement.

The discipline by the lash, when used on a 5-month pregnant inmate (who got pregnant while in prison) and who later died, prompted the state to construct a separate prison at **Mount Pleasant** for the women in 1839, the first women's prison in the United States (Rafter, 2009). Though it was close to the Sing Sing Prison and was in part administered by it, the Mount Pleasant Prison had its own buildings, staff, and administrator. This prison was built behind Sing Sing and overlooked the Hudson River. It was an Auburn-like building with Auburn- and Sing Sing–like sensibilities. It had a room for lectures and a chapel and a nursery. The matron's quarters were also in the prison.

Ohio's development of prisons facilities for women was similar to New York's (Rafter, 2009). At first they were held in less secure facilities along with the men. Then, in 1834, Ohio built an Auburn-like prison, called the Ohio Penitentiary, and in 1837 they housed the women in separate quarters from the men. However, the standards for prison operation in Ohio were even worse than New York's, and disease and corruption were rampant. They did build an annex specifically for the women in 1837, but this became crowded and fell into serious disrepair. The women had no matron and thus discipline was nonexistent; moreover, they were subject to sexual attack by male staff.

Among Tennessee prisons, when the first opened in Nashville in 1831, there was a progressive attitude toward standards and care, more like the New York model (Rafter, 2009). Women, however, were imprisoned in such small numbers before the Civil War that they were housed right with the men and worked with them in mines and on railroads. There were no matrons to protect them, and there were no separate accommodations for women in Tennessee prisons until the 1880s, and then they were placed in small overcrowded quarters in the Nashville Prison, with no room for work or exercise.

Race in Early Prisons

Maryland opened its first prison, the Maryland State Penitentiary, in 1811 in Baltimore (Young, 2001). From its inception, it housed the women in a congregate fashion, in the same prison with the men, until 1921, although they were separated. At first the races were not separated, but by the 1830s, black men and white men were separated at the prison and later so were the women. As Maryland was a "border state" separating the South with slavery before the Civil War, from the North without it, administrators of its penitentiary wrestled with issues of slavery and free and enslaved blacks. The research on incarcerated black women indicates that they were disproportionately incarcerated in the Northeast and Midwest before the Civil War, but that there were few blacks, male or female, incarcerated in the South before the war. After the war, however, black men and women were also disproportionately incarcerated in all prisons, but particularly Southern prisons where slavery-like treatment and work requirements were imposed (Oshinsky, 1996; Rafter, 2009; Young, 2001).

In her study of the Maryland State Penitentiary between the years 1812 and 1869, Young (2001) found that 72% of the incarcerated females were black and that as the Civil War drew to a close, the proportion of black women only increased. This was also true regarding the incarceration of black women around the Civil War, but especially during the antebellum stage, in Texas, Kansas, and Missouri. It is possible that a Maryland law, passed in 1858, that made black women who committed a larceny subject to "sale," rather than prison, resulted in less incarceration of black women before the Civil War in that state (Young, 2001).

Both black and white women in the United States of the 19th century tended to be incarcerated for property crimes, particularly larceny (Rafter, 2009; Young, 2001). Very few women were incarcerated for violent crimes, about 3% to 4% (Young, 2001). White women did tend to be incarcerated for "offenses against morality" more than black women were, perhaps because they were more "visible" to the police in white areas of town or the police were more attentive to them (Young, 2001). But there were also convictions for other felonies, miscellaneous offenses, and vagrancy. Young also found that black female inmates were required to serve a greater proportion of their sentence than were white female inmates, and they tended to be pardoned less and die more often, while incarcerated, than white female inmates.

Discipline in Women's Prisons

As with the male prisons during the 1800s, methods of discipline moved from the severe to the soft, depending on the availability of supervision, the facilities, the number of women incarcerated, and the inclinations of the keepers. Rafter (2009) reports that rarely was the lash used at the Mount Pleasant Prison for women, but the gag was used all of the time. At the Ohio prison, for instance, by the 1870s the discipline of women was quite severe, and women were beaten or placed in solitary confinement to enforce it (Rafter, 2009). By 1880, the "hummingbird" punishment was used in Ohio "[t]hat forced the naked offender to sit, blind-folded, in a tub of water while steam pipes were made to shriek and electric current was applied to the body" (Rafter, 2009, p. 53).

Hiring of Female Matrons

A serious problem for many of the first prisons was the absence of a female matron to supervise and, in some cases, protect the women. Writing in 1845 (reprinted in 1967), after visiting several primarily male facilities that housed women, the reformer Dorothea Dix noted that matrons had been hired in several prisons where women were housed (e.g., Connecticut Prison, Sing Sing, Eastern Penitentiary, Maryland Prison), but not in many county jails or prisons.

Houses of Refuge for Girls and Boys

Developed in tandem with the adult prisons were juvenile facilities in larger states for delinquent, neglected, abandoned, and abused children. **Houses of refuge** were part of the Jacksonian movement (named after President Andrew Jackson) of the early 1800s to use institutions as the solution for social problems. The first was opened in New York in 1825, the second in Boston in 1826, and the third in Philadelphia in 1828 (Beaumont & Tocqueville, 1833/1964, p. 136). Their stated purpose was to remove impressionable youth, mainly boys, but also girls, from the contamination that association with more hardened adult prisoners might bring. As Harris (1973) comments, such facilities for younger inmates had existed in Europe, particularly Holland, since the 17th century. The difference she notes between the American experiment with houses of refuge and the Dutch experiment was that they (the Dutch) were used only for the delinquent and those thought likely to become so without intervention, and that they were devised to achieve reform among their inmates.

From the first, the American house of refuge was a private institution. Such institutions were developed through private charity and subscription and operated by people hired by private subscribers. The states sanctioned their development this way and paid "some pecuniary assistance" to them, but they had no part in their administration (Beaumont & Tocqueville, (1833/1964, p. 137).

The early American houses of refuge were to be a mix of prison and school. The keepers in such houses were guardians, and such guardianship continued until the children reached the age of 20. In some of the houses, the children were separated at night and worked or went to school together during the day. In others, they were in congregate situations both at night and during the day. The requirement of silence, imposed on adult institutions of the time, was not visited on the children, as it was thought to be impossible to keep them completely quiet. In the New York and Philadelphia houses, the children worked in a workshop making shoes or cloth, and as carpenters, for 8 hours a day and spent another 4 hours in school. In the Boston house, they spent only 5-1/2 hours in the workshop, 4 hours in school, 1 hour in religious study, and there were a few hours for play (Beaumont & Tocqueville, 1833/1964, p. 142). The workshops in the houses were operated by private contractors. Notably, the girls did all of the domestic work around the houses including the cleaning, cooking, and sewing of clothes for themselves and the boys. The discipline used in the houses varied from deprivation of recreation, to solitary confinement, to restrictions on food and water, and sometimes the use of corporal punishment or the use of stripes (lashes with a leather belt).

Since the children placed in houses of refuge were not always sent there for punishment, it was thought that their time had to be indeterminate, as a magistrate could not tell at the beginning how long it would take to reform or correct the children. So it was left to those who operated the institutions to decide when a child was ready to leave them. Moreover, if he or she did not do as well as was hoped when released, and if still under 20, the staff at the institution were still his or her guardians and had the right to call the child back into it. Beaumont and Tocqueville (1833/1964)

Photo 10.1

The Virginia K. Johnson House of Refuge for Unmarried Mothers (circa 1913)

acknowledged that these absolute rights to deprive liberty might lead to abuse, but they point out that judges and parents do have some rights, to oversee, and protest, the incarceration of these children in the courts.

In a recidivism study of the New York House of Refuge, conducted by Beaumont and Tocqueville in 1831 (1833/1964), the authors found that of over 500 children released, more than 200 of the boys and girls had been "[s]aved from infallible ruin" (p. 151). As to the other 313 children released, they found that their behavior since release was either doubtful, "bad," or very "bad" (p. 151). Of course, lacking a control group for these releasees, it is difficult to know how to interpret these findings, but it was an admirable attempt by these French observers to try to find evidence as to the effectiveness of such early houses of refuge.

Dorothea Dix (1845/1967—whose own research is described more fully in Chapter 3) also visited houses of refuge in Boston, New York, Philadelphia, and Baltimore, and a farm school for children on Long Island. Her impression of these facilities, 14 years after Beaumont and Tocqueville visited some of them, was generally favorable. She liked that the children were employed in useful work, that the facilities were clean, and that the children were generally in good health. She noted that many houses of refuge provided schooling, training, and apprenticeships that would allow the children to succeed once they were able to leave. Some of the reports from the facilities that she reviewed indicated that children as young as 6 were incarcerated in these houses and that a possible offense that led to placement was being "stubborn" or "idle" along with other more common, but usually minor, criminal offending (Dix, 1945/1967, p. 91). Boys were often apprenticed to farmers and girls to domestic work once they reached their age of majority, and so were able to leave the institution.

Growth in Numbers of Women and Girls

As mentioned in the foregoing, the number of incarcerated and supervised women under the correctional umbrella has never been larger, but it was not always so. In the past, the number of women inmates and supervisees was proportionately smaller. For instance, we know from U.S. Census reports that women constituted 3% or 4% of state and federal prison populations from 1910 through the 1970s (Cahalan, 1986, p. 65). In 1980, that number had risen to 5% and has only increased since. If you add in reformatories, women and girls accounted for, on average, about 5% of those incarcerated in correctional facilities, from 1910 through 1959 (p. 66). Jail inmates were anywhere from 5% to 9% female from 1910 to 1983 (p. 91). Juvenile institutions averaged about 21.2% female residents (aged 15–19) from 1880 to 1980 (p. 130). Unfortunately, a gender breakdown of parolees and probationers is not available, but given the overall increase in the percentage of incarcerated women generally, and women and girls on probation or parole currently, it is likely that historically they were not as subject to the criminal justice system as they are today either.

The best explanation for the historically low number of female offenders in the U.S. criminal justice system (as compared to males) has been the fact that they commit fewer street crimes that would garner this distinction. Most murders, robberies, rapes, burglaries, and even larcenies are committed by men and boys (Federal Bureau of Investigation [FBI], 2007). Even among corporate/white-collar and environmental crimes, the more likely offender is male, if for no other reason than the fact that more males are in a position to commit such crimes than females. As mentioned in other chapters, the drug war of the last 30 years has brought more female offenders into the system, and this fact has resulted in their greater proportional growth among correctional populations, but even with this sort of offense, they are in the clear minority (FBI, 2007).

❖ Current Figures on the Number of Women and Girls in Corrections

Female Correctional Clients

More recently and by any measure, however, the number of women and girls as inmates or supervisees in corrections has grown exponentially over the last several years. In 2000, women comprised 11.4% of jail populations, but by 2009, that figure was 12.1% (a decrease from a peak of 12.6% in 2006) (Minton, 2010, p. 6). In 2000, women comprised 6.4% of prison populations, and by 2009, that figure was 6.9% (West, 2010a, p. 11). Girls confined in residential facilities increased from 13.6% of all juveniles in 1997 to 15.1% in 2003 (Sickmund, Sladky, & Kang, 2005). But the largest growth as far as correctional populations are concerned, has come in probation. Women constituted 22% of probation and 12% of parole caseloads in 2000, and by 2008 that percentage had increased to 24% for probation and remained the same for parole (Glaze & Bonczar, 2009). Though some of these percentage increases seem small, they represent increases, in the case of probation, of thousands of women on probation, as total caseloads in 2006 included 5 million people. For girls on probation, the increases do not even seem small. In 1985, girls constituted 19.3% of juveniles on probation, and by 2004 that percentage had grown to 27.2 (Puzzanchera & Kang, 2007).

Female Staff

The trajectory of employment of female correctional officers has not been as steep or as steady as it has been for those women and girls under correctional supervision. As already mentioned, women were employed as matrons, to a limited degree, to work with females in some of the earliest prisons and jails (Pollock, 2002b; Stohr, 2006; Zupan, 1992). However, they really did not make significant inroads into the correctional profession until the 1964 Civil Rights Act was amended in 1972 and they began using that law to sue in courts to gain employment in both female and male correctional institutions. According to reports from the Bureau of Justice Statistics (BJS), in 1999 (the latest data available), women occupied 28% of correctional officer positions in jails and in 2005 they constituted 26% of correctional officer positions in state prisons, 13% of correctional officers in federal prisons, and 48% of correctional officers in private prisons (Stephan, 2001, p. 8; 2008, p. 4). As was mentioned in Chapter 8, it is probably no coincidence that the prisons that pay the most (federal) employ the fewest women as officers and the prisons that pay the least (private) employ the most.

Staff demographic statistics regarding probation and parole officers for adults and children are not always readily available. According to a recent BJS report, 49% of all state-level parole agency staff are women, but this figure includes all staff, not just parole officers (Bonczar, 2008, p. 3). Unfortunately, the BJS does not even supply this level of information regarding probation officers at the state or federal level.

About 22% of correctional officers in Canadian minimum and medium security federal prisons are women (Correctional Service of Canada, 2003). Though these numbers are not particularly impressive, they do represent an increase since 1988 when Zupan (1992) reported that women constituted only 15% of correctional officers nationally and 10% in Canadian federal prisons in 1984.

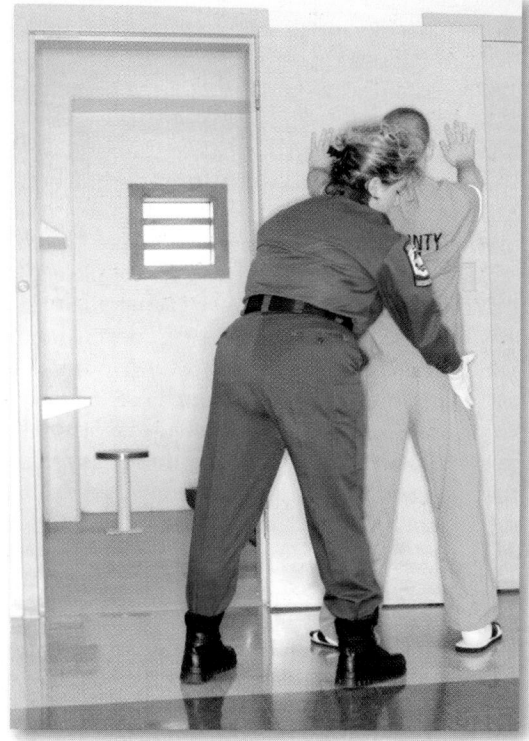

Photo 10.2

A female correctional officer pats down an inmate.

Feminism

Women staff would not be employed at the level they are and female inmates would not have the attention and programming they do, albeit usually less than men and boys, if not for the sustained efforts of feminist scholars and practitioners agitating for their rights and their needs (Pollock, 2002b; Smykla & Williams, 1996; Rafter, 1985; Stohr, 2006; Young, 1994; Zimmer, 1986, 1989; Zupan, 1992). As indicated by Rafter, the proponents of change in female corrections in the last half of the 1800s and first half of the 1900s tended to be of two minds, as represented by the moralists and the liberal feminists.

There were those *moralists,* who were sometimes social feminists, as Rafter terms them, who believed that women and girls involved in the criminal justice system were in effect morally impaired and therefore in need of religious and social remedies (prayers, efforts to keep them chaste, etc.). Women were crudely classified by these moralists as either "good," and thus acting in conformance with societal expectations for their gender role (labeled the *madonna*), or as "bad" and thus acting in opposition to their expected gender role (labeled the *whore*). This conceptualization and limited view of the possibilities of women and girls and its focus on sexuality was also shaped by social class and race/ethnicity. Those women who were of a higher class, and who were white, were believed to more closely approximate the madonna category, until, that is, they violated societal expectations that they be docile homemakers with nary a thought in their heads. Should they violate these expectations regarding their gender role and social and legal prohibitions against the commitment of crime, then they were *double deviants* (Rather than just being deviants, as men and boys who committed crime were, women and girls involved in crime were also deviants in terms of societal gender role expectations [Belknap, 2001]). Women of lower classes, but particularly women of color, were not expected to attain this madonna status. Women in the lower and even in the working classes—which described most women in the late 1800s and early 1900s—often worked outside of the home, because they had to in order to help support their families; thus, the belief—really a myth—that most women used to work only in the home, applied only to the middle-class and wealthier women, not to the majority of women. Women of color, who were disproportionately represented in the lower classes, then and now, were seen as more sexual in nature, perhaps as a justification for their exploitation in this manner, and so could not even aspire to a madonna status (Belknap, 2001).

There were others, those who espoused a *liberal feminist* perspective, who believed that the source of the crime problem for female offenders lay more with the social structure around these women and girls (e.g., poverty and lack of sufficient schooling or training, along with patriarchal beliefs) (Daly & Chesney-Lind, 1988). Liberal feminists believed the solution to female crime lay in preparing those inclined to engage in it for an alternate existence, for work, and sometimes this involved "traditional women's work," so that they would not turn to crime (Rafter, 1985). Some of these early feminists believed, as liberal feminists do today, that men and women are inherently equal and, as such, women and girls are entitled to the same rights, liberties, and considerations (e.g., in corrections this would be programming, quality of institutions, and equal employment as staff) as men and boys (Belknap, 2001; Daly & Chesney-Lind, 1988).

The moralists triumphed, though not completely, in the argument over what lay at the heart of female criminality. As a consequence, we have had over a century of correctional operation that has tended to be overly concerned with the sexuality of females (Giallombardo, 1966; Heffernan, 1972; Owen, 1998; Rafter, 1985). Another consequence of this triumph was that reform efforts were directed at training female inmates to be proper wives and mothers while forgetting that as members of the lower classes, they would need to make a living for themselves and their children once they reentered their communities. Despite this morals-of-the-fallen-woman focus—the soiled dove, if you will—feminist women and men were able to agitate for, and sometimes get, separate facilities for women and girls and other services (e.g., educational and job training) that were geared toward helping women and girls become independent and self-supporting in the free world (Hawkes, 1998; Yates, 2002).

One societal obstacle to achieving equal treatment in corrections has been *patriarchy*. **Patriarchy** involves the attitudes, beliefs, and behaviors that value men and boys over women and girls (Daly & Chesney-Lind, 1988). Members of patriarchal societies tend to believe that men and boys are worth more than women and girls. They also believe that women and girls, as well as men and boys, should have certain restricted roles to play and that those of the former are less important than those of the latter. Therefore, education, and work training that helps one make a living, and better pay, are more important to secure for men and boys than for women and girls who are best suited for more feminine—and by definition in a patriarchal society, less worthy—professions. Feminist scholars have determined that many cultures, even today, hold such beliefs and engage in the practices that derive from them.

In the United States, much effort has been expended over the centuries, by male and female feminists, to address the patriarchal belief system, and there has been some success in this regard (Dworkin, 1993; S. Martin & Jurik, 1996; Morash, 2006; Whittick, 1979). In terms of corrections, feminists have been instrumental in pushing for more and better programming for incarcerated and supervised women and girls, for the reduction in the incarceration of girls for status offenses, for the attention to the sexual abuse of women and girls while incarcerated, and for the greater employment of women in adult male and female correctional institutions.

❖ Females in Corrections: Needs, Programming, Abuse, and Adjustment

Needs and Programming

As a practical matter, then, if not just because women and girls have historically been valued less by this society (patriarchy), but perhaps because crime has generally been the purview of men and boys, correctional facilities and correctional practices have tended to focus on men. This focus led to disparate treatment that disadvantaged women and girls from the beginning and resulted in little concern for their needs (Muraskin, 2003).

Yet, women and girls are more likely to have mental and physical health problems than incarcerated men and boys. They are also more likely to have substance abuse problems than their male counterparts. Moreover, they have the same kinds of educational and job training deficits and needs as men and boys (Gray et al., 1995; Morash, Haarr, & Rucker, 1994; Owen & Bloom, 1995; Pollock, 2002b). Their need for gainful employment is likely as great as, if not greater than, that of men and boys, as they most often have to support themselves and their children, whereas fewer men had custody of their children before they were incarcerated (Owen, 2006); about 70% of the women had custody of their children at the time of their incarceration (Henriques, 1996, p. 77). Moreover, a greater percentage of women, perhaps as high as 60%, were the victims of sexual abuse in their past, and this is likely to negatively shape their self-concept and their relations with others, thus necessitating more programming (Belknap, 2001; Blackburn, Mullings, & Marquart, 2008; Comack, 2006; Morash, 2006; Pollock, 2002b).

Given all of these needs, and assuming that policy makers would not want women and girls to reenter the system, if for no other reason than that they cost much more to incarcerate (because of their needs and a reduction in economies of scale [separate female institutions house fewer inmates, but require almost the same number of administrative and support staff as much larger male institutions]), one would think that all of their needs would be met with adequate programming. Unfortunately, this has been far from the case in most jurisdictions. Though there has been some recognition by the federal government of the need to develop programming that fits the needs of women and girls, it is unclear how much this has spread to state and local facilities (Morash, 2006).

Though most of these needs are far from met in correctional environments, and far less than a majority of women are involved in meaningful programming, Pollock (2002b) notes that some states have made renewed efforts to address the needs of women and girls for educational, vocational, parenting, and substance abuse issues, and their histories of past victimizations. However, the numbers of these programs and their quality (very few are rigorously evaluated in terms of desired outcomes) leave much to be desired (Pollock, 2002a). The sad truth is that most women and girls who need programming in corrections are not able to access it, or if they are, it is sometimes of dubious worth.

Researchers find that women and girls have programmatic needs and styles that determine whether some rehabilitative approaches are more effective than others (Loper & Tuerk, 2006; Staton-Tindall et al., 2007; E. M. Wright, Salisbury, & Van Voorhis, 2007). One type of programming with particular relevance for women, given that most had physical custody of their children prior to incarceration, is parenting programs. Loper and Tuerk found in their research on such programming that they are delivered in several prisons, but their purported value in terms of helping mothers and fathers become better parents has not been rigorously studied (see also Pollock, 2002a, and Surratt, 2003). Craig (2009), in her historical-to-the-present review of mother and child programs in prisons, found that many states and localities have had programs where infants or very small children may stay with their mothers, at least initially. However, most correctional facilities where women are housed do not have such programs, and the qualifications for their use, even in states that have them, are widely varied.

In an interesting study of male and female inmates in 20 substance abuse treatment programs, Staton-Tindall and colleagues (2007) found that the females reported more psychosocial dysfunction (e.g., anxiety, depression), less criminal thinking (e.g., cold-heartedness, entitlement, irresponsibility), and greater involvement in programming (e.g., willingness to participate and receptivity to input) than did the males. The authors maintain that these findings support other research that indicates programming for women must be shaped to fit their abuse histories and mental health needs.

Relatedly, in a study of 272 incarcerated women offenders in Missouri, Emily Wright and her colleagues (2007) found that "gender-responsive" problems related to parenting, child care, and self-concept affect prison misconduct. Brown (2006), in her study of native and non-native Hawaiian women imprisoned in Hawaii, developed the alternate, but parallel idea of "pathways" women traverse that can lead to crime. A pathway strewn with violence, trauma, and addiction, coupled with discrimination based on race, gender, and class, is more likely to end in criminal engagement for women. Such a pathway, Brown explains, may be related to poorer treatment outcomes for incarcerated women.

The health care needs of incarcerated women are tied up in needs specific to gender and in the particular pathway that many poor women tread. Women in jails and prisons have numerous gynecological and obstetrical as well as psychological/psychiatric health ailments that are more specific to their gender, along with health problems that are common to both genders. As the health care needs of poor women and children in communities are going unmet, it is not surprising that similar circumstances apply to incarcerated women. Predictably, then, Moe and Ferraro (2003) found from their interviews of 30 women incarcerated in an Arizona jail that while basic needs were met in this jail, when the care required was of the long term, extensive, or individualized type, it was lacking.

Abuse

Unfortunately, abuse does not necessarily end at the corrections door. One of the primary reasons that women and girls were removed from facilities for men and boys in the 1800s and 1900s, and

Photo 10.3
Women inmates working, sewing American flags.

female staff were hired to supervise them, was that they were targets of sexual abuse by correctional staff and male inmates (Henriques & Gilbert, 2003). Though the separation from male inmates has reduced their abuse, the sexual abuse by male staff, though likely much less prevalent than it once was, in part because of the inclusion of more female staff, has not been eliminated.

One of the authors of this volume had occasion to serve as an expert witness for the plaintiffs in a civil suit in 2004 against a city in New Mexico whose judge and a few correctional officers for the local jail were involved in the sexual abuse of female inmates (*Salazar et al. v. City of Espanola et al.* [2004]). The male judge and a few male correctional staff had an arrangement whereby female offenders whom the judge found attractive would be placed in the jail (whether their alleged offense merited it or not), and then the judge would have access to them when they were sent over to "clean" his chambers. Inevitably, he would make passes at them, using the threat of more jail time, denied privileges, or a lengthened sentence, as a way to coerce them into sexual activity with him. Meanwhile, a few of the correctional staff were harassing the female inmates, such as watching and commenting on their bodies as they showered, making sexual advances toward them, and touching them inappropriately. Two male officers were even involved in removing a few females from their cells and having sex with them in the control room at night when no one else was around. There were no female staff on duty at the time of these sexual assaults and this abuse. After this kind of activity occurred for a period of time, and due to the concerted efforts of several ex-inmates and their attorneys, the judge was convicted of rape, and the judge, correctional staff, and city lost a million-dollar lawsuit (*Salazar et al. v. City of Espanola et al.* [2004]).

Unfortunately, the sexual abuse by male staff of female inmates is not limited to adult facilities. In 2003, the American Civil Liberties Union (ACLU) investigated reports of abuse by male staff of juvenile girls in the Hawaii Youth Correctional Facility and reported that male staff observed the girls using the toilet and showers, made comments about their bodies, threatened to rape them, and in fact several girls did have sex with the officers in exchange for cigarettes (Chesney-Lind & Irwin, 2006). A year later, one officer pled guilty to three counts of sexual assault and to threatening a female ward. A key circumstance that came out in the ACLU report was that there were no female officers on duty at night when much of the abuse of the girls took place.

Such abuse is particularly damaging when one considers that about half of incarcerated women and girls have experienced some form of sexual abuse in the past (Gray et al. 1995; Henriques & Gilbert, 2003). The Ninth Circuit Court, though not the Supreme Court, recognizing this fact, put some restrictions on body searches of female inmates by male staff, noting that such searches may serve to revictimize the women with sexual abuse histories (*Jordan v. Gardner* [1993]).

Efforts to reduce sexual abuse in correctional institutions have centered on ensuring that staff have the proper training and are supervised sufficiently to prevent abuse. Moreover, the value of disciplinary measures to reinforce appropriate practices cannot be overstated. Clearly, staff who violate the rights of their charges in a way that is as serious as sexual abuse, should be fired and prosecuted, and there is some evidence emanating from the reporting required by the Prison Rape Elimination Act that this may be occurring.

The hiring of more women officers to cover living units is another way that correctional agencies have worked to keep sexual predators from gaining access to relatively powerless female victims. There is no question, however, that lawsuits have been successful in spurring some of these needed changes in correctional practice. But the problem with lawsuits is that their application is hit or miss at best, and the success of plaintiffs is always iffy. Therefore, the best preventative measures are those that focus on hiring competent people, training them to behave professionally, rewarding them when they do, and punishing them (up to and including firing and prosecuting them) when they do not.

Adjustment and Pseudo Families

Women's adjustment in corrections is associated with their sometimes problematic personal relationships; separation from children; greater propensity for mental health problems; and the fact that in prisons, women tend to be charged with infractions for more minor offenses (Owen, 2006). As Van Tongeren and Klebe (2010) found in their study of female inmate adjustment in a maximum security prison in Colorado, adjustment is a multidimensional concept that encompasses an inmate's particular circumstances and environment as well as his or her criminal thinking and adoption of the prison subculture.

Regarding the circumstances that women prison inmates find themselves in, in most states, there is still only one women's prison, which is often located away from urban centers where most of the women came from. As a result, it is difficult to maintain familial relationships and friendships when a woman has a lengthy sentence. Poor families will find it much more difficult to visit their incarcerated family member, as they will not have reliable or inexpensive transportation available. Moreover, incarcerated women do not have access to cell phones or computers to contact family and friends. The pay phones they do have access to are expensive to use for poor families who must pay for the collect calls (even local calls). As many of these women are incapable of writing a letter, and their children may be unable to read or be unreachable to these women, the ability to maintain contact is further impaired. The separation from children is particularly acute, as the mothers lose control over their children's housing and care. Often the children are placed with family members who have a history of abuse or in the foster care system, which in most states is overwhelmed (S. F. Sharp & Marcus-Mendoza, 2001).

As with males, females incarcerated or supervised in corrections must grapple with the pains associated with that status and find some way to adjust to its strictures. Early researchers in women's prisons (e.g., see Giallombardo, 1966) reported on the formation of *pseudo families* as a way for women to meet their needs for companionship, support, and love, as well as sexual gratification. It was thought that women were importing these familial roles from traditional family structures and playing them out in the prison setting. In any given pseudo family, there were inmates who took on the roles of fathers, mothers, grandmothers, daughters, aunts, and cousins. More recent research has shown that some women do indeed form "families" while in prison, but the strength of these relationships is perhaps more casual than was first reported (Owen, 1998). Moreover, as might be expected, women incarcerated for longer periods of time and who are farther from their release date may be more likely to maintain their pseudo-familial relations than those who are not as immersed in the subculture due to a shorter incarceration.

❖ Female Correctional Officers

Overcoming Employment Obstacles

As with the accused and convicted in the system, women have always constituted a minority in terms of correctional staff (as discussed earlier in this chapter). Though one would expect that women might comprise a greater percentage of staff, given their representation in the larger community, the current figures actually represent a significant improvement over 30 or 40 years ago. At that time, women—with exceptions allowed for matrons in women's and girls' facilities working for lower pay than men working in male facilities—were prohibited by practice, tradition, or law from working in the more numerous men's and boys' correctional institutions or in probation and parole. As was mentioned earlier, it was not until the Civil Rights Act of 1964 was passed and amended in 1972 that women were given the legal weapon to sue for the right to work and be promoted in all prisons, jails, detention centers, juvenile facilities, halfway houses, and in community corrections. Many women did in fact sue; they had to if they wanted the same kinds of jobs and promotional opportunities then available only to men in corrections, policing, and law (Harrington, 2002; Hawkes, 1998; Stohr, 2006; Yates, 2002). As a result of this agitation and advocacy, and slowly, the available jobs and promotional opportunities became open to these pioneering women, resulting in the more diverse correctional workforce we see today.

Current Status: Equal Employment Versus Privacy Interests of Inmates, Qualifications for the Job, and Sexual and Gender Harassment

As the number of women employed in corrections has increased, three issues have been particularly problematic for them in the workplace:

1. Whether women's rights to equal employment in male correctional facilities are more important than male inmates' rights to privacy in those same facilities

2. Whether women are physically and mentally suited to do correctional work with men

3. How to deal with sexual and gender harassment—primarily from other staff—while on the job

As mentioned before in this section, women achieved the legal right to equal employment in corrections through law and lawsuits. Most of the jobs in institutional corrections or in communities are in dealing with male inmates or offenders. On the other hand, one can certainly understand the male inmate perspective that they would like some privacy when engaged in intimate bodily functions, like using the toilet or showering. The courts in this matter, however, have tended to side with the female employees', or prospective employees', right to equal protection over the male inmates' right to privacy (Farkas & Rand, 1999; Maschke, 1996). Their reasoning was likely as much influenced by the fact that inmates in the United States have very limited rights while incarcerated, with no real "right to privacy," as by the fact that correctional staff should be respectful and professional, no matter their gender, in their dealings with inmates.

The issue of whether women are physically and mentally qualified to work with male inmates has generally become a settled matter in most institutions, agencies, and states: They are. But when they were first making inroads into the correctional workplace, there were plenty of doubts about the ability of women to handle the work (Jurik, 1985b, 1988; Jurik & Halemba, 1984; Zimmer, 1986, 1989). Even as regards their propensity to use aggression in the course of their work, men and women correctional staff are similar (Tewksbury & Collins, 2006). Moreover, the Supreme Court has left open the possibility that if there is a bona fide job requirement that women could not fill in a male prison, they can be excluded from that work (Bennett, 1995; Maschke, 1996).

In a qualitative study of female parole agents in California, Ireland and Berg (2006) noted that these agents reported subtle harassment in the form of less desirable shifts or assignments because of their gender. They also felt that they were overlooked for promotions and an administrative career track. One 40-year veteran of the department observed that it was not the clients who harassed the women agents, but their colleagues who "often questioned her competence and treated her unprofessionally" (p. 140). In response to such bias and views, the women reported that they overcompensated, or did more than was expected of males on the job. Some female parole agents adapted by taking on stereotypical feminine roles such as acting helpless and in need of male assistance, or flirtatious, or maternal. Still another adaptation by these agents was to refuse to acknowledge that there was any bias at all in the workplace; this latter group of women did not think that considerations of gender or race/ethnicity had hampered their ability to advance in their career.

Clearly, because of their biology, most women are not as physically strong as most men, and sometimes strength is called for in dealing with an unruly inmate. However, the use of brute force is rather rare in most correctional institutions and in the community when functioning as a probation or parole agent (note the discussion of violence in corrections in an earlier chapter). Secondly, there are defensive (and offensive) tactics that give a trained and armed woman some advantage in a physical altercation over a male inmate. Third, there is some evidence that female staff may have a calming effect on male inmate aggression, because they are more inclined to use their interpersonal and communication skills and are less likely to be seen as a threat (Jurik, 1988; Jurik & Halemba, 1984; Lutze & Murphy, 1999; Zimmer, 1986).

More recent research has indicated that both male and female correctional officers value a service orientation over a security orientation to their work, so their work styles and preferences may be more similar than dissimilar (Farkas, 1999; Hemmens, Stohr, Schoeler, & Miller, 2002; Stohr, Lovrich, & Mays, 1997; Stohr, Lovrich, & Wood, 1996). In research on attitudes of 192 male and female jail officers in a southwestern state's jail system regarding conflict resolution with inmates, Hogan and her colleagues (Hogan, Lambert, Hepburn, Burton, & Cullen, 2004) also found that the genders reacted similarly in this area as well, though when the gender of the inmate was female, they were more likely to react aggressively, and they perceived a greater physical threat from male inmates.

In research by Kim and colleagues (Kim, DeValve, DeValve, & Jonson, 2003) on the attitudes of male and female wardens, they found that 90 of the female wardens (out of a total of 641 male and female wardens surveyed) were more inclined to value programming and amenities in their prisons that promoted the health, education, and programming for inmates, than were their male colleagues. However, these researchers also found that there were many more similarities than differences between the two genders as they viewed and appreciated their work.

Taken in total, what all of this research on women's ability to do the work and on the differences and similarities in work styles between men and women indicate is that men and women mostly view and do correctional work similarly and that some women, perhaps more than some men, can calm an agitated inmate and that some men, perhaps more than some women, are better at physically containing an agitated inmate.

Unfortunately, the problem with gender and sexual harassment of female staff by male staff has not become a settled matter. This is not to say, of course, that male staff are not harassed by female staff. This does happen and can be as debilitating for the male employee as it is for the female. But a number of studies have shown, over a period of years, that females are much more likely to be the victims of male harassment by bosses and coworkers in the workplace, and that when men are victims, they are as likely to be harassed by other men as by women (Firestone & Harris, 1994, 1999; Mueller, De Coster, & Estes, 2001; O'Donohue, Downs, & Yeater, 1998; Pryor & Stoller, 1994). Male institutions in particular, with their smaller percentages of women employees and managers, and their traditions aligned with male power in the workplace, are more susceptible to this kind of behavior than would be other correctional workplaces (Lawrence & Mahan, 1998; Lutze & Murphy, 1999; Pogrebin & Poole, 1997). The types of harassment that occurs can be of the **quid pro quo** type (something for something, as in you give me sexual favors and you get to keep your job) or the less serious, but still workplace stultifying, **hostile environment** harassment (when the workplace is sexualized with jokes, pictures, or in other ways that are offensive to one gender).

Thankfully, there are remedies, imperfect and cumbersome though they might be, that can be employed to stop, or at least significantly reduce, such harassment. Initially, women had to sue in order to stop the harassment, because many managers of their workplaces simply would not do anything to stop it. In one mid-1990s case, one of the authors of this book served as an expert witness for the plaintiff against the State of California's San Quentin Prison. The female victim won over a million dollars for enduring harassment by several male staff, and one inmate, that started in the 1970s and ended when she quit in frustration in the 1990s, and was never stopped by the prison administration (*Pulido v. State of California, et al.* [1994]). As successful as this case was, there was incontrovertible evidence of the harassment (as provided by memos and diaries and staffing logs and witnesses), evidence that is usually not available to support most victims' stories. Moreover, the female victim, an African American woman, lost her job and had to endure almost 2 years of an uncertain legal battle, before the case was tried and the judge ruled. Even then, the State of California appealed, and it took another year before the matter was finally settled in the plaintiff's favor.

What this story illustrates is that there are few true "winners" when sexual and gender harassment cases go to trial. Most such cases fail, as there is not sufficient evidence of the abuse, beyond a he said/she said scenario. Victims of such abuse suffer untold harm, in terms of their psychological and physical well-being, both during the abuse and as they relive it during the legal process. Even when cases are successful at trial, taxpayers (not just the instigators of the abuse, who often do not have the "deep pockets" of their governmental employer) have to pay for the illegal practices of their own governmental entities and actors.

In other words, there has got to be a better way—and there is! Researchers and correctional practitioners have agreed that there are proactive steps that managers and other employees can take to prevent or stop sexual and gender harassment in the correctional workplace. Such steps would involve hiring, training, firing, and promoting based on respectful treatment of other staff and clients. Training in particular can reinforce the message of a "no tolerance" policy regarding harassment. But to be effective, employees need to see that people are rewarded when they do, or punished when they do not, adhere to the policy.

As a discussion of the current status of female staff working in corrections would indicate, women have made some significant advances in these workplaces. They have not just made gains in employment, but female supervisors and managers are also no longer anomalies in most states. Though nowhere near matching the numbers of men as staff and management in corrections, and while still grappling with the pernicious problem of sexual and gender harassment, nonetheless they have come a long way since the days of working as matrons, with lower pay and respect, attributes that typified their work for most of the history of corrections.

Summary

The study of women and girls in corrections was not always a priority for scholars (Flavin & Desautels, 2006; Goodstein, 2006). Patriarchal perceptions and beliefs, along with women's status as numerical minorities, have served to shape organizational and scholarly priorities in a way that favors men and boys. Since the 1970s, however, there has been more scholarly focus on the reality of women and girls who work, live, or are supervised under the correctional umbrella.

- Part of this shift in focus has occurred as the result of feminist work to equalize the work for women and the living and supervision arrangements for women and girls under correctional supervision.
- Recent research on women and girls under correctional supervision has highlighted the outstanding needs they have for educational, substance abuse, work training, parenting, and surviving abuse programming; Unfortunately, it has also shown that little programming is provided in either jails or prisons, or in the communities, to meet these needs.
- Female correctional officers have also faced a number of legal and institutional barriers to their full and equal employment in corrections.
- For the most part, many of these formal barriers have been removed as female officers have demonstrated their competencies in handling correctional work.

- Some researchers have even found a feminine style to officer work that employs the successful use of interpersonal communication skills to address inmate needs.
- However, there is still evidence that sexual and gender harassment, and the sexual abuse of female inmates, continues in some correctional environments. Though organizational remedies exist to "deal" with such abuse, they are not always employed by managers.
- Taken in tandem, the research presented in this chapter should shift our perspective from the much more numerous and normative study of males, to females. By shifting our gaze to the female side, we are as likely to see the sameness of the genders as the contrasts that distinguish them (Rodriguez, 2007; Smith & Smith, 2005). We might also see that the life course of a woman or girl entangled in the criminal justice system (e.g., see Brown [2006] and E. M. Wright et al. [2007]) forms a predictable pattern that might be fruitfully addressed if we only had the will to do so.

Key Terms

Hostile environment

Houses of refuge

Mount Pleasant

Patriarchy

Quid pro quo

Discussion Questions

1. In what ways have women and girls occupied a minority status in corrections? How has that status affected how they are treated in the system?

2. What is feminism and what do feminists advocate? How have they had an effect on the work of women and their (and girls') experience with incarceration?

3. What is patriarchy? What kind of effect did and does patriarchy have on corrections for women and girls? How might it negatively affect the experience of men and boys?

4. What sorts of factors are likely to lead to the greater abuse of female inmates in correctional institutions? How might such abuse be prevented?

5. What sorts of factors are likely to lead to sexual and gender harassment in correctional work? How might such harassment be stopped or prevented?

6. How might female officers' supervision styles differ from those of male officers? What might be the advantages of hiring women to work with men and boys in corrections? What might be the advantages of hiring men to work with women and girls in corrections?

Useful Internet Sites

American Correctional Association: www.aca.org

American Jail Association: www.corrections.com/aja/

American Probation and Parole Association: www.appa-net.org

Bureau of Justice Statistics (information available on all manner of criminal justice topics): http://bjs.ojp.usdoj.gov/

National Organization for Women: www.now.org

CHAPTER 11

Minorities and Corrections

❖ Introduction

The race and ethnicity of America's population have shaped its law and practice from the beginning. At the very writing and ratification of the Constitution, full citizenship was denied to those who were not white and, for several decades, male and in possession of significant amounts of property. The institution of slavery, the forcible seizure of American Indian lands, the limitations on the immigration of nonwhites, and their rights while in the United States, have all marked and marred this country. Accordingly, police agencies, courts and correctional institutions and programs, and their actors, have historically treated people differently based on their race and ethnicity.

Minority group members were more likely in some parts of the country to be incarcerated when they were innocent or sentenced for periods that were longer than their white brothers and sisters. Once in the correctional system, minority group members were sometimes segregated into separate institutions, sections of institutions, and programs. At times, they were given less desirable jobs and housing in jails and prisons. Whether such discriminatory treatment continues today is a matter of some debate, but there are indications that some laws, police, courts, and correctional practices have the effect of maintaining a separate and unequal system for minority group members. In this chapter, we briefly discuss this history and use it as a context for current practices of, and experiences in, corrections.

❖ Defining Race, Ethnicity, Disparity, and Discrimination

Race is a term that refers to the skin color and features of a group of people. It is based on biology. The extent to which different racial groups truly differ biologically is still being determined by scientists. Scientists are still putting together the collective pieces of our human history. However, the genotyping of the whole human race indicates that our species likely originated in Africa (Diamond, 1997; McAuliffe, 2010). Waves of migration then occurred, beginning at least 50 thousand years ago, and continuing over thousands of years, to Europe and to Asia resulting in a variation of skin and other features of racial groupings, who in turn migrated to other continents and islands (Mann, 2006; McAuliffe, 2010). It is worth noting that even as these racial distinctions developed, there has been much intermingling, both historically and currently, among races, resulting in populations that are substantially mixed, rather than distinct in their "racial" heritage. For this reason, using racial designations such as white or black or Asian might be necessary to ensure that one group is not advantaged over another, but we should recognize that they can be somewhat arbitrary designations, being that true racial differences, though visible to the eye, may be measured more in gradations rather than in clear distinctions, particularly in racially mixed societies.

Ethnicity, on the other hand, refers to the differences between groups of people based on culture. An ethnic group will often have a distinct language, as well as distinct values, religion, history, and traditions. Ethnic groups may be made up of several races and have a diverse national heritage. For instance, the term *Hispanic* is applied to an ethnic grouping in the United States that includes white, black, and Asian racial groupings whose ancestors may hail from Cuba, Puerto Rico, Mexico, or Central or South America. Italians, Irish, French, German, and other ancestral ethnic Europeans who have immigrated to the United States are usually racially white, but not always, as while in Europe, or after immigrating to the states, those groups may have intermingled with Africans and Asians. For instance, people known as Creoles are both ethnically and racially differentiated by their white and African racial background and the French ethnic cultural influences in Louisiana. And who are "black Irish" Americans, but primarily white ethnic Irish people who intermingled with Spanish Moorish people while in Europe (who were at least partially racially black) and who then immigrated as Irish to the United States. Among black people in the United States, there are distinct ethnic differences between those whose ancestors have been in the country for hundreds of years either as free people or as those forcibly brought here through slavery, and those whose families are more recent immigrants from Africa or predominantly racially black areas of the world (e.g., immigrants from the Caribbean Islands, Haiti). More recent immigrants from the Sudan or Nigeria or Kenya are different ethnically; that is to say, they have a distinct culture, as well as nationality, from each other and from those blacks whose families have been in the United States for generations.

Photo 11.1

California inmates in Folsom Prison sitting in mixed-race groups.

Disparity and Discrimination

Clear et al. (2011) define **disparity** as "The unequal treatment of one group by the criminal justice system, compared with the treatment accorded other groups" (p. 527). In turn, they define **discrimination** as "Differential treatment of an individual or group without reference to the behavior or qualifications of the same" (p. 527). We would add that disparity can happen in many organizations and entities and is not just restricted to the criminal justice system, and that often discrimination is linked in law to classes of people distinguished by race, ethnicity, gender, age, disability, religion, nationality, sexual orientation, and income.

❖ A Legacy of Racism: African Americans, American Indians, Hispanics, Asian Americans

The legacy of *racism* (or discriminatory attitudes, beliefs, and practices directed at one race by another) runs long and deep in the United States. Notably, sometimes the term racism is also applied when one ethnic grouping holds discriminatory attitudes or beliefs, or engages in such practices, against another ethnic group. Correctional institutions and programs as social institutions are products of their larger social/political and economic environments, and therefore the legacy of racism has affected, and continues to affect, their operation.

African Americans

Slavery, or the involuntary servitude of black Africans by white Europeans in the Americas, was practiced almost from the settling of the United States (K. C. Davis, 2008). Many of the founding fathers were slave owners, and the practice of slavery was protected in the Constitution (through the three-fifths designation of slaves in Article I [the worth that slaves had for states who wanted to count them for representation in Congress] and Article IV [which caused fugitive slaves to be returned to the slave owner]).

Slavery was a lucrative business for ship owners in the colonial United States, both Northern and Southern, and for plantation owners in the South, as it provided the back-breaking agricultural labor that built the Southern economy. Though slavery officially ended with the Civil War between the Northern and the Southern states, and the subsequent adoption of the Thirteenth Amendment in 1865, it lived on in civil society and law for a hundred years through discriminatory laws and practices (see the discussion of Jim Crow laws later in the chapter).

In Focus 11.1

The Scottsboro Case

Nine African American teenage boys were hoboing in 1931 on a freight train headed to Memphis, Tennessee. The train was stopped, and they were arrested and accused of the rape of two white girls on the train. The case was first tried in Scottsboro, Alabama, where the boys received little representation and the trial was rushed. All of the boys in this first trial, except the youngest (a 12-year-old), were convicted of rape and sentenced to death.

The case was appealed and made it all the way to the Supreme Court. In the famous decision *Powell v. Alabama* (1932), the Supreme Court ruled that the due process rights of the accused—in this case their right to counsel in capital cases, particularly as these teenagers were indigent and illiterate—was violated. The Court reversed their convictions and sent their case back for retrial.

Photo 11.2

The Scottsboro Boys meeting with their lawyer, in jail

(Continued)

(Continued)

In the second trial, 7 of the 8 convictions were upheld by all-white juries (Black voters in Alabama were purposefully excluded from lists for juries.). The reconvictions happened despite the fact that the case was moved to Decatur, Alabama, for retrials, and one of the two victims recanted her story, claiming that the story was made up and the boys never touched either of them.

The Supreme Court, in 1935, again reheard the case in light of the all-white jury composition; the Court reversed again. The Alabama judge set aside the verdict and scheduled a new trial where the boys were again found guilty.

Eventually, charges were dropped for 4 of the 9 defendants, but the others received sentences of 75 years to death, and 3 of those 5 served prison time. The one who was sentenced to death was eventually pardoned, in 1976 (Walker, Spohn, & DeLone, 1996). This case is widely regarded by legal scholars as a gross miscarriage of justice, and as emblematic of the way African Americans were treated in racist sectors of this country well after slavery was abolished.

Correctional institutions, particularly in the South following the Civil War, were devised to maintain the slavery system with newly freed, and often unemployed blacks, incarcerated for minor or trumped-up charges, and leased out to Southern farmers for work on the same plantations on which they or their brethren had been slaves (Oshinsky, 1996; Young, 2001). In the same time period, in the North and Midwest, African American inmates were sometimes segregated from whites in prisons and jails, and given substandard housing and the least desirable work assignments (Hawkes, 1998; Joseph & Taylor, 2003).

The Scottsboro case exemplified the racist attitudes of communities and how those attitudes were translated into discriminatory practices by law enforcement, courts, and corrections (Walker et al., 1996; see In Focus 11.1). Lynchings of black men, fueled by mob rule and widespread Ku Klux Klan hate group activity, was also practiced in many states and communities following the Civil War and well into the 1900s (Keil & Vito, 2009). Lynching reinforced a culture of fear that prevented African Americans from achieving an equal and decent footing in communities. The Klan's avowed purpose was to target and persecute Catholics, Jews, and nonwhites, but especially blacks, particularly in the South and midwestern states. Membership was widespread among public and criminal justice officials in the first half of the 1900s and even included those who rose to such lofty heights as the Supreme Court (e.g., Supreme Court Justice Hugo Black was a member in the 1920s) or members of Congress (e.g., Senator Robert Byrd of West Virginia was a member and defender of the Klan well into the 1950s).

There is little doubt that up until the civil rights movement and the implementation of laws and practices that reduced racism in public and private organizations, there was *institutional racism,* or racism practiced by many, if not most, institutional members, in criminal justice and other organizations. In correctional institutions, it was not until the civil rights movement morphed into the *prisoner rights movement* in jails and prisons, that these practices were changed and African American and white inmates were treated more similarly, or were legally required to be so treated, in correctional institutions (see In Focus 11.2) (Belbot & Hemmens, 2010).

In Focus 11.2

Muslim Inmates Agitate for Their Rights

As the civil right movement, fueled in part by Malcolm X and his Nation of Islam, helped raise the political consciousness of African Americans, black prisoners were affected. Correctional officials were threatened by the language and attitudes of such inmates and banned the practice of Islam in prisons (Belbot & Hemmens, 2010). Muslim inmates could not worship together or receive religious materials from the outside community. In 1962, Thomas Cooper sued the officials at the Stateville Prison in Illinois, claiming that he was denied benefit of religious clergy and religious literature, including the Koran, access that he claimed was protected by the First Amendment to the Constitution. Though he initially represented himself in court, the Nation of Islam took up his cause, and in 1964 the United States Supreme Court decided the case in Cooper's favor. This case, *Cooper v. Pate* (1964), set a precedent for the expansion not just of religious rights of all prisoners, but of other rights as well (see Chapter 13).

In Focus 11.3

Fourteen Examples of Racism in the Criminal Justice System

By Bill Quigley*

The biggest crime in the U.S. criminal justice system is that it is a race-based institution where African Americans are directly targeted and punished in a much more aggressive way than white people.

Saying the U.S. criminal system is racist may be politically controversial in some circles. But the facts are overwhelming. No real debate about that. Below I set out numerous examples of these facts.

The question is—are these facts the mistakes of an otherwise good system, or are they evidence that the racist criminal justice system is working exactly as intended? Is the US criminal justice system operated to marginalize and control millions of African Americans?

Information on race is available for each step of the criminal justice system—from the use of drugs, police stops, arrests, getting out on bail, legal representation, jury selection, trial, sentencing, prison, parole and freedom. Look what these facts show.

One. The U.S. has seen a surge in arrests and putting people in jail over the last four decades. Most of the reason is the war on drugs. Yet whites and blacks engage in drug offenses, possession and sales, at roughly comparable rates—according to a report on race and drug enforcement published by Human Rights Watch in May 2008. While African Americans comprise 13% of the US population and 14% of monthly drug users they are 37% of the people arrested for drug offenses—according to 2009 Congressional testimony by Marc Mauer of The Sentencing Project.

Two. The police stop blacks and Latinos at rates that are much higher than whites. In New York City, where people of color make up about half of the population, 80% of the NYPD stops were of blacks and Latinos. When whites were stopped, only 8% were frisked. When blacks and Latinos are stopped 85% were frisked according to information provided by the NYPD. The same is true most other places as well. In a California study, the ACLU found blacks are three times more likely to be stopped than whites.

Three. Since 1970, drug arrests have skyrocketed rising from 320,000 to close to 1.6 million according to the Bureau of Justice Statistics of the U.S. Department of Justice.

African Americans are arrested for drug offenses at rates 2 to 11 times higher than the rate for whites— according to a May 2009 report on disparity in drug arrests by Human Rights Watch.

(Continued)

(Continued)

Four.   Once arrested, blacks are more likely to remain in prison awaiting trial than whites. For example, the New York State Division of Criminal Justice did a 1995 review of disparities in processing felony arrests and found that in some parts of New York blacks are 33% more likely to be detained awaiting felony trials than whites facing felony trials.

Five.   Once arrested, 80% of the people in the criminal justice system get a public defender for their lawyer. Race plays a big role here as well. Stop in any urban courtroom and look at the color of the people who are waiting for public defenders. Despite often heroic efforts by public defenders the system gives them much more work and much less money than the prosecution. The American Bar Association, not a radical bunch, reviewed the US public defender system in 2004 and concluded, "All too often, defendants plead guilty, even if they are innocent, without really understanding their legal rights or what is occurring. . . . The fundamental right to a lawyer that America assumes applies to everyone accused of criminal conduct effectively does not exist in practice for countless people across the US."

Six.   African Americans are frequently illegally excluded from criminal jury service according to a June 2010 study released by the Equal Justice Initiative. For example in Houston County, Alabama, 8 out of 10 African Americans qualified for jury service have been struck by prosecutors from serving on death penalty cases.

Seven.   Trials are rare. Only 3 to 5 percent of criminal cases go to trial—the rest are plea bargained. Most African American defendants never get a trial. Most plea bargains consist of promise of a longer sentence if a person exercises their constitutional right to trial. As a result, people caught up in the system, as the American Bar Association points out, plead guilty even when innocent. Why? As one young man told me recently, "Who wouldn't rather do three years for a crime they didn't commit than risk twenty-five years for a crime they didn't do?"

Eight.   The United States Sentencing Commission reported in March 2010 that in the federal system black offenders receive sentences that are 10% longer than white offenders for the same crimes. Marc Mauer of The Sentencing Project reports African Americans are 21% more likely to receive mandatory minimum sentences than white defendants and 20% more like to be sentenced to prison than white drug defendants.

Nine.   The longer the sentence, the more likely it is that non-white people will be the ones getting it. A July 2009 report by The Sentencing Project found that two-thirds of the people in the US with life sentences are non-white. In New York, it is 83%.

Ten.   As a result, African Americans, who are 13% of the population and 14% of drug users, are not only 37% of the people arrested for drugs but 56% of the people in state prisons for drug offenses. Marc Mauer, May 2009 Congressional Testimony for The Sentencing Project.

Eleven. The US Bureau of Justice Statistics concludes that the chance of a black male born in 2001 of going to jail is 32% or 1 in 3. Latino males have a 17% chance and white males have a 6% chance. Thus black boys are five times and Latino boys nearly three times as likely as white boys to go to jail.

Twelve. So, while African American juvenile youth [are] but 16% of the population, they are 28% of juvenile arrests, 37% of the youth in juvenile jails and 58% of the youth sent to adult prisons. 2009 Criminal Justice Primer, The Sentencing Project.

Thirteen. Remember that the US leads the world in putting our own people into jail and prison. The New York Times reported in 2008 that the US has 5 percent of the world's population but a quarter of the world's prisoners, over 2.3 million people behind bars, dwarfing other nations. The US rate of incarceration is five to eight times higher than other highly developed countries and black males are the largest percentage of inmates according to ABC News.

Fourteen. Even when released from prison, race continues to dominate. A study by Professor Devah Pager of the University of Wisconsin found that 17% of white job applicants with criminal records received call backs from employers while only 5% of black job applicants with criminal records received call backs. Race is so prominent in that study that whites with criminal records actually received better treatment than blacks without criminal records!

So, what conclusions do these facts lead to? The criminal justice system, from start to finish, is seriously racist.

*Authors' Note: Bill Quigley is Legal Director for the Center for Constitutional Rights and a law professor at Loyola New Orleans. This excerpt was taken from an article he wrote for the *Huffington Post* (www.huffingtonpost.com/bill-quigley/fourteen-examples-of-raci_b_658947.html). Reprinted with permission.

Native Americans or American Indians

American Indians are another group of people who have been victims of racism in this country. Note, the terms *Native American* and *American Indian* are both used to describe the peoples who were here when Columbus landed in 1492. Columbus mistakenly thought he was in India and thus dubbed the native peoples "Indians." The name stuck, giving rise to the more recent use of the name "Native Americans" by those not wishing to associate these native peoples with Columbus. The problem is that sometimes people who are not Indians have adopted the Native American term, as they were born in the United States. However, both names are used by natives and non-natives, and they will be used interchangeably in this book (Mann, 2006).

At the time of the arrival of the first of Columbus's ships on what later became United States shores, there were reportedly as many as 20 million native peoples residing in North America (Colbert, 1997; K. C. Davis, 2008; Diamond, 1997; Mann, 2006). Emerging archeological evidence has established that complex cities and agriculture flourished in the Americas, particularly in South and Central America, thousands of years before this wave of Europeans arrived (There are theories and some evidence that Africans, other Europeans, and Asians all made trips to the Americas, and many times, over the millennia and well before this latter foray by the Spaniards and Columbus—see History Channel, 2010; Mann, 2006.). Within a few short decades, those populations had been decimated by disease (smallpox mostly), wars, and massacres. Over the course of a few hundred years, only a small percentage of those original peoples survived, and they were overwhelmed by the influx of European immigrants who through wars and treaties relocated American Indians, often forcibly, off their lands and onto reservations.

Such reservations, at least initially, were in essence forms of correctional institutions whose purpose it was to incarcerate a whole people on a piece of land, by restricting their movement away from the reservation. Such land was usually less desirable than the land the tribe originally resided on and often inadequate to support the survival of that tribe. As a consequence, American Indian reservations of the 1800s and 1900s were populated by poor, underfed, and undereducated peoples with few prospects for regaining their land, wealth, or status (Blalock, 1967; Kitano, 1997; Stannard, 1992). Federal policy regarding American Indian tribes has shifted over time from efforts to segregate them from white communities, to efforts to integrate tribal members into the larger society, to more current efforts to respect their identity and cultures.

As a result of this complicated history, the interplay of tribal and federal and state law is complex and depends on the time period, and the state and tribe involved. Currently, there are 565 federally recognized tribes in the United States, and there are a number of tribes that have not received or sought this recognition (Bureau of Indian Affairs, 2011). On large reservations, more minor criminal offending by tribal members falls under the jurisdiction of that tribe, while felony offenses or off-the-reservation criminal activity by tribal members might be handled by the tribe or the state or the federal government, depending. Larger reservations maintain their own jails for tribal members accused of crimes, for minor offenders and those with shorter sentences for incarceration. Despite the existence of these separate legal and correctional systems on larger reservations, at least as regards less serious offending, the number of Native Americans in federal and state prisons is often disproportionate to their representation in the larger population of that state (Perry, 2004).

As reported in a recent Bureau of Justice Statistics publication regarding Indian jails (Minton, 2011), there were 80 jails in Indian country in 2009, and the number of inmates confined in those jails has been increasing. Nationwide, the number of accused and offenders who are Native American and under correctional supervision generally has also increased recently, from 75,400 in 2008 to 79,600 in 2009 (Almost two-thirds of the offenders were on probation or parole.) (p. 1).

Hispanics/Latinos/Latinas

As mentioned previously, the term *Hispanic* is used to designate an ethnic group that spans many races and nations of origin, to the point where it may not be descriptive (Martinez, 2004). For this reason, other monikers are often used to describe Hispanics that may better represent who they are, such as the more general Latinos/Latinas (which can be used to describe those who originally hailed from Latin America or whose ancestors did), or national heritage–specific Mexican Americans or Cuban Americans, and so forth. Each of these groups of people has a history, with a distinct American experience. Sometimes that history has included discrimination by criminal justice actors during incarceration.

The history of Mexican Americans, the largest subgroup of Hispanics or Latinos/Latinas, has been one in which they and their land were forcibly made part of the American Southwest. As a result of the Mexican–American War, which lasted from 1846 to 1848, Mexico lost almost half of its land, the area that has become the American Southwest, from Texas to California and all of the states in between. However, there is evidence that at least some of the Mexicans in these territories were willing to become citizens of the United States (PBS, 2011).

Today in border states, the number of Mexican Americans and Cuban Americans is so high, and their assimilation into the culture so thorough (e.g., in New Mexico, Arizona, parts of Texas, California, and Florida), that the existence of a clear racial or ethnic majority group has disappeared or has become the Hispanic/Latino/Latina group itself. The increased numbers of Mexican Americans in these states and the immigrants crossing over the southern border into the United States, have sparked the political debate over recent Mexican immigrants and whether or not they should be accorded citizenship rights. At the center of the debate is the passage by the State of Arizona of a new immigration law that allows law enforcement there to demand papers from any person whom they *suspect* might be in the country illegally, without further cause (Archibold, 2010). Civil libertarians and civil rights groups allege that this law will result in discrimination against Hispanics in Arizona and the potential to fill jails, if not prisons, in that state. Whether in Arizona or in other states, however, and as with American Indians and African Americans, the representation of

Hispanics in American prisons and jails is already disproportionate to their representation in the general population.

Asian Americans

As with most immigrants to America of the 1800s and 1900s, Japanese and Chinese immigrants (who collectively represent the largest groupings of Asian Americans, but certainly not the only groupings [space prevents us from sufficiently exploring the Korean, Cambodian, Vietnamese, Pacific Islanders, or East Asian experiences, for example]) were looking for a better life for themselves and their families. Though to some extent they found such a life in varying degrees, their experience, like that of the other ethnic and racial minorities mentioned in this chapter, was tinged with racism. Originally settling primarily in western states in the 1800s and early 1900s, Chinese and Japanese immigrants were heavily involved in mining and agriculture in pioneer communities.

Chinese labor was crucial to the construction of the first transcontinental railroad (1863–1869). Later barred from owning property in some states and from voting in others, they made do by engaging in service professions (e.g., laundries, restaurants, herb shops) and settling together in parts of cities for both comfort and safety (Moyers, 2008; Wei, 1999). When economies soured in some of those cities or states, Asian immigrants were blamed for taking jobs from poor whites—much like the way blacks were blamed by poor whites in the South after the Civil War, or recent Mexican immigrants are blamed by poor whites in the West today—and they were often run out of town. They were literally placed on ships and sent home, even though they and their families may have lived in the states for decades, if not generations.

The first restrictive immigration law in the country, the Chinese Exclusion Act of 1882, was directed at reducing immigration from China (Wei, 1999). Some of the first drug laws, laws against opium dens dating from the 1870s onward, were passed because Chinese immigrants were thought to be corrupting the white population by spreading the use of the drug; such laws were ironic, as opium was first introduced to China by Westerners (Moyers, 2008).

Much like the Chinese immigrants, the Japanese immigrants provided cheap labor as they were employed in the construction of railroads, as well as agriculture, restaurants, and many other businesses, primarily in the American West. In fact, when Chinese immigrants were excluded, the Japanese immigrants filled the gap, beginning in the 1880s, until their own immigration was also restricted in 1908. Barred from owning their own land, many Japanese

Photo 11.3

A cartoon that appeared in the *Harper's Weekly* regarding the "Chinese Question," which depicts Lady Justice defending a Chinese man being persecuted by a racist mob.

Americans earned their living in the late 1800s and early 1900s by leasing land and growing beets in Oregon and Idaho, for instance. As their economic strength grew, however, they were regarded as a threat by the local white population, and there were numerous instances where they were forcibly run off their land and out of town (Mercier, 2010).

> Despite the Issei's [another word for the first Japanese immigrants] hard work in the early twentieth century, envy and racial discrimination led to increasing anti-Japanese attitudes on the West Coast, much as the sentiment had developed against perceived Chinese competition. Residents of Mountain Home, Nampa, and Caldwell, Idaho drove out Japanese workers, and white mobs near Coeur d'Alene and in Portland threatened Japanese railroad workers. Tensions led to the so-called "Gentleman's Agreement" between the U.S. and Japan that effectively limited after 1908 the numbers of laborers that could emigrate from Japan. Instead, the two governments allowed wives and brides to join earlier male immigrants in the United States, changing the character of the immigrant community. (Mercier, 2010, p. 10)

The internment of 120,000 Japanese Americans in 1942 in 10 inland concentration camps during the Second World War, along with the confiscation of their property, was not based on the actual threat they presented to the safety of western states, or at least no more, say, than the German Americans who were scattered all over the United States at the time, and who were not incarcerated (Mercier, 2010, p. 1). The internment of whole Japanese American families in prison camps was instead based on the racist-tinged beliefs about who could be trusted and ignorance regarding the allegiance that such citizens felt for this, their country.

As far as incarceration of most Asian Americans these days goes, they tend to be underrepresented in correctional organizations in relation to their representation in the general population. It is not clear why such underrepresentation exists, but it is likely related to their tight-knit and supportive families and communities and the value those cultures have placed on education and achievement, resulting in higher incomes and education for many Japanese and Chinese American citizens (Mercier, 2010). Notably, successful integration into American society, as measured by economic and educational achievements, is not uniform across all Asian Americans. Those emigrating from war-torn Cambodia and Vietnam in the latter half of the 20th century were not always as "successful" or able to stay out of the criminal justice system and its correctional institutions.

❖ The Connection Between Class and Race/Ethnicity

Americans are often averse to recognizing the existence of a class system in the United States. In part, this dislike of class labels springs from our history of revolution, which was spurred in part by a desire to separate ourselves from the rigidity of the class system in England and Europe. Also, our economic, political, and social systems have allowed people in lower classes to advance through ingenuity, education, or drive, or some mix of those, to the middle or upper classes. However, this upward mobility is hampered in any number of ways by poverty and related ills such as poor nutrition and schools, limited access to health care, and parents who are absent or neglectful. When poverty is combined with long-term and systematic discrimination against a people, such that their families are destroyed, as occurred with the social institution of slavery and the continued discrimination against African Americans, recovery of communities can take generations. Not surprisingly, illegal drug use catches on in such poor communities, as does other forms of involvement in street criminality.

Certain racial and ethnic minorities are more likely to be poor and thus caught up in the criminal justice system and overrepresented in correctional institutions and programs (see Table 11.1). Race and traditions of discrimination regarding African Americans has stymied

Table 11.1

Percentage of People in Poverty by Different Poverty Measures, 2009

	Number* (in thousands)	Official*	Std error	Research SPM (percent below poverty threshold)	Std error
All People	304,280	14.5	0.2	15.7	0.2
Children	75,040	21.2	0.3	18.0	0.3
Non-elderly adults	190,627	13.0	0.2	14.8	0.2
Elderly	38,613	8.9	0.3	16.1	0.4
In married couple family	183,532	7.1	0.2	9.7	0.2
In female householder family	58,949	27.4	0.5	28.4	0.5
In male householder family	31,599	17.3	0.4	22.4	0.5
In new SPM family groups	30,199	31.0	0.6	20.6	0.7
White, not Hispanic	197,436	9.5	0.2	10.7	0.2
Black, not Hispanic	38,624	25.7	0.5	24.0	0.6
Other	23,252	16.5	0.7	19.1	0.6
Hispanic origin	48,901	25.4	0.5	28.7	0.6
Nativity					
Native born	266,674	13.8	0.2	14.3	0.2
Foreign born	37,605	19.1	0.4	26.2	0.5
Naturalized citizen	16,024	10.8	0.4	17.7	0.6
Not a citizen	24,581	25.2	0.6	32.5	0.7

Source: U.S. Census Bureau, Current Population Survey, 2010 Annual Social and Economic Supplement.

For information on confidentiality protection, sampling error, nonsampling error, and definitions,

see www.census.gov/hhes/www/p60_238sa.pdf

*Includes unrelated individuals less than 15 years of age.

their ability to assimilate. Language barriers and discrimination regarding race have also prevented some Hispanics, American Indians, and Asians from moving to the middle and upper classes. Cultural differences have created a similar barrier for these groups. The drug war, which is discussed more fully in the following, has tended to target illegal drugs and their use and has had a disparate impact on minority groups such as Hispanics and African Americans. The drug war has led to the phenomenon of disproportionate representation by these minority groups in correctional organizations.

❖ Minorities: Policies and Practices That Have Resulted in Increased Incarceration

As was mentioned in this chapter and in Chapters 5, 6, and 7, African Americans and Hispanics, particularly, but also American Indians, are disproportionately represented as the accused or convicted in jails, prisons, and community corrections in the United States. Asian Americans are overrepresented in federal prisons. As already mentioned, most of these minority groups are also overrepresented among the poor in the United States and among those accused or convicted of street crimes (see Table 11.1). In the 1950s, an estimated 70% of the inmates in America's prisons were white, but by the year 2009, almost 60% of inmates of prisons and jails were African American (39.4%) or Hispanic (20.6%) (Belbot & Hemmens, 2010, p. 5; West, 2010a, p. 19). Yet, according to the 2009 United States Census, only 12.9% of the population as a whole were black or African American (39,641,060) and 15.8% were Hispanic or Latino of any race (48,419,324), while 79.6% were white (with other races constituting Asian [4.6%], Native Hawaiian or Pacific Islander [.1%], or two or more races) [7.9%]) (U.S. Census Bureau, 2010b, p. DP-1).

Stated another way, among the largest racial and ethnic groups in 2009, black, non-Hispanic males had an incarceration rate per 100,000 U.S. residents that was 6 times higher than white, non-Hispanic males and 2.6 times higher than Hispanic males (West, 2010a, p. 2). Likewise, 1 in 300 black, non-Hispanic females was incarcerated in a prison or a jail, as compared to 1 in 1,099 white, non-Hispanic females and 1 in 704 Hispanic females (West, 2010, p. 2). All of this means that, no matter how one views the data, blacks or African Americans and Hispanics/Latinos are disproportionately incarcerated in the United States when compared to their population composition.

In Focus 11.4

Harsh Justice and the Scott Sisters

In 1993, two sisters, Jamie Scott, 22, and her pregnant 19-year-old sister Gladys, were convicted of using three teenage boys to set up the armed robbery of two men (Pitts, 2010). The Scott sisters supplied the shotgun to the teenagers. Eleven dollars was stolen during this robbery, and the victims were unharmed. For this crime, the sisters, who had no prior criminal history, were given a double life sentence and as of November 2010, had served 16 years of it.

The teenage boys, two of whom testified against the sisters as part of their plea bargains, received 2-year sentences, which they completed years ago. The Scott sisters claimed and still claim they are innocent. The mother of the sisters argues that the harsh sentences were revenge for the family's willingness to testify against a corrupt sheriff (Pitts, 2010). As news columnist Leonard Pitts explains,

> Whatever the proximate cause of this ridiculous sentence, the larger cause is neon clear: The Scott sisters are black women in the poorest state in the union. And as report after report has testified, if you are poor or black (and God help you if you are both), the American justice system has long had this terrible tendency to throw you away like garbage. Historically, this has been especially true in the South. . . . How many other Scott sisters and brothers are languishing behind bars for no good reason, doing undeserved hard time on non-existent evidence, perjured testimony, prosecutorial misconduct or sheer racial or class bias.

The Scott sisters did, finally, get some relief from their sentence. Due to the advocacy of Pitts, and others such as the NAACP and the original prosecutor of the sisters, the governor of Mississippi, Haley Barbour, suspended the sisters' sentences as long as Gladys donates a kidney to her sister Jamie, whose kidneys have failed (Diaz-Duran, 2010). They were released from prison in January of 2010.

The Drug War: The New Jim Crow?

The rhetoric for the modern drug war was initiated by President Richard Nixon. He ran on a hard-line law enforcement platform for president, and as a consequence was interested in implementing "tough on crime" policies and practices. His efforts, however, were stymied by the fact that law enforcement was (and still is) primarily a responsibility of the states and their counties and cities. Ronald Reagan was the next president interested in enlarging the reach of the federal government into the states' business regarding law enforcement. His administration was responsible for declaring a "war on drugs" and for asking Congress to allocate money for prisons and law enforcement. Therefore, President Reagan is often credited (or blamed, depending on one's perspective) for starting the modern drug war.

Riding this popular "tough on crime" rhetoric of the 1980s and 1990s, Presidents Bush (H. W.), Clinton, Bush (G. W.), and Obama each continued to fund, and at times expand, the reach of the federal drug war. The practical effect of this modern war, if not the intent of its architects, has been to incarcerate unprecedented hundreds of thousands of minority men and women, primarily African Americans and Hispanics, who would otherwise not be incarcerated in the correctional system (Lurigio & Loose, 2008).

Michelle Alexander, in her book *The New Jim Crow: Mass Incarceration in the Age of Colorblindness* (2010), asserts that the modern drug war has been focused on the poor and the minority, while ignoring the fact that most drug users and drug sellers are white. She notes that in 2004, an estimated 75% of those incarcerated for drug offenses were black or Latino, while the majority of the drug users and sellers were white (p. 97). She argues the case that the drug war, as executed, has the practical effect of reinstating Jim Crow laws in the United States. She maintains this is so because of the police sweeps of poor and minority neighborhoods; the law enforcement focus on small-time marijuana possession offenders; and laws' nonsensical emphasis on crack cocaine over powder cocaine, though they are similar in addiction and pharmacologically the same (see the following discussion of this topic and Chapter 4 on Sentencing). Moreover, the implementation of the drug war has led to the erosion of civil liberties protections regarding search and evidence.

Jim Crow laws were devised by Southern states following the Civil War, starting in the 1870s and lasting until 1965 and the civil rights movement, to prevent African Americans from fully participating in social/economic and civic life. These laws restricted the rights and liberties of black citizens in employment, housing, education, travel, and voting. Interestingly enough, voter disenfranchisement, or preventing African Americans from voting, was a key part of the Jim Crow laws back then (Alexander, 2010). Today, a felony offense, gained through even a relatively minor drug possession conviction, can mean the loss of employability, access to public housing or food stamps, and voter disenfranchisement, or much the same effect as the Jim Crow laws of a century ago.

Crack Versus Powder Cocaine

The concern over crack cocaine started in the 1980s. The sentencing disparity that occurred when crack cocaine possession was treated as 100 times worse than possession of powder cocaine in federal law, was tied to the race and class of the persons associated with each drug (Alexander, 2010; The Sentencing Project, 2011). Poorer and disproportionately black and Hispanic people tended to use the cheaper crack cocaine, while richer and disproportionately white people tended to use the more expensive powder cocaine. Though there was never any real evidence that crack was more harmful or addictive than powder cocaine, there were a number of news stories sensationalizing news of "crack babies" and mothers—portrayed as *black* babies and their mothers—in the 1980s when the Reagan administration promoted the disparate sentencing. Alexander (2010) reports that the Reagan administration used the emergence of crack as a means of justifying the drug war and its focus on poor and minority people:

> They hired staff whose job it was to find reports of inner-city crack users, crack dealers, crack babies, and crack whores and to feed those stories to the media. The media saturation coverage of crack was no accident. It was a deliberate campaign that fueled the race to incarcerate. Legislators began passing ever harsher mandatory-minimum sentences in response to the media frenzy. (as quoted in Cooper, 2011, p. 7)

The harsher sentencing for crack cocaine possession is another example of a current criminal justice policy that even the United States Sentencing Commission concedes has had the practical and discriminatory effect of vastly increasing the incarceration of African Americans and Hispanics. Though the federal law was changed in 2010, crack cocaine sentences at the federal level are still much harsher than powder cocaine, by a factor of 18 to 1 (rather than 100 to 1 as they were under the 1986 law). Even so, the United States Sentencing Commission (2011) estimates that 12,811 federal inmates will be affected by the retroactive application of the reduced sentences for crack cocaine and that 85% of those affected will be African Americans (p. 19). However, state laws may still treat crack cocaine use more harshly than powder cocaine, which results in disproportionate incarceration of minority group members for this drug offense in state prisons.

Racial Profiling and DWB

In addition to the drug war and its effect on increasing minority involvement as the accused or convicted in corrections, scholars note that racial profiling by the police can have a similar effect. DWB, which stands for **Driving While Black or Brown,** refers to the police practice of focusing law enforcement attention on black- or brown-skinned drivers. The research in this area has been mixed, with some researchers finding that this practice affects arrests, while others were unable to establish the existence of this practice (Rice, Reitzel, & Piquero, 2005).

Police officers will tend to stop older vehicles, and such cars are often owned by poorer and minority group members. Having said this, Langan and colleagues (Langan, Greenfield, Smith, Durose, & Levin, 2001) found in a review of Bureau of Justice Statistics data from a police and public national contact survey, that blacks and Hispanics were more likely to report being stopped by the police than whites were. They also found that minority group members were more likely to report negative criminal justice outcomes for themselves, such as being ticketed, arrested, handcuffed, or searched, or subjected to the use of force by officers when stopped. Rice and colleagues (2005) found in their study of the perceptions of 700 randomly selected young adult (aged 18–26) New Yorkers, that the nuances of these stops might hinge on what shade one's skin is (p. 63). They found that blacks and black Hispanics were more likely to report that racial profiling was widespread and that they were racially profiled than were whites or non-black Hispanics.

In a study of drug arrests in Seattle, Washington, Beckett and colleagues (Beckett, Nyrop, Pfingst, & Bowen, 2005) found that the disparity in arrests between minorities and whites can be explained by racialized justice. The drug problem there was seen as a dangerous crack problem, which in turn was seen as a problem of use by blacks and Hispanics, despite comparable use of illegal drugs by whites. More minorities were therefore stopped by the police, as they were seen as more involved in illegal drug use.

Of course, the more such stops one is subjected to, the more likely that one is to run afoul of the law and to enter a correctional institution such as a jail or to find oneself on probation (Hawkins, 2005). Relatedly, these experiences are also more likely to result in the building of a record, which later, should one become entangled in the system again, might be used to justify a conviction or a more severe sentence.

In Focus 11.5

Los Angeles Police Department (LAPD) and Racial Profiling

The U.S. Department of Justice was critical of the Los Angeles Police Department's efforts to end racial profiling by its officers. Of particular concern for the Department of Justice was a conversation between two officers and a supervisor when the officers were unaware that they were being recorded.

"So, what?" one said, when told that other officers had been accused of stopping a motorist because of his race. The second officer is heard twice saying that he "couldn't do [his] job without racially profiling." The officers' comments, Justice officials found, spoke to a "perception and attitude of some LAPD officers on the street" and suggested "a culture that is inimical to race-neutral policing." (Rubin, 2010, p. 1)

❖ Minorities: Adjustment to Incarceration

Victor Hassine (2009), a writer and inmate doing life, since 1980, in Pennsylvania prisons for a capital offense, comments that race was and is an integral part of the prison life he has and does experience. Segregation in housing and by gangs (both voluntary), and racial bias in treatment by staff, were common in the Graterford Prison where he was an inmate during the 1980s. Most of the inmates in this prison were black, while most of the staff were white (Notably in the 1980s in the Graterford Prison, the only choices for self-identifying inmate race *or* ethnicity were white or black.). Most of the staff in this prison identified as Christian, while a sizable proportion of the black inmates were Muslim. Added to these differences based on race and religion was their place of origin: Many inmates tended to come from urban areas, while many staff were raised in more rural settings. Such differences between staff and inmates led to a difficult adjustment for minority inmates (see the following discussion of minority staff) and were cited as one of the complaints by inmates in the 1971 riot at the Attica Prison in New York.

Walter Rideau (2010), in his first-person account of incarceration in Louisiana prisons, describes the setting for his third trial, and the racial politics of the day in Baton Rouge, Louisiana, in this way:

> In 1970, at the time of my third trial, the Klan was using the kind of intimidation for which it was famous. It invaded North Baton Rouge—the black part of town—and plastered the utility poles and other upright surfaces with signs showing a rearing white-hooded horse carrying a hooded white rider, his left hand holding aloft a fiery cross. Beneath the horse's feet was the Klan's motto: FOR GOD AND COUNTRY. The poster was dominated by the horse and rider and by the big, bold print in the upper left corner that read SAVE OUR LAND, and beneath the picture it read JOIN THE KLAN. (p. 61)

Rideau (2010) encountered racism from some staff and inmates over the course of his long incarceration, but he noted that it lessened in degree and frequency as the years went on. Today, the racial mix of staff is more likely to reflect that of the community where inmates come from, which has tended to reduce race as a source of conflict between staff and inmates. However, Ross and Richards (2002) note that a "color line" still divides prison inmates into at least these groups: blacks, whites, and Hispanics. Between and among these groups, there are different styles of living and means of surviving.

Victimization by Race and Ethnicity

As regards victimization in prisons, Wolff, Shi, and Blitz (2008) found that African Americans were more likely to report sexual or physical violence from staff than from other inmates; non-Hispanic whites were more likely to report victimization by other inmates than by staff, and Hispanics had above-average reporting of victimization by both staff and other inmates. When both types of victimization were accounted for, though, all three groups reported about the same amount of victimization, just from different sources.

Probation or Prison?

Some research indicates that black offenders may prefer prison over community alternative sentencing, whereas white offenders express the opposite preference. In a study by Wood and May (2003), the authors note that blacks and whites "differed in their willingness to participate in alternative sanctions, in their preference for prison over alternatives, and in the amount of these alternatives they were willing to serve" (p. 624), with blacks less willing to participate in alternatives or the amount of alternatives and more likely to prefer prison over alternatives. There are several explanations for these differences discovered by these and other researchers.

Crouch (1993) argued that blacks might be more able to accept prison, and adjust to it, over alternatives because they were more likely to find people they know housed there, and are less likely to be threatened by prison life than whites, given that they have suffered the violence and deprivations of the cities already. Wood and May (2003) add that it is possible that blacks may also prefer prison because the alternatives to it in the community may subject them to abuse and harassment and ultimate revocation of their probation anyway. Therefore, it is not likely true that blacks or whites "prefer" prisons or the alternatives (e.g., probation or other programming); they just disagree about which is the lesser evil.

❖ Minorities Working in Corrections

As with women, the employment of minority group men and women in correctional organizations did not increase until the Civil Rights Act of 1964 was passed and affirmative action plans were developed to encourage their employment. Today, however, the number of minorities employed in corrections, though not always reflecting their representation in the community, particularly as this regards minority group women, has increased substantially. Though data in this area are not always consistent or up to date, we do know from the *Sourcebook of Criminal Justice Statistics* (2000 [Table 1.104], 2003 [Table 1.101], 2004 [Table 1.107]) that black non-Hispanics in 1999 accounted for 23.7% of local jail correctional officer employees (when their representation in the general populations was at about 12.2% [according to the U.S. Census for 2001]), 19.5% of all employees in state and federal and private prisons, and 24.3% of correctional officers

Photo 11.4

A correctional officer fingerprints an incoming inmate.

in federal prisons in 2004. On the other hand, non-Hispanic whites and Hispanics are underrepresented among staff, when compared to their representation in those communities. In 2001, whites constituted about 69.0% of the U.S. population and Hispanics about 12.9% [U.S. Census Bureau, 2001]), in these same data, although whites were still the majority racial group employee (59.3%, 63.3%, and 60.6%, consecutively) in these jails and prisons, and Hispanics constituted a substantial ethnic minority (7.7%, 7.3%, and 12.4%, consecutively).

Corrections: The Essentials

Comparative Perspective: Aborigine Inmates in the Victoria (Australia) Prison System

The Aborigine, or original people, of Australia have resided there for some 30,000 years. They lost their land to encroaching white settlers in the 1800s and 1900s in much the same way as the American Indians did in the United States. As indicated in a Victoria Department of Justice publication (2010), as of June 2010, about 6% of inmates in Victoria's prisons identified themselves as Aboriginal or Torres Strait Islanders (another minority group in Australia). This represents an increase over 2006 when they comprised about 5.5% of the population. Given the relatively small numbers of Aborigines in Australia (they comprise about 2% of the general population), these numbers represent an astounding imprisonment rate per 100,000 Indigenous adults of 2,482.5 for males (pp. 18–22).

Females who identified themselves as Aboriginal or Torres Strait Islanders rose from 15 in 2006 to 27 in 2010. The female indigenous imprisonment rate was 245.9. When compared with imprisonment rates in the United States, which hover around 750, these numbers for the males do not appear so outlandish. However, when compared to the rates of imprisonment in Victoria for all adult males (200) or for all adult females (14.3), these indigenous imprisonment rates in Australia seem as out of sync as they are for African Americans in the United States (Victoria Department of Justice, 2010, pp. 18–22).

Summary

- American history encompasses a racist past, which has affected the operation of correctional entities and the criminal justice system generally.
- Those who fall below the poverty line in the United States are also more likely to be enmeshed in street criminality. Some racial and ethnic groups who are more likely to be poor (e.g., African Americans and Hispanics) are also more likely to be engaged in street crime.
- Police, court, and correctional practices have had the effect of increasing the disproportionate incarceration of minority group members. Driving While Black or Brown, the drug war generally, and the harsh sentencing for crack cocaine specifically, along with the disenfranchisement that comes with a felony conviction (and in some states stays with a

felony conviction) all serve to reinforce disparity in treatment by the criminal justice system of racial and ethnic minorities.
- Physical and sexual victimization in prisons varies by type of victimization and by race and ethnicity, though the total amount of such victimization appears to be similar for these groups.
- The numbers of racial and ethnic minorities working in corrections has increased substantially over the years, and for African Americans at least, it appears that they mirror their numbers in the community.
- Disproportionate incarceration of minority group members is not isolated to just the United States. Aboriginal inmates in the Victoria, Australia, prison system are also incarcerated at a much higher rate than whites in that country.

Key Terms

Discrimination	Driving While Black or Brown	Jim Crow laws
Disparity	Ethnicity	Race

Discussion Questions

1. What sorts of criteria differentiate race and ethnicity? Why might it not always be clear what race or ethnicity a person is? Are there reasons to make such distinctions?

2. What evidence is there of disparity and discrimination against racial and ethnic minorities in the United States in the past?

3. What evidence is there of disparity and discrimination against racial and ethnic minorities in the United States currently?

4. How and why is adjustment in corrections affected by one's race and ethnicity?

5. Discuss how we might reduce the amount of disparity and discrimination against minorities in the United States? What specific steps can be taken in this direction? What are the likely barriers to accomplishing these changes?

Useful Internet Sites

American Civil Liberties Union (ACLU): www.aclu.org

National Association for the Advancement of Colored People (NAACP): www.naacp.org

The Sentencing Project: www.sentencingproject.org

Southern Poverty Law Center: www.splcenter.org

United States Sentencing Commission: www.ussc.gov

Juveniles and Corrections

❖ Introduction: Delinquency and Status Offending

The juvenile justice system generally falls under the broad umbrella of the civil law rather than criminal law. This placement emphasizes the distinction that the law makes between adults and juveniles who commit the same illegal acts. Juveniles who commit acts that are criminal when committed by adults are called **delinquents** rather than criminals, conveying the notion that the juvenile has *not* done something he or she *was* supposed to do (behave lawfully) rather than *done* something he or she *was not* supposed to do (behave unlawfully). This difference is a subtle one that reflects the rehabilitative, rather than punitive, philosophy of American juvenile justice.

Juveniles are subject to laws that make certain actions that are legal for adults such as smoking, drinking, not obeying parents, staying out at night to all hours, and not going to school, illegal for them. These acts are called **status offenses** because they apply only to individuals having the status of a juvenile, and they exist because the law assumes that juveniles lack the maturity to appreciate the long-term consequences of their behavior. Many of these acts can jeopardize juveniles' future acquisition of suitable social roles because they may lead to defiance of all authority, inadequate education, addiction, and teenage parenthood (Binder, Geis, & Bruce, 2001). If parents are unwilling or unable to shield their children from harm, the juvenile justice system becomes a substitute parent. Status offenses constitute the vast majority of juvenile offenses, and consume an inordinate amount of juvenile court time and resources (J. Bynum & Thompson, 1999). Because of this, some states have relinquished court jurisdiction over status offenses to other social service agencies where terms such as "child in need of supervision" (CHINS) or "person in need of supervision" (PINS) are used to differentiate status offenders from juveniles who have committed acts that are crimes when committed by adults. In this chapter, we will discuss the extent of juvenile delinquency and status offending, the likely causes of it, the history of dealing with children in corrections, and current processing of delinquents in the system.

Photo 12.1

Immaturity of adolescent behavior is matched by the immaturity of the adolescent brain.

❖ The Extent of Delinquency

Figure 12.1 shows the juvenile proportion of all reported arrests in 2008. Juveniles ("youths under 18") accounted for 16% of all violent crime arrests and 26% of all property crime arrests. According to the U.S. Census Bureau (2006), the percentage of the population between the age of responsibility (the age at which juveniles can be held accountable for their actions) and adulthood averaged across all states was about 11.5% in 2005. Juveniles are thus overrepresented in most of the crime categories in Figure 12.1.

Figures such as these are troubling, but antisocial behavior is normative (although not welcome) for juveniles; juveniles who do *not* engage in it are statistically abnormal (Moffitt & Walsh, 2003). Adolescence is a time when youths are "feeling their oats" and temporarily fracturing parental bonds in their own personal declaration of independence. Looking at data from 12 different countries, Junger-Tas (1996) concluded that delinquent behavior is a part of growing up, and that the peak ages for different types of crimes were similar across all countries (16–17 for property crimes and 18–20 for violent crimes). Biologists tell us that adolescent rebellion is an evolutionary design feature of all social primates. Fighting with parents and seeking out age peers with whom to affiliate "all help the adolescent away from the home territory" (Powell, 2006, p. 867). As Caspi and Moffitt (1995) put it, "every curfew broken, car stolen, joint smoked, or baby conceived is a statement of independence" (p. 500). The juvenile courts are thus dealing with individuals at a time in their lives when they are most susceptible to antisocial behavior.

Figure 12.1

Juvenile Proportion of Arrests by Offense, 2008

Source: Puzzanchera (2009).

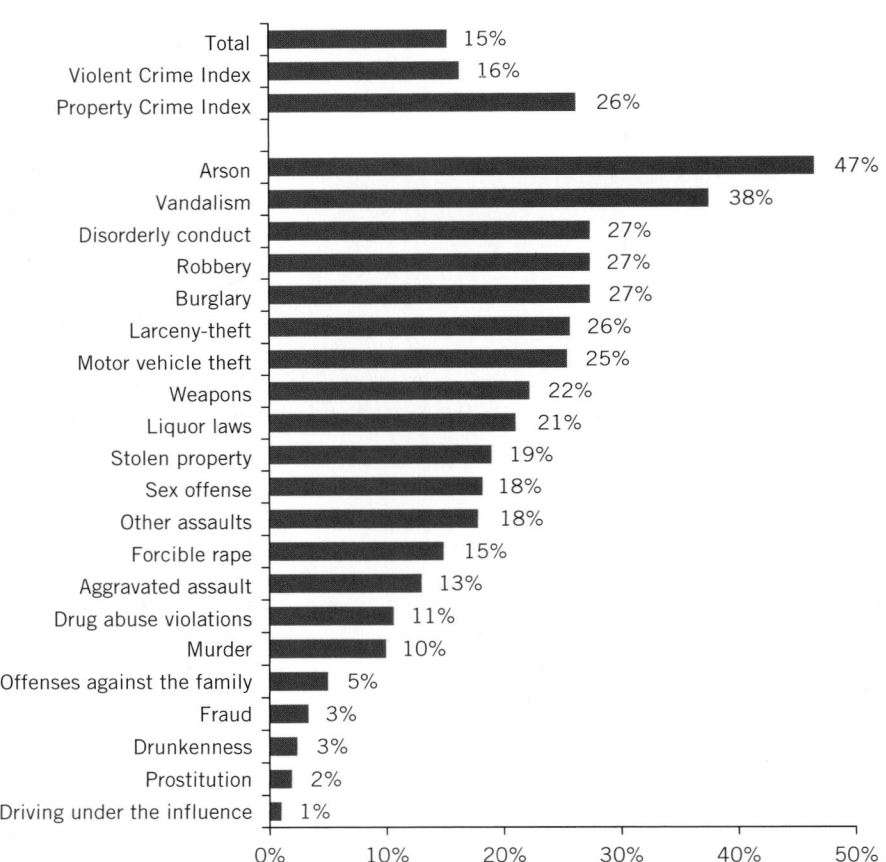

❖ The Juvenile Brain and Juvenile Behavior

Neuroscience research over the past 15–20 years has thrown much light on why there is a sharp rise in antisocial behavior in adolescence across time and cultures, and some very important court decisions in juvenile justice (such as the abolition of the juvenile death penalty) have been influenced by this research (Garland & Frankel, 2006). What has emerged from this research is that the immaturity of adolescent behavior is matched by the immaturity of the adolescent brain, as Aaron White's (2004, p. 4) summation of the key messages from the 2003 conference of the New York Academy of Sciences (NYAS) makes clear:

1. Much of the behavior characterizing adolescence is rooted in biology intermingling with environmental influences to cause teens to conflict with their parents, take more risks, and experience wide swings in emotion.

2. The lack of synchrony between a physically mature body and a still maturing nervous system may explain these behaviors.

3. Adolescents' sensitivities to rewards appear to be different than in adults, prompting them to seek higher levels of novelty and stimulation to achieve the same feeling of pleasure.

4. With the right dose of guidance and understanding, adolescence can be a relatively smooth transition.

The onset of puberty also brings with it a 10- to 20-fold increase in testosterone in males, a hormone linked to aggression and dominance-seeking (L. Ellis, 2003), and brain chemicals that excite behavior increase in adolescence while chemicals that inhibit it decrease (Collins, 2004; Walker, 2002). Many other events are reshaping the adolescent's body and brain during this period that lead to the conclusion that there are *physical* reasons why adolescents often fail to exercise rational judgment and why they tend to attribute erroneous intentions to others. When the brain reaches its adult state, a more adult-like personality emerges, with greater self-control and conscientiousness (Blonigen, 2010). It is important to understand these biological processes, and it is especially important to note the last of the NYAS's messages: "With the right dose of guidance and understanding, adolescence can be a relatively smooth transition."

❖ History and Philosophy of Juvenile Justice

Up until about 300 years ago, the concept of childhood was not recognized; children were considered not much different from property, and no special allowances for children were recognized in matters of determining culpability and punishment. The minimum legal age of criminal responsibility was defined in early English common law as 7 (In the modern United States, it ranges from 6 in North Carolina to 10 in Arkansas, Colorado, Kansas, Pennsylvania, and Wisconsin, as it is in modern England.) (Snyder, Espiritu, Huizinga, Loeber, & Petechuck, 2003). Under the increasing influence of Christianity, English courts in the Middle Ages began to exempt children below the age of 7 from criminal responsibility, and children between the ages of 7 and 14 could only be held criminally responsible if it could be shown that they were fully aware of the consequences of their actions. Fourteen was the cut-off age between childhood and adulthood for the purpose of assigning adult criminal responsibility because individuals were considered rational and responsible enough at this age to marry (Springer, 1987).

Ever since the formation of the English chancery courts in the 13th century, there has been movement toward greater state involvement in children's lives. Chancery courts adopted the doctrine of **parens patriae**, which means "state as parent." *Parens patriae* gave the state the right to intercede and act in the best interest of the child or any other legally incapacitated person such as someone who is mentally ill. This meant that the state and not the parents had the ultimate authority over children, and that children could be removed from their families if they were being delinquent and placed in the custody of the state (Hemmens, Steiner, & Mueller, 2003).

Despite *parens patriae,* the family was still considered the optimal setting for children to be reared in, and as such, orphans or children with inadequate parents were assigned to foster families through a system known as binding out. Children whose parents could not control them or who were too poor to provide for them were apprenticed to richer families who used them for domestic or farm labor. This period saw the establishment of the first laws directed specifically at children, including laws that condemned begging and vagrancy (P. Sharp & Hancock, 1995). The concern over vagrancy led to the creation of workhouses in which "habits of industry" were to be instilled. The first one, called *Bridewell,* was opened in 1555, and in 1576 the English Parliament passed a law establishing **bridewells,** or workhouses (also discussed in Chapter 2), in every English county (Whitehead & Lab, 1996). These places were generally dank, harsh, and abusive, but the idea behind them was that if vagrant youths were removed from the negative influences of street life, they could be reformed by discipline, hard work, and religious instruction.

❖ Childhood in the United States

American notions of childhood and how to deal with childhood misconduct were imported whole from England. Based on the Bridewell model, the New York House of Refuge was established in 1825 to house orphans, beggars, vagrants, and juvenile offenders. Several other cities, counties, and states soon established their own homes for "the perishing and dangerous classes" as they were thought of (Binder, Geis, & Bruce, 2001, p. 202). Children in houses of refuge (discussed in Chapter 10) lived highly disciplined lives and were required to work at jobs that brought income to the institution. The indeterminate nature of children's residence allowed the institutions a great deal of latitude in their treatment. Children were required to work long hours, often received little or no training, and were frequently mistreated (Whitehead & Lab, 1996).

It was a frequent practice for poor parents to place their children in residence for idle and disorderly behavior, making it clear that the courts would have to create standards for admission. The courts did this in *Ex Parte Crouse* (1838). The subject of the case was a child named Mary Ann Crouse who was placed in the Pennsylvania House of Refuge by her mother against the wishes of her father. Mary's father argued that it was unconstitutional to incarcerate a child without a jury trial, but the Pennsylvania Supreme Court ruled that parental rights are superseded by the *parens patriae* doctrine. This landmark decision established *parens patriae* as settled law in American juvenile jurisprudence (del Carmen, Parker, & Reddington, 1998).

❖ The Beginning of the Juvenile Courts

Greater concern for children's welfare in the 19th century created an impetus for change in the way juvenile offenders were handled as it became increasingly obvious that adult criminal courts were not equipped to apply the spirit of the *parens patriae* doctrine. In 1899, Cook County, Illinois, enacted legislation providing for a separate court system for juveniles, and by 1945 every state in the union had established juvenile court systems (Hemmens et al., 2003). These courts combined the authority of social control with the sympathy of social welfare in a single institution and afforded judges a great deal of latitude in determining how "the best interests of the child" could be realized. The creation of a separate system of justice for juveniles brought with it a set of terms describing the processing of children accused of committing delinquent acts that differentiated it from the adult system. These terms reflect the protective and rehabilitative nature of the juvenile system in contrast with the punitive nature of the adult system. Table 12.1 lists the terms used to describe the procedure or event from initial contact with authorities to the last in both the adult and juvenile justice systems today.

Table 12.1

Comparing Procedural/Event Terminology in Adult and Juvenile Court Systems

Procedure or Event	Adult System	Juvenile System
Police take custody of offender	Placed under arrest	Taken into custody
Official who makes initial decisions about entry into the court system	Magistrate	Intake officer
Place accused may be held pending further processing	Jail	Detention
Document charging the accused with specific act	Indictment or information	Petition
Person charged with illegal act	Defendant	Respondent
Accused appears to respond to charge(s)	Arraignment	Hearing
Accused verbally responds	Enters a plea of guilty, not guilty, or no contest	Admits or denies
Court proceeding to determine if accused committed the offense	Public jury trial	Adjudicatory hearing
Decision of the court as to whether accused committed offense	Verdict of jury	No jury; not public; adjudication by judge
Standard of proof required	Beyond a reasonable doubt	Beyond a reasonable doubt
Court proceeding to determine what to do with person found to have committed offense	Sentencing hearing	Dispositional hearing
Institutional confinement	Prison	Juvenile correctional facility
Community supervision	Probation; parole if had been imprisoned	Probation; aftercare if had been confined to juvenile correctional facility

❖ Processing Juvenile Offenders

Figure 12.2 illustrates the flow of juvenile cases through the juvenile courts in the United States in 2002 (Snyder & Sickmund, 2006). We see that only 58% of the juveniles taken into custody ("arrested") or otherwise referred to juvenile court were petitioned (formally charged). Among those not petitioned, most had their cases dismissed, some are placed on informal probation (probation without a formal adjudication of delinquency, sometimes known as diversion), and "other sanctions"—this could be something as minor as a written apology to something as serious as placement in a mental institution.

Figure 12.2

Juvenile Court Case Processing

Source: Snyder & Sickmund (2006).

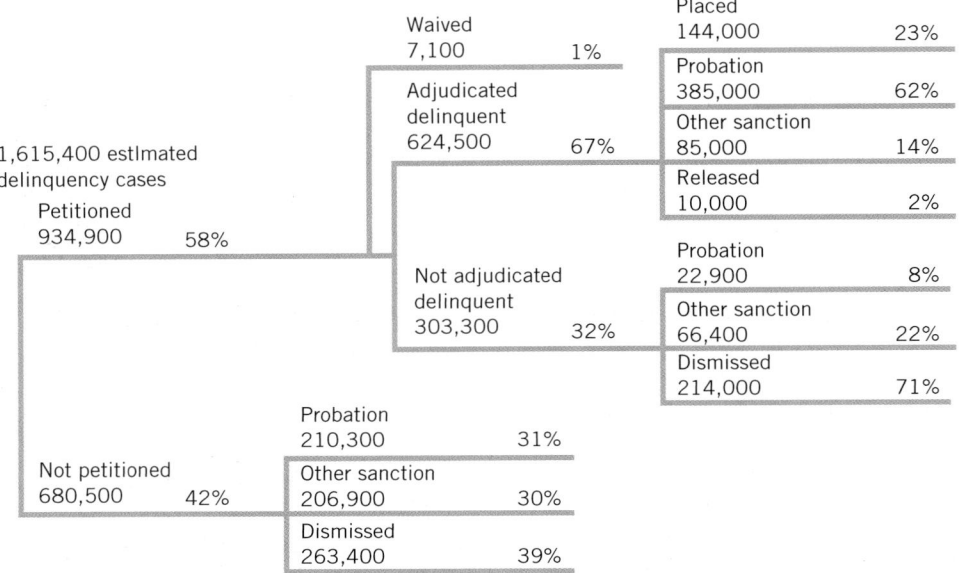

In 2002, the most severe sanction ordered in 85,000 adjudicated delinquency cases (14%) was something other than residential placement or probation, such as restitution or community service

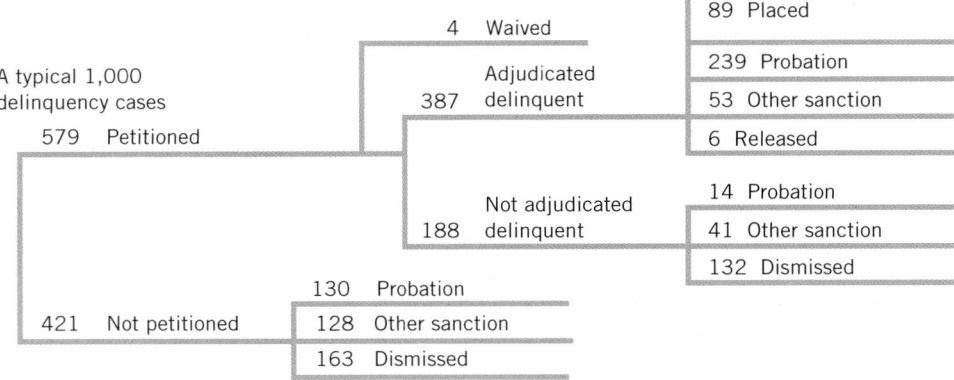

Adjudicated cases receiving sanctions other than residential placement or probation accounted for 53 out of 1,000 delinquency cases processed during the year

When a petition has been filed by the juvenile court prosecutor, the court has to decide if it should take jurisdiction of the case. In about 99% of the cases, the court does accept jurisdiction, and about 66% of the time the juvenile is adjudicated delinquent (found guilty). Note that in 32% of the cases without a finding of adjudication, 30% of those cases were not outright dismissed. In adult court, a finding of "not guilty" always means that the defendant is now a free person. Under the principle of *parens patriae,* however, the juvenile court has the power to intervene in a child's life as a proactive measure even though he or she has been found not guilty of any wrongdoing. If the juvenile court does not accept jurisdiction, it means that the case is waived to adult court, which is one of the most controversial issues in juvenile justice.

Juveniles Waived to Criminal Court

As noted earlier, juveniles can sometimes be *waived* (transferred) to adult criminal court where they lose their status as minors and become legally culpable for their alleged crime and are subject to criminal prosecution and punishments. A transfer to adult court is called a **waiver** because the juvenile court waives (relinquishes) its jurisdiction over the child to the adult system. Waivers are designed to allow the juvenile courts to transfer juveniles over a certain age who have committed particularly serious crimes, or who have exhausted the juvenile system's resources in trying to rehabilitate chronic offenders, to a more punitive system. Juveniles become increasingly more likely to be waived if they are chronic offenders approaching the upper age limit of their state's juvenile court's jurisdiction. Note from

Photo 12.2
A 10-year-old juvenile, charged with murder, listens to an explanation of his rights by his public defender.

Figure 12.2 that only about 1.0% of juvenile cases nationwide are waived to the criminal courts.

There are three primary (non–mutually exclusive) ways in which juveniles can be waived to criminal court:

(1) **Judicial waiver:** A judicial waiver involves a juvenile judge deciding after a "full inquiry" that the juvenile should be waived. (Forty-eight states at present use this judicial discretionary model.) In some states, there are mandatory waivers for some offenses, but juvenile judges are involved in determining if the criteria for a mandatory waiver are met. Twelve states use a system of presumptive waivers in which the burden of proof is on juveniles to prove that they are amenable to treatment and therefore should not be waived, not on the prosecutor to prove they should.

(2) **Prosecutorial discretion:** This model allows prosecutors to file some cases in either adult or juvenile court. In such cases (usually limited by age and seriousness of the offense), the prosecutor can file the case directly with the adult court and bypass the juvenile court altogether. Fourteen states and the District of Columbia allow prosecutorial discretion waivers.

(3) **Statutory exclusion:** These are waivers in cases in which state legislatures have statutorily excluded certain serious offenses from the juvenile courts for juveniles over a certain age, which varies from state to state. These automatic waivers are found in 31 states.

Studies have shown that juveniles waived to adult courts are more likely to recidivate than youths adjudicated for similar crimes in juvenile court, although remember that only the most delinquent-prone youths are waived (Butts & Mitchell, 2000). Neither does a waiver necessarily guarantee a more punitive disposition. Waived juveniles who commit violent crimes are likely to be incarcerated, but juveniles waived for property and drug offenses often receive more lenient sentences than they would have received in juvenile courts (Butts & Mitchell, 2000).

Extending Due Process to Juveniles

Contrary to the "best interests of the child" philosophy, juvenile courts often punished children in arbitrary ways that would not be tolerated in the adult system, as illustrated in some famous juvenile cases presented below. Critics argued that the *parens patriae* doctrine allowed too much latitude for courts to restrict the rights of juveniles, and that because the courts could remove juvenile rights to liberty, juveniles should be afforded the same due process protections as adults. Supporters of *parens patriae* countered that it was suitable and proper for the treatment of children, and that any problems concerning juvenile court operation were problems of implementation, not philosophy (Whitehead & Lab, 1996).

The United States Supreme Court maintained a "hands off" policy with regard to the operation of the juvenile courts until 1966 when it agreed to hear *Kent v. United States*. In 1961, 16-year-old Morris Kent broke into a woman's apartment, raped her, and stole her wallet. Because of Kent's chronic delinquency and the seriousness of the offense, he was waived to adult court. The adult court found Kent guilty of six counts of housebreaking and robbery, for which he was sentenced to 30 to 90 years in prison. Had Kent remained in juvenile court, he could have been sentenced to a maximum of 5 years (the remainder of his minority, which was until age 21 at the time). Kent appealed, arguing that the waiver process had not included a "full investigation," and that his counsel had been denied access to the court files.

The Supreme Court remanded Kent's case back to district court, with Justice Abe Fortas commenting that

> there is no place in our system of law for reaching a result of such tremendous consequences without ceremony—without hearing, without effective assistance of counsel, without a statement of reasons. . . . [T]he admonition to function in a "parental" relationship is not an invitation to procedural arbitrariness. (n.p.)

Justice Fortas also noted that under the *parens patriae* philosophy, the child receives the worst of both worlds: "he gets neither the protections accorded to adults nor the solicitous care and regenerative treatment postulated for children" (n.p.). The *Kent* decision determined that juveniles must be afforded certain constitutional rights, and thus began the process of formalizing the juvenile system into something akin to the adult system (Hemmens et al., 2003).

The Supreme Court heard a second case concerning juvenile rights one year later in *In re Gault* (1967). In 1964, 15-year-old Gerald Gault was adjudicated delinquent for making obscene phone calls and sentenced to 6 years in the State Industrial School. An adult convicted of the same offense would have faced a $5 to $50 fine and a maximum of 60 days in jail. The Supreme Court used this case to establish five basic constitutional due process rights for juveniles: (1) the right to proper notification of charges, (2) the right to legal counsel, (3) the right to confront witnesses, (4) the right to privilege against self-incrimination, and (5) the right to appellate review, all of which had been denied to Gault.

A third significant juvenile case is *In re Winship* (1970). In 1967, 12-year-old Samuel Winship was accused of stealing $112 from a woman's purse taken from a locker. Winship was adjudicated delinquent based on the civil law's "preponderance of evidence" standard of proof (i.e., is it more likely than not that the person committed the act he or she is charged with?) and was sent to a state training school. Upon appeal, the Supreme Court ruled that when the possibility of commitment to a secure facility is a possibility, the "beyond a reasonable doubt" standard of proof must extend to juvenile adjudication hearings.

In *McKeiver v. Pennsylvania* (1971), the sole issue before the Court was "Do juveniles have the right to a jury trial during adjudication hearings?" The Supreme Court ruled that they do not. The Court did not rule that the states cannot provide juveniles with this due process right, only that they are not constitutionally required to do so. And in *Breed v. Jones* (1975), the Supreme Court ruled that the prohibition against double jeopardy applied to juveniles once they have had an adjudicatory hearing (which is a civil process and not technically a trial). Breed had an adjudicatory hearing, and was subsequently waived to adult court. The Court ruled that he had been subjected to the burden of two trials for the same offense and therefore the double jeopardy clause of the Fifth Amendment had been violated.

In the case of *Schall v. Martin* (1984), the issue before the Supreme Court was whether the preventative detention of a juvenile charged with a delinquent act is constitutional. The Court ruled that it was permissible because it serves a legitimate state interest in protecting both society and the juvenile from the risk of further crimes committed by the person being detained while awaiting his or her hearing. This ruling established that juveniles do not enjoy the right to bail consideration and reasserted the *parens patriae* interests of the state.

The last major juvenile case is *Graham v. Florida* (2010) in which 17-year-old Terrance Jamar Graham was sentenced to life in prison without possibility of parole. Graham had been convicted of robbery when he was 16 and placed on probation. He was subsequently arrested for an armed home invasion and also admitted to several other armed robberies. The sentencing judge considered him to be highly dangerous and beyond rehabilitation. The U.S. Supreme Court overturned Graham's sentence, stating,

> The Constitution prohibits the imposition of a life without parole sentence on a juvenile offender who did not commit homicide. A State need not guarantee the offender eventual release, but if it imposes a sentence of life it must provide him or her with some realistic opportunity to obtain release before the end of that term. (n.p.)

The issue of the juvenile death penalty will be addressed in Chapter 13 and will not be discussed here except as part of Figure 12.3, which presents a summary of important Supreme Court cases regarding juveniles' due process rights from *Kent* (1966) to *Graham* (2010). Taken as a whole, what these cases essentially mean is an erosion of the distinction between juvenile and criminal courts. On the positive side, these rulings have helped to create a juvenile court system that more closely reflects the procedural guidelines established in adult criminal courts. On the negative side, they have in effect criminalized juvenile courts. In order to gain due process rights enjoyed by adults, juveniles have surrendered some benefits such as the informality of solicitous treatment they nominally enjoyed previously. Only time will tell if this convergence of systems results in more just outcomes for juveniles than they received under unmodified *parens patriae.*

Figure 12.3

Supreme Court Cases Altering Juvenile Justice, 1966–2010

Kent v. United States (1966)	In re Gault (1967)	In re Winship (1970)	McKeiver v. Pennsylvania (1971)	Eddings v. Oklahoma (1982)
Courts must provide essentials of due process when waiving juveniles to adult system.	In hearings that could result in institutional commitment, juveniles have four basic constitutional rights.	The state must prove guilt beyond a reasonable doubt in delinquency matters.	Jury trials are not required in juvenile court hearings.	All mitigating factors should be considered in deciding to apply the death sentence to juveniles.

Schall v. Martin (1984)	Sanford v. Kentucky (1989)	Roper v. Simmons (2005)	Graham v. Florida (2010)
Pretrial preventive detention of juveniles is permissible under certain circumstances.	It is constitutionally permissible to impose the death penalty on 16- and 17-year-olds.	Death penalty for juveniles is unconstitutional.	Life without possibility of parole is unconstitutional for juveniles.

❖ Juvenile Community Corrections

As seen from Figure 12.4, juvenile corrections mirrors the adult system in that the majority of adjudicated delinquents are placed into some form of community-based corrections and less than a quarter are sent to residential facilities. Juvenile community corrections offers a wide variety of options, all ostensibly designed to implement the three-pronged goal of the juvenile justice system: (1) to protect the community, (2) to hold delinquent youths accountable, and (3) to provide treatment and positive role models for youths. This is known as the **balanced approach** to corrections (K. Carter, 2006).

When juveniles are taken into custody, a complicated process of determining how to best deal with them with the above goals in mind is initiated. Juveniles may be released to their parents or detained in a detention center until this determination is made. The most lenient disposition of a case is known as **deferred adjudication**. Depending on the jurisdiction, a deferred adjudication decision can be made by the police, the prosecutor, a juvenile probation officer, or a juvenile magistrate/judge. A deferred adjudication means an agreement is reached between the youth and a juvenile probation officer, without any formal court appearance, that the youth will follow certain probation conditions. This form of disposition is only used for status offenses or minor property offenses, and as long as the juvenile has no further charges he or she is discharged from probation within a short time period. No formal record of the proceedings of the case is made in deferred adjudications.

Other juveniles may be placed on formal probation after adjudication in court by a juvenile judge. In such cases, there are records of the proceedings and probationers are more strictly monitored. As in the adult courts, juvenile judges typically make their dispositional decision based on recommendations made by probation officers. Juvenile probation officers write **predisposition reports** (analogous to the adult presentence investigation report) and

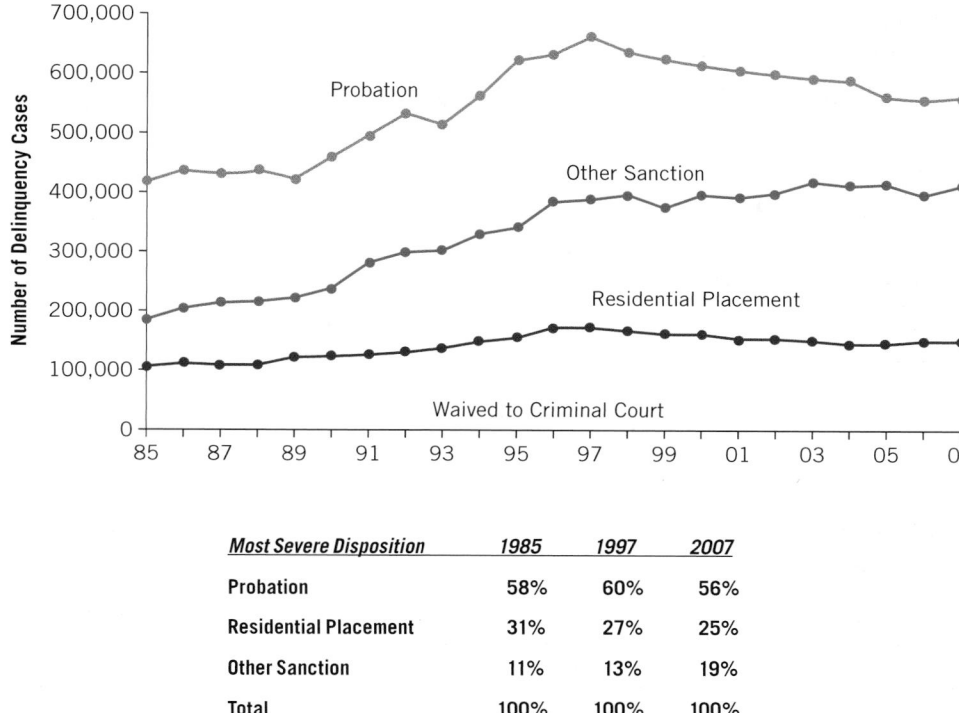

Figure 12.4

Juvenile Delinquency Probation Caseload, 1985–2007

Source: Livsey (2010).

Most Severe Disposition	1985	1997	2007
Probation	58%	60%	56%
Residential Placement	31%	27%	25%
Other Sanction	11%	13%	19%
Total	100%	100%	100%

will have a variety of classification instruments very similar to those used in the adult system and discussed in Chapters 6 and 14, to help them to formulate their recommendations.

Once a youth is placed on probation, under the doctrine of *parens patriae,* the probation officer becomes a surrogate parent to the youth. But with probation officers being saddled with a nationwide average of around 42 cases (Taylor, Fritsch, & Caeti, 2007), they can do very little "parenting." Probation officers may see their charges only once a month for perhaps 30 minutes, whereas the juveniles' natural parents see them (or should) every day. Juvenile probation officers therefore insist on parental support in working with their children, because parental involvement in the rehabilitative effort of juveniles is considered a "must" (Balazs, 2006). It is a must because while probation serves the positive goal of keeping youths in the community and thus avoiding the stigma of institutionalization and the exposure to other seriously delinquent youths, the potential danger is that the probationer may view the disposition as a slap on the wrist and return with more confidence to the old ways that led to his or her adjudication.

If the child comes from an antisocial family rife with substance abuse and criminality, officers are not likely to get any sort of positive support. Even if juvenile probationers come from prosocial two-parent families, there is often resistance by parents to juvenile authorities "poking their noses" into family affairs and "picking on" their children, who of course are victims of "bad company" (Walsh & Stohr, 2010, p. 457). If parents are eager to help their children, however, there are some excellent parent effectiveness training programs out there. The relatively short-term *Prosocial Family Therapy System* described by Bleckman and Vryan (2000) is a good comprehensive system with some very encouraging results reported.

❖ Intensive Probation

There will always be some juveniles who require more extensive supervision and treatment than others. To meet their needs and the needs of community protection, a variety of methods have been devised. One such method is **intensive supervision probation (ISP)**, described in Chapter 6. ISP is usually imposed on youths as a last chance before incarceration. Juvenile probation officers with an ISP caseload typically supervise only 15–20 juveniles and may carry a gun (Taylor et al., 2007). Officers may make daily contact with their charges, visiting them at home, school, and work to monitor their behavior and progress in these settings. Officers will also enlist the help of other agencies (both public and private) that can provide probationers with more specialized and concrete help of the kind outside the purview of the juvenile court. These agencies will include mental health clinics, substance abuse centers, educational and vocational guidance centers, and welfare agencies (to help juveniles' families). ISP officers know that they cannot possibly provide all the needs of their probationers themselves, and that efficient case management consists of them delivering services by using networks of collaborative providers. Delany, Fletcher, and Shields (2003) point out the importance of collaborative efforts to assist youths with multiple problems: "Without some level of collaboration among agencies, the odds of relapse and recidivism, which often leads to repeated institutionalization, are high" (p. 66).

Other forms of more intense supervision include electronic monitoring and/or house arrest. These sanctions were discussed in Chapters 6 and 8, and since they operate for juveniles exactly as they do for adults, they will not be discussed again here.

Youths who commit property crimes are frequently made to pay restitution to their victims to compensate for the victims' losses. This both compensates the victim and holds the youth accountable for his or her actions. To compensate the community as a whole, adjudicated delinquents may receive a **community service order**, which is part of a disposition requiring the probationer to work a certain number of hours doing some kinds of tasks to help their communities. This work can range from cleaning graffiti from walls to picking up trash along highways or in parks. Restitution and community service orders can go a long way to help juveniles to develop a sense of responsibility and the ability to accept the consequences of their actions without rancor. For these reasons, community service and restitution have been called "integral components of the restorative justice philosophy" (Walsh & Stohr, 2010, p. 455).

Restorative justice (also discussed in Chapter 6) may be defined as "every action that is primarily oriented toward justice by repairing the harm that has been caused by the act" and "usually means face-to-face confrontation between victim and perpetrator, where a mutually agreeable restorative solution is proposed and agreed upon" (Champion, 2005, p. 154). Restorative justice defines delinquency as an offense committed by one person against another rather than against the state, and by doing so personalizes justice by engaging the victim, the offender, and the community in a process of *restoring* the situation to its pre-offense status. Restorative justice thus gives equal weight to the needs of offenders, victims, and the community and focuses equally on each of these rather than being driven solely by offenders ("What do we do with them now?") (K. Carter, 2006). Victims or their representative are included in the justice process with the sentencing procedure addressing the needs of the victims, including their need to be heard and to be restored to wholeness again as far as possible.

Just as the retributive model reemerged after the alleged failure of the medical model, the restorative justice model has emerged with the apparent failure of "get tough" programs (Welch, 1996). However, the restorative model may not suit all victims because many victims understandably feel that things cannot be "put right" so easily and want the offender punished. However, it would be a mistake to see it as a New Age "touchy-feely" approach to corrections. It holds offenders fully accountable for their actions by applying appropriate punishment and adds the additional dimensions by requiring offenders to accept responsibility for taking action to repair the harm done (Bazemore, 2000).

A meta-analysis found that restorative justice programs had a weak-to-moderate positive effect on victim satisfaction, a weak positive effect on offender satisfaction and recidivism, and a moderate effect on restitution compliance (Latimer, Dowden, & Muise, 2005). These findings should be viewed in light of the fact that both victims and offenders select themselves into such programs.

Comparative Perspective: Juvenile Justice Philosophies

There are a variety of philosophies regarding juvenile justice around the world, which Reichel (2005) classifies into four families or models: welfare, legalistic, corporatist, and participatory models. The models reflect broad generalities, and there is much overlap among the countries used to exemplify each. The *welfare* model reflects a concern for the well-being of children at the expense of legalities in which "troubled youths" are funneled through a series of non-judicial agencies designed to address the problem. Police cautions (a warning by a senior police officer) and restorative justice constitute a big part of this model, which Reichel sees the Australian and New Zealand systems as exemplifying. Sweden is even deeper into the welfare model since it does not even have a juvenile court system. The police turn over offenders under age 20 to social boards that proceed very informally and only rarely refer a case to the criminal courts. The Swedish government has even abolished imprisonment for youth. All this is designed to reduce stigma and support treatment, but it has led to youths knowing that they can commit criminal acts almost with impunity (Terrill, 2003).

The *legalistic* model emphasizes the law over treatment, although this does not necessarily mean that the model is less humanistic. Reichel uses Italy as an example of this model. Legalism comes in with the realization that there is a need to treat juveniles differently in the criminal code (Individuals cannot be held responsible for criminal actions in Italy until they reach age 14.). When an individual over age 14 commits a criminal act, he or she is treated procedurally just like an adult except for the purposes of punishment. In other words, all due process rights afforded adults are also afforded juveniles. Reichel sees the welfare and legalistic models as opposites in which the problem with the former is the lack of legal protections and the problem with the latter is it lacks compassion and flexibility.

Reichel sees the *corporatist* model, exemplified by England, as being a compromise between the extremes of the welfare and legalistic models. This model seeks the middle ground between realizing the lack of maturity of juveniles and thus wanting to "treat" them, and at the same time asserting the need to hold them responsible for their actions, and thus wanting to "punish" them. Juvenile justice in the UK is primarily the responsibility of Youth Justice Boards (YJBs), which are locally managed social institutions set up in 1998 that operate semi-autonomously from the government. There is a great reliance on restorative justice programs in YJBs, but every dispositional option available in the United States, including incarceration, is available in the UK. Despite huge sums being poured into juvenile crime prevention in the UK, a 10-year evaluation of the progress made by YJBs found that they have had zero to minimal impact on youth offending (E. Solomon & Garside, 2009).

Whereas the welfare, legalistic, and corporatist models are all examples of efforts to control antisocial behavior through *formal* social control, the final model—the *participatory* model—seeks to control behavior more via informal social control. This model assumes that reform is best achieved if youthful offenders are dealt with outside the formal court system, and enlists the aid of family, school, and various neighborhood "committees" to control juvenile behavior. This system is popular in communist/socialist countries that (theoretically) see no division between the state and the people. Such a model would obviously work best in traditional societies in which there was relatively little geographic mobility and relatively little racial/ethnic diversity. In other words, it would work best in a community in which everyone more or less knew everyone else and had for a long time, and in which everyone held the same values and attitudes. It does not, therefore, seem like a model that would work in modern Western societies.

❖ Residential and Institutional Juvenile Corrections

One of the dispositional sanctions that can be imposed on adjudicated delinquents is placement in some sort of program in a residential facility. A *residential facility* or *residential treatment center* is not analogous to an adult prison, but rather more like a halfway house. Boot camps, discussed in the chapter on adult probation, are one example of such facilities. Another one is *wilderness* or *survival programs.* These are more self-discipline programs designed to test delinquents' characters and coping skill by providing them with structured challenges. Overcoming these challenges is said to build youth's confidence and self-esteem and show them they are capable and not simply victims of circumstance. In such programs, there are no drill instructors bawling and spitting in their faces and belittling them; rather, there are guides who set adventurous challenges for them and provide encouragement. Wilderness programs do seem to be better than boot camps at reducing recidivism, although it is probably true that on the whole, fewer serious offenders are assigned to wilderness programs than to boot camps. Nevertheless, one review of 22 studies of wilderness effectiveness found that wilderness participants recidivated at a rate of 29% versus 37% for comparison subjects (S. Wilson & Lipsey, 2000).

Photo 12.3

Juveniles in the cafeteria of a secure detention facility.

Another alternative is a group home. *Group homes* are typically operated by private organizations that contract with juvenile authorities. These group homes tend to specialize in some form of programming such as drug treatment or treatment for "troubled girls." Youth in these homes remain in their communities and attend school and all the other normal functions of school-age children, but live with perhaps 10 to 30 other youth at the home.

A commitment to a juvenile institutional corrections facility is a serious matter and is typically the disposition reserved for juveniles who have committed violent offenses or for chronic repeat offenders. There are two broad categories of institutional correctional facilities: long-term and short-term. Short-term facilities include reception and detention centers (the equivalent of adult jails) where children may be held while awaiting release to parents or court adjudication, or youth shelters. Long-term facilities are those used for housing juveniles after adjudication. They include secure detention centers/training schools (the equivalent of an adult prison) and boot camps, and less secure youth centers/ranches and adventure forestry camps.

Juveniles sent to long-term secure correctional facilities tend to have committed very serious delinquent acts or be chronic offenders. A study of juveniles sent to long-term secure facilities found that 35% were committed for violent offenses and the remaining 65% for property, drug, or status offenses (Sickmund & Wan, 2003). Minority youths are even more overrepresented in secure juvenile correctional facilities than minority adults are in adult prisons. Gus Martin (2005) reports that whereas there are about 204 white juveniles per 100,000 in secure facilities, there are 1,018 per 100,000 African American juveniles, and that about 70% held in custody for violent offenses are minorities (p. 247).

Steiner and Giacomazzi (2007) examined recidivism among juveniles waived to adult court and placed into a boot camp program compared with a control group of juveniles who were also waived to adult court but placed on probation rather than in a boot camp. They found no difference between the boot camp and control groups on rates of recidivism, but boot camp juveniles were significantly less likely to be reconvicted, which may be one bright spot in an otherwise dark performance of boot camps.

Other important differences between juvenile and adult facilities are that juvenile facilities are almost always much smaller (rarely more than 250 juveniles), that the costs associated with incarceration are considerably higher, and much more money is spent on programming relative to security (Taylor et al., 2007). For instance, the California Youth Authority spends 52% of its budget on academic and vocational training, case planning, counseling, and skills training as opposed to only 13% on custody and security (Taylor et al., 2007). Nevertheless, many of the same problems seen in adult prisons are also seen in juvenile facilities, especially in the larger institutions with a low staff-to-resident ratio. As in adult prisons, gangs form along racial/ethnic and neighborhood lines, and there is always the danger of violence and sexual assault against the unaffiliated (G. Martin, 2005).

Summary

- The juvenile justice system in the United States is based on civil law and deals with status offenses (those applicable only to juveniles) and delinquency (crimes if committed by adults).

- Juveniles commit a disproportionate number of both property and violent crimes, and this has always been true across time and cultures. Recent scientific evidence relates this situation to the hormonal surges of puberty and a brain undergoing numerous changes. Although most adolescents commit antisocial acts, only a small proportion continue to do so after brain maturation is completed.

- The history of juvenile justice has three distinct periods. Originally, Western culture relied heavily on parents to control children. As society has changed, so have the expectations regarding juvenile delinquency. Institutional control of wayward youth was the model from the mid-1500s until the inception of the juvenile courts in the United States in the late 1800s/early 1900s. The juvenile court follows the doctrine of *parens patriae,* but recently there has been a movement away from the broad discretion formerly accorded to juvenile courts to a model more closely reflecting the constitutional protections afforded adult offenders. Much of this change has issued from the increased waivers of juveniles to adult courts and from the often arbitrary control juvenile justice authorities have exercised over juveniles.

■ Much of what constitutes juvenile corrections mirrors what we have written about in other sections in this book; thus we have only briefly highlighted differences between the juvenile and adult systems. Major differences include a greater emphasis on rehabilitation as exemplified by the ratio of programming to security spent in juvenile correctional facilities and the lesser likelihood of juveniles being sent to secure facilities relative to adults.

Key Terms

Balanced approach

Bridewells

Community service order

Deferred adjudication

Delinquents

Intensive supervision probation (ISP)

Judicial waiver

Parens patriae

Predisposition reports

Prosecutorial discretion

Restorative justice

Status offenses

Statutory exclusion

Waiver

Discussion Questions

1. Discuss the development of the concept of childhood in Western culture.

2. Discuss the doctrine of *parens patriae* in relation to the development of the juvenile court system in the United States.

3. Do you think that restorative justice is workable? In what circumstances would it be or not be?

4. Which of the models of juvenile justice outlined by Reichel do you favor? Give your reasons.

5. Do you think that a highly dangerous person such as Terrance Jamar Graham should ever be released back into the community just because he committed his crimes as a juvenile?

Useful Internet Site

The Office *of* Juvenile Justice and Delinquency Prevention (offers many wonderful publications pertinent to the juvenile justice system): www.ojjdp .gov/

CHAPTER 13

Legal Issues in Corrections

❖ Introduction: Historical Background

Prisons are not nice places, but they were never meant to be. This does not mean that society is ever justified in treating prisoners in less than humane ways. Winston Churchill (and perhaps many others before him) once said that a civilization could be judged by the way it treated its prisoners (Morris, 2002). We do not treat them very well and never have, but many members of the public see them getting better treatment than they deserve, as summed up in the line, "If you can't do the time, don't do the crime." Convicted criminals are perhaps the most despised group of people in any society since they are generally viewed as evil misfits who prey on decent people, so why worry about what happens to them while in custody, paying their debt to society? With an estimated 1 in every 100 adults in jails or prisons in the United States in 2008 (Robertson, 2010), however, issues relating to how these people are treated while incarcerated becomes more pressing. The only system capable of monitoring the treatment of people under criminal justice supervision, other than the criminal justice system itself, is the court system. The courts have changed their attitudes toward the treatment of prisoners over the history of the United States. These attitudes have ranged from complete indifference to almost attempting to micromanage a state's entire prison system. In this chapter, we review how inmates in correctional institutions were legally viewed in the past up to how they are legally protected today. Though much as changed in this regard, the trend in recent years has been to remove some of those legal protections and access to courts that inmates gained in the latter half of the 20th century.

The Hands-Off Period: 1866–1963 and Prisoners as Slaves of the State

The general public's attitude of indifference has also been the general attitude of the courts throughout much of American history, as is evident in their so-called **hands-off doctrine**. This doctrine basically articulated the reluctance of the judiciary to interfere with the management and administration of prisons—to keep their "hand off." The doctrine rested primarily on the status of prisoners who suffered a kind of legal and civil death upon conviction. Most states had **civil death statutes**, which meant that those convicted of crimes lost all citizenship rights such as the right to vote, hold public office, and—in some jurisdictions—the right to marry. In affirming the ruling of a lower court in *Ruffin v. Commonwealth* (1871), the Virginia Supreme Court made it plain what the status of a convicted offender was:

> For the time being, during his [the convict's] term of service in the penitentiary, he is in a state of penal servitude to the State. He has, as a consequence of his crime, not only forfeited his liberty, but all his personal rights except those which the law in its humanity accords to him. He is for the time being the slave of the State. He is *civiliter mortuus* [civilly dead]; and his estate, if he has any, is administered like that of a dead man. (n.p.)

Ruffin was a state case, and thus not binding on other states, but the case was consistent with the earlier U.S. Supreme Court case *Pervear v. Massachusetts* (1866). It is *Pervear* rather than *Ruffin* that first clearly enunciated the lack of concern for prisoners' rights contained in the hands-off doctrine. In this case, the Supreme Court made plain the slave-like status of prisoners, ruling that they did not even enjoy the protections of the Eighth Amendment, which forbids cruel and unusual punishment. Convicted felons thus found themselves totally at the mercy of prison officials and fellow prisoners, without any kind of constitutional protection provided by judicial oversight.

The hands-off doctrine also prevailed because the courts viewed correctional agencies as part of the executive branch of government and did not wish to violate the Constitution's separation of powers doctrine. Correctional officials were considered quite capable of administering to the needs of prisoners in a humane way, without having to deal with the complicating intrusions of another branch of government. Besides, if prisoners have been stripped of any rights under civil death statutes, there is nothing that the courts have to monitor and protect.

The Prisoners' Rights Period: 1964–1978

As part of a growing trend toward an overall greater respect for individual rights in the mid-20th century, as African Americans, women, gays and lesbians, and other disadvantaged groups strongly agitated for them, the courts began to enter into the area of prisoners' rights. The major issue in prisoner litigation has been the conditions of confinement, but the first significant case was *Ex parte Hull* (1941), which dealt with the denial by prison officials of a Michigan inmate's petition for an appeal of the legality of his confinement (the term *ex parte* refers to situations in which only one party appears before the court). Although Hull's petition was denied (Hull had committed a statutory sexual offense, violated his parole, and was returned to prison to serve out his original sentence), the United States Supreme Court ruled that inmates had the right to unrestricted access to federal courts to challenge the legality of their confinement. This ruling was the beginning of the end of the hands-off doctrine, although as we shall see there is growing evidence of its return (Federman, 2004).

The technical term for a challenge to the legality of confinement is a writ of habeas corpus. **Habeas corpus** is a Latin term that literally means "you have the body," and is basically a court order requiring that an arrested person be brought before it to determine the legality of his or her detention. Habeas corpus is a very important concept in common law, which precedes even the Magna Carta of 1215, although its precise origins are unknown. It has been called the "Great Writ" and was formally codified into English common law by the Habeas Corpus Act of 1679. Indicative of the respect in which habeas corpus was held by the Founding Fathers is that it is only one of three individual rights mentioned in the United States Constitution (the other two are the prohibition of bills of attainder—imposing punishment without trial—and the prohibition of ex post facto laws—legislation making some acts criminal after the fact). The other individual rights that Americans enjoy were formalized in the first 10 amendments to the Constitution (the **Bill of Rights**), almost as an afterthought. A writ of habeas corpus is not a direct appeal of a conviction, but rather an indirect appeal regarding the legality of a person's confinement or the conditions of confinement.

Two cases signaled the end of the hands-off period: *Jones v. Cunningham* (1963) and *Cooper v. Pate* (1964). In *Jones,* the Supreme Court went further than it did in *Hull* and ruled that prisoners could use a writ of habeas corpus to challenge the conditions of their confinement as well as the legality of their confinement. This went beyond the original meaning of habeas corpus, which was only meant to address the pre-conviction issue of the legality of a petitioner's detainment. In *Cooper,* the Court went even further and ruled that state prison inmates could sue state officials in federal courts under the Civil Rights Act of 1871, which was initially enacted to protect Southern blacks from state officials. This act is now codified and known as 42 USC § 1983, or simply as "section 1983," and any deprivation of rights grievances filed under it is called a **civil rights claim**. The relevant part of the act reads as follows:

> Every person who under color of any statute, ordinance, regulation, custom, or usage of any state or territory, subjects or causes to be subject, any citizen of the United States or other person within the jurisdiction thereof to the deprivation of any rights, privileges, or immunities secured by the Constitution and laws, shall be liable to the party injured in an action at law. (n.p.)

What was a mere trickle of habeas petitions before *Pate* quickly became a flood that threatened to drown the federal courts with grievances ranging from the petty to the deadly serious. The most serious petition led to a federal appeals judge declaring the entire prison system of Arkansas unconstitutional and a "dark and evil world" when he placed it under federal supervision (*Holt v. Sarver,* 1969). This case involved the torture of inmates and severe deprivation of basic rights and gave birth to what has come to be known as a "conditions of confinement lawsuit." From then on, the federal courts became very much involved in the monitoring and operation of entire prison systems. The vast majority of habeas corpus grievances filed today are about the conditions of confinement, not the legality of an inmate's confinement. Inmates filing a petition challenging their confinement face an uphill battle because the state's defense against such a claim is based on inmates' convictions, which is the obvious legal basis for their confinement!

The Deference Period: 1997–Present

Convicted criminals are no longer considered civilly dead while under correctional supervision. However, after a short time in which prisoners' rights were granted and extended, there began an era that correctional scholars have called the *deference period*. This time is a partial return to the hands-off period, and basically refers to the courts' willingness to defer to the expertise and needs of prison authorities. Some courts have come to the conclusion that it is necessary to place restrictions on prisoners' rights that do not apply to non-offenders because of the need to balance the rights of offenders and the legitimate needs and concerns of correctional authorities, particularly the safety and security needs of prisons and jails. The basic stance of the U.S. Supreme Court on this matter has not changed since it was first enunciated in *Bell v. Wolfish* (1979), the case widely considered to be the one that signaled the onset of the deference period:

> Simply because prison inmates retain certain constitutional rights does not mean that these rights are not subject to restrictions and limitations. There must be a "mutual accommodation between institutional needs and objectives and the provisions of the Constitution that are of general application." Maintaining institutional security and preserving internal order and discipline are essential goals that may require limitation or retraction of the retained constitutional rights of both convicted prisoners and pretrial detainees. (n.p.)

We begin by looking at certain fundamental rights guaranteed by the First, Fourth, and Fourteenth Amendments to the Constitution and how they apply to convicted felons.

❖ First Amendment

The **First Amendment** guarantees freedom of religion, speech, press, and assembly, but in a prison setting these freedoms do not extend to activities and materials that jeopardize prison safety or security. The *Cooper v. Pate* (1964) case discussed earlier was essentially a First Amendment issue because Cooper, a Black Muslim, alleged that he was denied certain religious publications solely on the basis of his religion. Prison authorities claimed that Black Muslim literature was dangerous and jeopardized the safety and security of the prison because it preached violent revolution and sought to recruit new members. The Supreme Court acknowledged that such literature may have an incendiary effect, but ruled that Cooper's right to free exercise of his religion trumps what *might* result in security problems.

Religious freedom cannot extend to demanding alcohol or exotic foods to satisfy real or invented religious requirements. But the law is more an ideological exercise than a science. As if to prove our point, two federal circuit courts came to opposite conclusions in the same year (2008) regarding Muslim inmates' right to a *halal* diet (a strict diet in which only the meat of ritually slaughtered animals, excluding pigs, can be eaten). The Eighth Circuit Court ruled that the non-provision of a *halal* diet did not place an undue burden on the inmate and that there were alternative means of obtaining such a diet. The more liberal Ninth Circuit Court ruled in a different case that the prison's refusal to supply a Muslim inmate with a *halal* diet impinged on the free exercise of religion clause of the First Amendment (Robertson, 2010).

Photo 13.1

An inmate attending a prison church service. Inmates retain some First Amendment rights, including the right to practice their religious beliefs if doing so does not unduly interfere with prison security.

Restrictions on inmates' rights to free speech can exceed those necessary to ensure safety and security. In *Smith v. Mosley* (2008), the plaintiff made a statement in a grievance that prison authorities saw as insubordinate and false for which he received disciplinary sanctions. The inmate sought relief in the federal court claiming that he had been punished for exercising his right to free speech. The Court disagreed, ruling that while filing a grievance is considered protected speech, the statements made within it were not, and therefore the imposition of sanctions was constitutionally permissible. Freedom of speech or expression can also be limited on moral or ethical grounds. For instance, inmates can write and publish their thoughts or sell personal memorabilia, but "notoriety-for-profit" statutes enacted by the federal government and most states forbid inmates from profiting monetarily from those activities (Walsh & Hemmens, 2011).

The right of assembly allows for attendance at religious services and for visitation from family and friends, but it obviously cannot be construed as allowing inmates to assemble at a tattoo conference outside the prison walls. Federal courts have also ruled that while Black Muslim groups had the right to assemble for worship, their right to hold religious services could be denied if prison administrators considered such services to constitute potential breaches of security (Inciardi, 2007).

❖ Fourth Amendment

The **Fourth Amendment** guarantees the right to be free from unreasonable searches and seizures. What is reasonable inside prison walls is, of course, quite different from what is reasonable outside them. For all practical purposes, inmates have no Fourth Amendment protections since their prison cells are not "homes" of personal sanctuary deserving of privacy (*Hudson v. Palmer,* 1984). The one area in which Fourth Amendment rights have not been completely extinguished for inmates is that involving opposite-sex body searches. The courts have had to wrestle with conflicting claims on this issue. One is the equal employment claim of female corrections officers who want to work in male institutions where, because of their size and scope, promotion prospects are greater than they are in female prisons. Working in all-male prisons necessarily means that women officers will occasionally view inmates undressed or using toilet facilities, and sometimes they may be required to perform pat downs and visual body cavity searches (Physical searches of body cavities may only be performed by medical personnel.). A frequent inmate claim is that cross-gender searches are "unreasonable" within the meaning of the Fourth Amendment.

Bennett (1995) notes that the great majority of cross-gender search complaints are filed by males, which is not surprising since males constitute about 94% of all state prison inmates (Bohm & Haley, 2007). On the other hand, it is surprising, given the frequent complaints about female officers, that some male inmates seem to take every opportunity they can to behave in a sexual manner in the presence of female officers (Cowburn, 1998). Of course, this does not mean that many inmates are not genuinely embarrassed and offended by having to bear the indignity of female officers observing them using the toilet. However, ever since the doors were opened in *Hull,* the filing of all sorts of complaints has become a sort of inmate hobby for some that serves the purpose of relieving boredom, getting "one over" on prison authorities and possibly a ride or two into town to attend court (McNeese, 2010). There are many legitimate prisoner complaints, but Dilworth (1995) reports that 75% of prisoner petitions are dismissed by the court's own evaluation, 20% are dismissed in a grant of a state's motion, and about 2% result in a trial in which less than half result in a favorable verdict for the prisoner. Thus, less than 1% of prisoner complaints are considered legitimate. Examples of some frivolous and malicious petitions are given in In Focus 13.1.

In *Turner v. Safley* (1987), the Supreme Court enunciated what has come to be known as the *balancing test,* which means that the courts must balance the rights of inmates against the interests of penological concerns of security and order. In ruling that lower courts were wrong in applying the strict scrutiny standard of review (a standard of review used by the courts if a "fundamental right"—anything in the Bill of Rights—or a "suspect classification"—race, religion, or national origin—is involved) to inmates' constitutional complaints, the Court ruled that these cases require a lesser standard and involve the issue of whether a prison regulation that impinges on inmates' constitutional rights is "reasonably related" to legitimate penological interests. This "reasonableness" revolves around a number of factors, including whether there is a valid and rational connection between the regulation and a legitimate government interest that is justified in the name of staff and inmate safety and security.

According to the balancing test, then, the viewing of opposite-sex inmates is constitutionally valid "if it is reasonably related to legitimate penal interests." Bennett (1995) tells us how lower courts have interpreted "reasonableness" and concludes that while female officers conducting or observing strip searches of male inmates are tolerated in emergency situations, similar observation and searches by male officers of female inmates are considered unreasonable. This double standard has been justified on two grounds: (1) Males do not experience loss of job opportunities if they are forbidden to frisk female inmates, and (2) intimate touching of a female inmate by a male officer may cause psychological trauma because many female inmates have histories of sexual abuse.

Farkas and Rand (1999) also support gender-specific standards for cross-gender searches on the basis of prior sexual abuse that an inmate may have suffered and state, "Cross-gender searches have the very real potential to replicate that suffering in prison" (p. 53), and further

In Focus 13.1

Any Complaints This Morning?

"Any complaints this morning?" was the drill sergeant's daily cynical question to newly drafted soldiers during World War II. Of course, none of the draftees confined in the sweltering barracks was ever bold enough to make any complaint, although living conditions were much less comfortable than they are in many of our prisons today. We wonder what these old soldiers would say about the following.

In dismissing a lawsuit in one case, the Supreme Court noted that the majority of prisoner petitions are frivolous and/or malicious, cost taxpayers millions of dollars, and waste precious court time. Florida Attorney General Bob Butterworth (1995) asserts that "[m]y office spends nearly $2 million a year defending that state against inmate suits, most of which contain ridiculous charges or demands" (p. 1). Similarly, Idaho Deputy Attorney General Timothy McNeese (2010) notes that fully 27% of all litigation in Idaho's Federal District Courts involve inmate petitions, most of which are "meritless," "downright frivolous," and "no doubt are filed by inmates out of frustration and anger and desire to get even with a correctional employee or the 'system'" (p. 321). Presented below are some examples of frivolous and/or malicious petitions gleaned from Butterworth and McNeese that judges must wade through to get to prisoner petitions that are really deserving of attention. The sheer audacity of these examples may raise a smile on you face, but they are no laughing matter to the courts, prison administrators, the taxpayers, or the prisoners with real grievances whose petitions are lost in the pile:

- Prisoner starts a riot, shatters glass in his cell, and files an Eighth Amendment suit claiming cruel and unusual punishment because he cut his foot on the glass.
- Prisoner sends for information about prison security and locks and sues because the warden refused to give him the mail containing the information.
- Prisoner sues because his ice cream was half melted when it was served to him.
- Prisoner sues over unsatisfactory haircut.
- Prisoner sues because jailers cut her sausage into small pieces because she had been caught previously masturbating with a whole sausage.
- Prisoner sues because he was required to eat off a paper plate.
- Prisoner files over 140 actions in state and federal court over finding gristle in his turkey leg.
- Prisoner sues to receive fruit juice at meals and an extra pancake at breakfast.
- Prisoner who murdered five people sues because he had to watch network TV programs after lightning knocked out the prison's satellite dish. These programs contained violence and other material that this multiple murderer said was objectionable.
- Prisoner sues over the inferior brand of sneakers issued to him.
- Prisoner loses a suit claiming his rights as a Muslim were violated because the prison put "essence of swine" in his food, then converts to Satanism and demands tarot cards and "doves' blood."
- Prisoner sues because the disciplinary cell he was placed in had no electrical outlet for his TV.

abuse would constitute cruel and unusual punishment. This raises the question of the possible legal validity of complaints about same-sex body searches if the complainant can show prior sexual abuse by a same-sex person. For instance, will such a person (or gays and lesbians) then be in a position to demand an opposite-sex body search?

❖ Eighth Amendment

Cruel and unusual punishment is forbidden by the **Eighth Amendment**. What constitutes cruel and unusual punishment is punishment applied "maliciously and sadistically for the very purpose of causing harm" (*Hudson v. McMillian,* 1992), although the inmate is responsible for proving that the punishment was so applied. Eighth Amendment protections are denied not only by prison officials doing something to inmates that they should not, but also if they fail to do something that they have a duty to do. Prison officials must provide inmates with the basic amenities of life such as food and medical attention, and they must provide them protection from the physical and sexual predations of other inmates, many of whom have histories of "maliciously and sadistically" causing harm to others.

Liability attaches to prison officials for inmate–inmate assaults if officials display *deliberate indifference* to an inmate's needs (Vaughn & Del Carmen, 1995). The courts have struggled to make plain what deliberate indifference means, but basically it occurs when prison officials know of, but disregard, an obvious risk to an inmate's health or safety (*Wilson v. Seiter,* 1991). In other words, prison officials must not turn a blind eye to situations that obviously imperil the health or safety of inmates entrusted to their care. Purposely placing a slightly built and effeminate young male in a cell with a known aggressive sexual predator is an example of a violation of the deliberate indifference standard for which prison authorities would be liable for any injuries suffered. For inmates to prevail in suits involving deliberate indifference claims, they must prove that (1) they suffered an objectively serious deprivation or harm, and (2) prison officials were aware of the risk that caused the alleged harm and they failed to take reasonable steps to prevent it. *Wilson* is seen as a key decision favoring correctional agencies because of these stringent proof requirements.

Despite movie depictions and public perceptions of jails and prisons as places where sexual assault is rife, the rate of sexual assault in prisons is 0.40 per 1,000 inmates, which is much lower than sexual violence rates on the outside (Beck & Harrison, 2007). The forcible rape rate in 2008 in the free community was at 0.69 per 1,000 females (FBI, 2009), and this does not include all other sexual misconduct offenses (nor rapes of males) of the kind included in the Beck and Harrison report. The Beck and Harrison report also noted that 38% of the allegations involved staff misconduct, sometimes referred to as "romance" between staff and inmate. The majority of forceful sexual assault was inmate on inmate (Robertson, 2010).

Inmate medical care is also covered by the concept of deliberate indifference. Cohen (2008) writes,

> Our jails and prisons have increasingly become the de facto clinical depositories for hundreds of thousands of inmates who are very sick and who require all manner of specialty medical, dental, and mental health care. Prisons are not only the new mental asylums; they are the community hospitals and emergency wards for certain segments of the poor. (p. 5)

The medical needs of inmates in today's prisons are as well addressed as those of the average free person of roughly similar class background presenting with similar health problems. Indeed, inmates are the only group of people in the United States with a constitutional right to medical care. According to a Bureau of Justice Statistics report on inmate mortality in state prisons, prisoners between 15 and 64 years of age had a mortality rate 19% lower than that of the U.S. general population (Mumola, 2007). The majority of the difference was attributable to African American male inmates under 45 who had a mortality rate 57% lower than the rate of black males of similar age in the general population. Of course, not all of this difference is attributable to the medical care inmates receive, and no one claims that such care is better than, or even equal to, the average level of medical care available to most people on the outside. The lower mortality rate is most likely due to the fact that incarceration lowers the probability of being murdered, being exposed to drugs, and having access to alcohol and tobacco and other risks to a person's health and safety that those who pursue a criminal lifestyle face every day on the outside.

❖ Fourteenth Amendment

The legal basis for granting limited procedural rights to individuals under correctional supervision is the **Fourteenth Amendment**, which reads in part, "No state shall make or enforce any law which shall abridge the privileges and immunities of citizens of the United States; nor shall any state deprive any person of life, liberty, or property, without due process of law." The due process clause of the Fourteenth Amendment was first applied to inmates facing disciplinary action for infractions of prison rules in *Wolff v. McDonnell* (1974). In *Wolff*, the Supreme Court declared that while inmates are not entitled to the same due process rights as an accused but unconvicted person on the outside, they are entitled to some. These rights are (1) to receive written notice of an alleged infraction, (2) to be given sufficient time (usually 24 hours) to prepare a defense, (3) to have time to produce evidence and witnesses on their behalf, (4) to have the assistance of non-legal counsel, and (5) to have a written statement outlining the disciplinary committee's findings.

In *Sandin v. Connor* (1995), the Supreme Court clarified and trimmed back inmate rights. The Court declared that the above due process rights are only triggered by any disciplinary action that may result in the loss of "good time," which amounts to an extension of an inmate's sentence. Conner had been given 30 days of punitive segregation for using foul and abusive comments to an officer while being subjected to a strip search. The Court ruled that due process rights are not triggered by actions that result in temporary placement in a disciplinary segregation unit, which does not amount to an extension of sentence. The Court also concluded that disciplinary segregation is not an atypical hardship relative to the ordinary hardships of imprisonment.

❖ The Civil Commitment of Sex Offenders

The idea that people who engage in socially disapproved behavior are sick and require treatment is seen as particularly applicable to sex offenders. In 1997, the Supreme Court upheld Kansas's Sexually Violent Predator Act (SVPA), which allowed for keeping sex offenders in state custody under civil commitment laws *after* they have served their full prison terms if they demonstrate "mental abnormality" or are said to have a "personality disorder" (*Kansas v. Hendricks,* 1997). The federal government and many other states have since passed similar involuntary confinement laws for sex offenders. Prior to *Hendricks*, civil commitments were limited to individuals suffering from mental illness, but seemingly for the express purpose of covering sex offenders, states have loosened the mental illness criterion in favor of the "mental abnormality" criterion (R. Alexander, 2004). The term *mental abnormality* can of course be used to cover almost anything society may disapprove of and serves as a justification for imprisonment in ways "reminiscent of Soviet policies that institutionalized dissidents" (Grinfield, 2005, p. 2). There is no doubt that Leroy Hendricks was a repeat predatory pedophile and a thoroughly nasty piece of work, but while many applaud his incapacitation, what concerns civil libertarians is that the *Hendricks* decision created a special category of individuals defined as "abnormal" who may be punished indefinitely for what they *might do* if released.

Hendricks appealed his confinement on double jeopardy (no person can be prosecuted twice for the same offense) and ex post facto (a person cannot be punished for acts that were not crimes at the time he or she committed them) grounds. The Supreme Court ruled that Hendricks's confinement was not double jeopardy because the commitment proceedings of the SVPA under which he was confined were civil rather than criminal, and thus did not constitute a second prosecution. Likewise, Hendricks's ex post facto claim was rejected because the ex post facto clause of the U.S. Constitution relates only to criminal statutes. The Court declared that because Hendricks was committed under a civil act, his commitment did not constitute punishment. The majority of Supreme Court justices seemed to have believed that putting a different name on state confinement against the will of the confined changes how that person experiences it.

In his analysis of civil commitment of sex offenders, R. Alexander (2004) lists a number of ways in which he believes the Supreme Court erred in its rulings in *Hendricks* and subsequent cases dealing with the same issue. The gist of most of these criticisms is that the ruling offends (Alexander's) ideas of social justice and does not accomplish any of the goals of state confinement (deterrence and rehabilitation) other than retribution and incapacitation. Farkas and Stichman (2002) also argue that what they call the "culture of fear" generated by atypically brutal sex offenses has resulted in laws that are constitutionally questionable and that have negative consequences for the treatment of sex offenders, the criminal justice system, and society in general. The popularity of the laws with the general public, and the extremely negative view of sex offenders that it holds, makes it unlikely that these laws will be changed in the near future. It should be noted that only between 1% and 2% percent of sex offenders are confined under civil commitment orders (R. Alexander, 2004).

❖ Curtailing Prisoner Petitions

According to Federman (2004), two congressional acts signed into law in 1996—the Prison Litigation Reform Act (PLRA) and the Antiterrorism and Effective Death Penalty Act (AEDPA)—have severely curtailed prisoner access to the courts. Both acts were passed in part to reduce the thousands of lawsuits filed by inmates that clog the federal courts. In the year of the passage of these acts, inmates filed 68,235 civil rights lawsuits in the federal courts compared to less than 2,000 in the early 1960s (Alvarado, 2009). By 2000, the number of such lawsuits dropped to 24,519 (Seiter, 2005), a 64% decrease from the 1996 figure. Despite claims that these acts are "silencing the cells" (Vogel, 2004), the number of lawsuits filed is still more than 12 times what it was in the 1960s. Figure 13.1 is from a Bureau of Justice Statistics report (Scalia, 2002). The report indicates that civil rights petitions fell from 37 to 19 per 1,000 inmates from 1965 to 2000. However, it is noted that habeas corpus petitions actually increased slightly from 19 to 23 per 1,000 inmates. Inmates in New Mexico prisons filed the most petitions (77 per 1,000) while Alaska and Utah inmates filed the least (6 per 1,000 each).

The primary intention of the PLRA was to free prisons and jails from federal court supervision as well as limit prisoners' access to the federal courts. Both intentions have largely succeeded. Among the requirements of the PLRA is one that state inmates cannot bring a **section 1983 suit** (civil rights lawsuit) in federal court unless they first exhaust all available administrative remedies such as filing a written grievance with the warden. The PLRA also states that inmates claiming to be unable to afford the required filing fee for the lawsuit may still have to pay a partial fee, which will be collected whenever money appears in their inmate accounts. This provision may limit the airing of genuine grievances in federal court because of financial difficulties.

As the name implies, the AEDPA is mostly about antiterrorism and the death penalty rather than an act specifically designed to limit habeas corpus proceedings. It was passed in response to the bombing of the Murrah Federal Building in Oklahoma City, with the reform of habeas corpus law being a rider to it. The AEDPA does not eliminate inmates' rights to habeas corpus, but it does restrict its availability (Alvarado, 2009). It does so by limiting successive petitions and judicial review of evidence, and may now apply only to inmates who have sought, but have been denied, state court remedies available to them. The AEDPA thus takes habeas corpus partially back along the road to again becoming the pre-conviction remedy against unlawful imprisonment that it was initially.

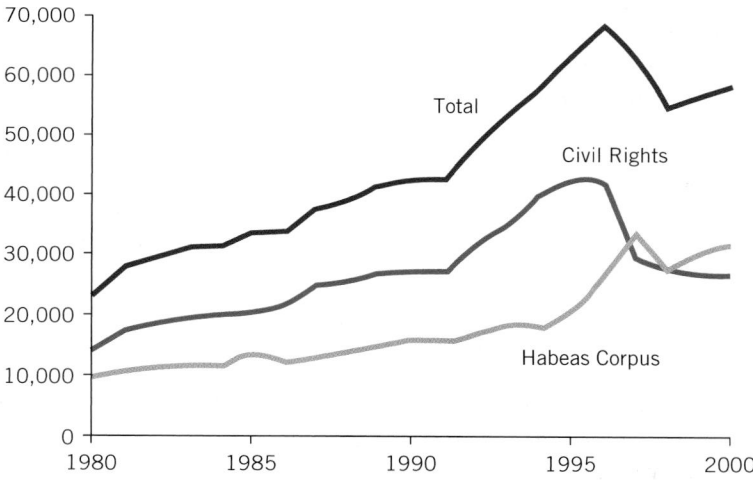

From 1995 to 2000 The Number of Civil Rights Petitions Filed by Prison Inmates Decreased 39%, as the Number of Habeas Corpus Petitions Increased 50%

Figure 13.1

Highlights of Study Assessing Changes in Prisoners' Petitions, 1980–2000

Source: Scalia (2002).

- During 2000, 58,257 prisoner petitions were filed in U.S. district courts—80% by State prison inmates and 20% by Federal inmates.

- The majority of petitions filed during 2000 were habeas corpus petitions (43%) or petitions by Federal inmates challenging the constitutionality of an imposed sentence (11%); 44% alleged civil rights violations; and 2% were mandamus actions.

- The 1996 Prison Litigation Reform Act appears to have resulted in a decrease in the number of civil rights petitions filed by State and Federal prison inmates. They filed 41,679 petitions during 1995 compared to 25,504 during 2000.

- Between 1995 and 2000 the rate at which Federal and State prison inmates filed civil rights petitions decreased from 37 to 19 per 1,000 inmates.

- The 1996 Antiterrorism and Effective Death Penalty Act appears to have resulted in an increase in the number of habeas corpus petitions filed by State prison inmates. State prison inmates filed 50% more habeas corpus petitions during 2000 (21,345) than during 1995 (13,627).

- Between 1995 and 2000 the rate at which State prison inmates field habeas corpus petitions increased from 13 to 17 per 1,000 inmates.

A review of Supreme Court decisions on habeas corpus since the AEDPA found that they have upheld the reforms largely as intended by Congress (Scheidegger, 2006). Although proponents of the statutes argue that they free up the federal courts and save the taxpayer literally millions of dollars, critics are concerned that states may return to the "bad old days" of decrepit institutions and abusive staff.

❖ The Death Penalty: Legal Challenges to the Ultimate Sanction

Given the emotional and philosophical issues surrounding the death penalty and particularly its finality (there is no way to put right an execution of someone later found to be innocent), it is understandable that the death penalty has been subjected to intense legal scrutiny in recent times. This has not always been the case, however. Throughout much of our history, the death penalty has been considered a legitimate, appropriate, and necessary form of punishment, and a clear majority of the American public still favors its retention, although if life without the possibility of parole is provided as an option, the percentage favoring the death penalty falls to about 52 (Radelet & Borg, 2000).

Photo 13.2

The electric chair

Legal challenges to the death penalty have revolved around the Eighth Amendment's prohibition of cruel and unusual punishment—usually about the constitutionality of the method of execution, not the penalty per se. The first case to challenge the penalty itself was *Furman v. Georgia* (1972). Furman had shot and killed a homeowner during the course of a burglary and was sentenced to death. Furman challenged the constitutionality of his sentence, and the Supreme Court in a 5-to-4 vote agreed. The Court decided that the death penalty per se was not unconstitutional, but rather the arbitrary and discriminatory way in which it was imposed was. The Court argued that because the death penalty is so infrequently imposed, it serves no useful purpose, and that when it is imposed, judges and juries have unbridled discretion in making life versus death decisions.

Because the way the death penalty decision, rather than the penalty itself, was found to be unconstitutional, states began the process of changing their sentencing procedures. Some states introduced bifurcated (two-step) hearings, the first to determine guilt (the trial), and the second to impose the sentence after hearing aggravating and mitigating circumstances to determine if death was warranted. Other states removed sentencing discretion (since this seemed to be the Supreme Court's problem with it) and made the death penalty mandatory for certain types of murder.

Georgia revised its statute and opted for the bifurcated hearing. Using this process, Troy Gregg was sentenced to death for two counts of murder and two counts of armed robbery.

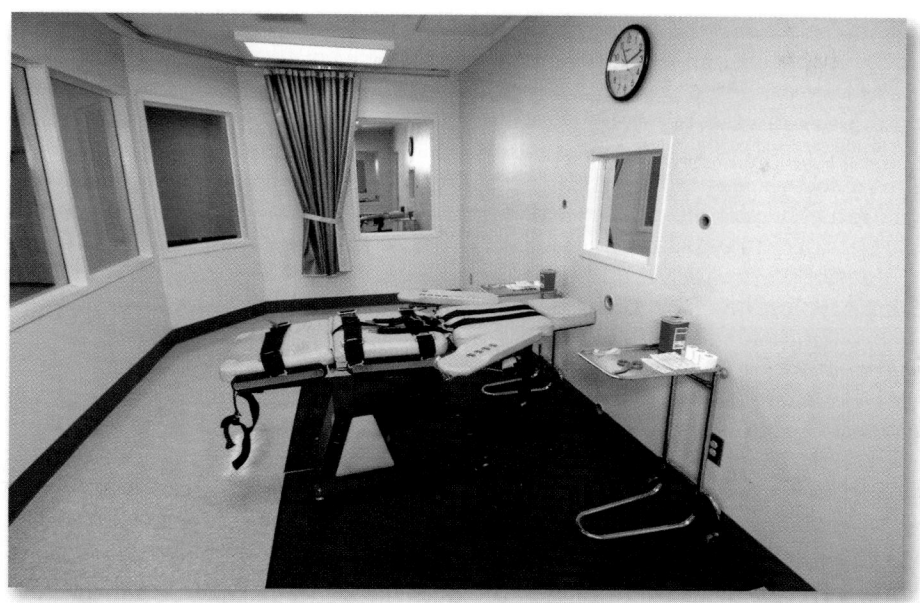

Photo 13.3

The execution chamber in the Walls Unit in Huntsville, Texas. Texas has executed more than 460 inmates since the death penalty was reinstated in 1982.

In *Gregg v. Georgia* (1976), the Supreme Court upheld the constitutionality of the bifurcated hearing and thus of Gregg's death sentence. On the same day, the Court also decided against mandatory death sentences in *Woodson v. North Carolina* (1976). The Court rejected as excessive and unduly rigid the North Carolina statute that mandated that all persons convicted of first-degree murder (Woodson had killed a convenience store cashier and seriously wounded a customer in the course of an armed robbery) should receive the death penalty.

In *Coker v. Georgia* (1977), the Supreme Court struck down the death penalty for rape. Coker had escaped from prison where he was serving time for murder, rape, and kidnapping and promptly proceeded to commit another rape and kidnapping. Nevertheless, the Court struck down the Georgia statute authorizing death for rape under certain circumstances as "grossly disproportionate" and thus repugnant to the Eighth Amendment.

Another concern is whether the death penalty is applied in a racially discriminatory fashion. Warren McCleskey, an African American man, was sentenced to death for killing a white police officer in the course of a robbery. In challenging his sentence in *McCleskey v. Kemp* (1987), McCleskey offered as evidence a statistical study purporting to show that racial disparity existed in Georgia (the state where the crime occurred) in that defendants who killed white victims were more likely to be sentenced to death than defendants who murdered black victims. In ruling against McCleskey's claim, the Court ruled that the statistical risk indicated in the study represented averages and does not establish that an individual's death sentence violates the Eighth Amendment. In other words, a study of past cases indicating average outcomes in no way serves as evidence that McCleskey was denied due process individually.

Further challenges to the death penalty have involved the constitutionality of imposing it on mentally disabled persons and on juveniles. In 1979, paroled rapist Johnny Penry was sentenced to death for rape and murder in Texas. Penry appealed to the Supreme Court on Eighth Amendment grounds, claiming that the jury was not instructed that it could consider his low IQ (between 50 and 63) as mitigating evidence against imposing the death penalty. The Court held in *Penry v. Lynaugh* (1989) that the constitution does not prohibit the execution of a mentally disabled person.

The Court overruled itself in this regard in (*Atkins v. Virginia* (2002). The Court concluded that there was a national consensus against executing the mentally disabled, and that since they are less capable of evaluating the consequences of their crimes, they are less culpable than the average offender. The Court also noted that mentally disabled individuals are more prone to confess crimes that they did not commit and therefore more prone to wrongful execution. Six of the justices concluded that the overwhelming disapproval of the world community must be considered a relevant factor in determining the imposition of capital punishment on mentally disabled individuals.

In *Panetti v. Quarterman* (2007), the Supreme Court was confronted with the issue of the constitutionality of the execution of the mentally ill (not the mentally disabled as in *Atkins,* but rather those individuals suffering from recognized psychiatric problems of an organic nature such as schizophrenia). Panetti, who killed his mother-in-law and father-in-law in 1992, has a long history of mental illness for which he was taking medication. Despite his undisputed mental illness and claim of mental incompetence, he was found competent to stand trial and to waive legal counsel (he defended himself). While the Court recognized that the Eighth Amendment forbids the execution of the insane, it declined to overturn Panetti's death sentence or to offer guidance for setting up standards for determining mental competence. Instead, it ordered a stay of execution and remanded the case back to the Texas courts so that they could more fully evaluate Panetti's incompetence claim.

Comparative Perspective: The Death Penalty

The death penalty as the ultimate criminal sanction was universally applied until quite recently. Some countries retain the death penalty in principle but not in practice, and as it stands today just over half of the world's countries have abolished it (Dammer & Fairchild, 2006). According to Amnesty International, in the nations that retain and use the penalty, at least 2,390 people were executed in 25 countries and 8,864 people were sentenced to death in 52 countries in 2008. Amnesty International also points out that 95% of all known executions were carried out in just six countries: China, Iran, Saudi Arabia, the United States, Pakistan, and Iraq. These figures should be considered minimums in all listed countries besides the United States. The number of *reported* executions in each of these countries as compiled by Amnesty International is given below (rates per million population computed by the present authors):

Country	Number of Executions	Rate per Million
1. China	1,718	1.29
2. Iran	346	4.66
3. Saudi Arabia	102	3.97
4. United States	37	0.12
5. Pakistan	36	0.21
6. Iraq	34	1.11

What is notable about this listing is that all the countries except the United States are strongly authoritarian states; China is a one-party communist state, and Iran, Iraq, Pakistan, and Saudi Arabia are Islamic countries with (to varying extents) theocratic legal values. Note that in terms of rates (the only valid way to make comparisons). Iran and Saudi Arabia are by far the leaders in the proportion of their populations that they execute (although China is suspected of greatly underreporting its executions). For instance, the Iranian rate is almost 39 times the American rate. Under Islamic law (known as *Shari'a*—"the path to follow"), quite a number of offenses are eligible for the death penalty, including apostasy (converting from Islam to some other religion), adultery, and homosexual fornication; Chinese law allows the death penalty for over 70 different offenses (Walsh & Hemmens, 2011).

All modern secular democracies except Japan and the United States have abolished the death penalty. The Netherlands abolished the penalty in 1870, Italy in 1905, Sweden in 1921, New Zealand in 1941, the United Kingdom in 1964, and France in 1981 (Fairchild & Dammer, 2001).

The justices have also had to wrestle with the moral issue of imposing the death penalty on individuals who committed their crimes as juveniles. From 1973 to 2003, a total of 22 such offenders were executed in the United States (Streib, 2003). The death penalty has only been applied to juveniles who have murdered in particularly heinous and depraved ways.

In 1977, 16-year-old Monty Lee Eddings and several companions stole an automobile. The car was stopped by an Oklahoma Highway Patrol Officer, and when the officer approached the car, Eddings shot and killed him. At Eddings's sentencing hearing, the state presented three aggravating circumstances to warrant the death penalty, but the judge only allowed Eddings's age in mitigation. In *Eddings v. Oklahoma* (1982), the Supreme Court vacated Eddings's death sentence, ruling that in death penalty cases the courts must consider all mitigating factors (e.g., Eddings had been a victim of abusive treatment at home) when considering a death sentence.

Thompson v. Oklahoma (1988) involved 15-year-old William Thompson, who was one of four young men charged with the murder of his former brother-in-law. All four were found guilty and sentenced to death. Thompson appealed to the Supreme Court, claiming that a sentence of death for a crime committed by a 15-year-old is cruel and unusual punishment. The Court agreed, and using the "evolving standards of decency" principle, it drew the line at age 16 under which execution was not constitutionally permissible.

Sixteen years later, in *Roper v. Simmons* (2005), the Court redrew the age line at 18 under which it was constitutionally impermissible to execute anyone. Christopher Simmons was 17 when he and two younger accomplices broke into a home, kidnapped the owner, beat her, and threw her alive from a high bridge into the river below where she drowned. Simmons had told many of his friends before the crime that he wanted to commit a murder, and he bragged about it to them afterward. His crime was a "classical" death penalty case—premeditated, deliberate, cruel, and he was totally unremorseful. Nevertheless, his sentence drew condemnation from around the world, with **amicus curiae** ("friend of the court" briefs presented to the Court arguing in support of one side or the other by interested parties not directly involved with the case) being filed in favor of Simmons by the European Union, the American and British Bar Associations, the American Medical and Psychological Associations, along with 15 Nobel Prize winners, among others.

In a 5–4 opinion, Justice Kennedy noted that the United States was the only country in the world that gives official sanction to the juvenile death penalty. He also noted the growing body of evidence from neuroscience about the immaturity of the adolescent brain. The majority opinion also cited *Atkins v. Virginia* (2002). In noting that the Court in *Atkins* had

Figure 13.2

Executions in the
United States,
1976–2010

Source: Death Penalty
Information Center (2010).

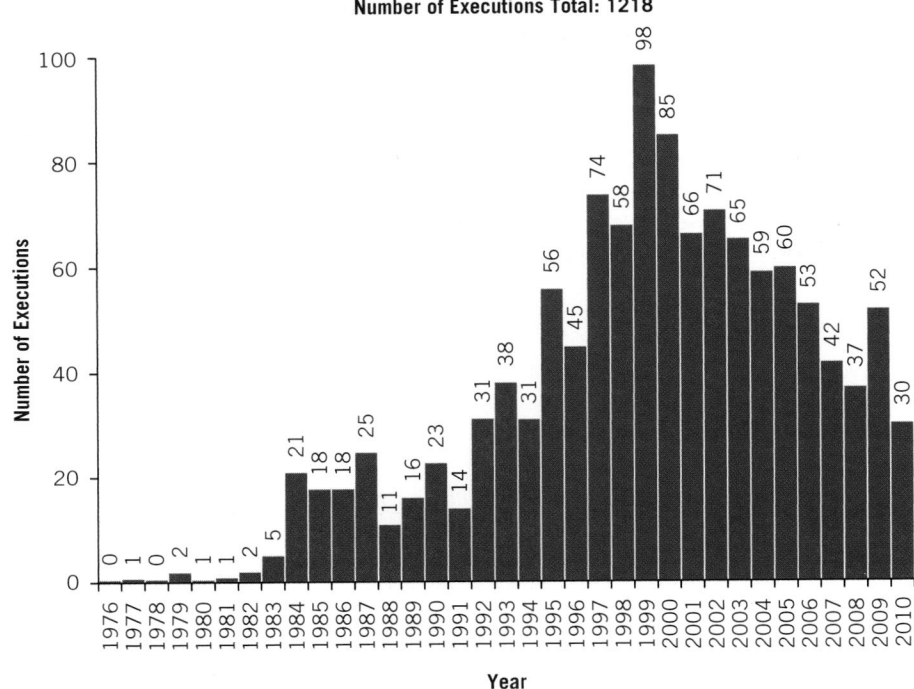

Number of Executions Total: 1218

Table 13.1

Number of Persons Executed by Race, Hispanic Origin, and Method of Execution, 1977–2007

	Number of Persons Executed				
Method of execution	White*	Black*	Hispanic	American Indian*	Asian*
Total	631	373	81	8	6
Lethal injection	536	301	79	7	6
Electrocution	82	69	2	1	0
Lethal gas	8	3	0	0	0
Hanging	3	0	0	0	0
Firing squad	2	0	0	0	0

*Excludes persons of Hispanic origin.

Source: Snell (2008).

ruled the execution of the mentally disabled to be cruel and unusual punishment because of the lesser degree of culpability attached to the mentally challenged, it reasoned that such logic should be applied to juveniles. The Court also pointed out that the plurality of states (30) either bar execution for juveniles or have banned the death penalty altogether, thus citing state legislation as part of the impetus behind its decision.

Figure 13.2 presents a graph of the annual number of executions in the United States from the mid-1970s, when states were coming to terms with *Furman,* to 2010. By 2010, the number of executions declined by 69% from their peak in 1999. In 2010, there were 3,263 prisoners under a sentence of death, and 43 (1.3%) were executed. Of those under sentence of death in 2006, 44% were white, 42% black, 12% Hispanic, and 2% other races. Only 1.6% of death row inmates were women (Death Penalty Information Center, 2010). Table 13.1 shows the racial/ethnic breakdown and method of execution of all those executed in the United States from 1977 to 2007.

Summary

- The courts have moved through three general periods with respect to inmates' rights: the hands-off period, a short period of extending many rights to prisoners, and the current retreat to a limited hands-off policy.
- During the hands-off period, prisoners were considered slaves of the state and had no rights at all. During the period of extending prisoners' rights, the federal courts extended a number of First, Fourth, Eighth, and Fourteenth Amendment rights to them, although these rights were obviously not as extensive as they would be outside prison walls (however, inmates are the only group of Americans with a constitutional right to medical treatment).
- Because of the granting of these rights, the federal courts became clogged with section 1983 suits challenging the conditions of inmates' confinement; some of these suits were demonstrably frivolous.
- The U.S. Congress passed the PLRA in 1996, limiting prisoner access to federal courts and loosening the grip of the courts on state correctional systems

because of these excessive suits. Congress also passed the AEDPA in the same year with a rider limiting inmates' habeas corpus rights.

- The courts have been active since the 1960s in providing rights to offenders under community supervision as well. The Supreme Court has ruled that probationers have a right to counsel at a deferred sentencing hearing, and that probationers and parolees have a right to minimal due process rights (a fair hearing to establish cause) at revocation hearings. The Court has also extended the greater search powers of probation/parole officers to police officers under certain circumstances, as we saw in the probation chapter.
- Because the United States stands almost alone among democracies in retaining the death penalty, the issue has generated much debate and numerous cases questioning its constitutionality. Two classifications of individuals (the mentally disabled and juveniles) were removed from execution eligibility in the early 2000s based on culpability issues among others.

Key Terms

Amicus curiae	*Habeas corpus*	Fourth Amendment
Bill of Rights	Hands-off doctrine	Fourteenth Amedment
Civil death statutes	Eighth Amendment	Section 1983 suits
Civil rights claim	First Amendment	

Discussion Questions

1. What were the two main reasons or justifications behind the hands-off doctrine?

2. Why does the concept of habeas corpus have such a revered place in common law?

3. In what way did the Court in *Jones v. Cunningham* go beyond the original meaning of habeas corpus?

4. Do you think that the majority of male inmate complaints about female officers frisk searching them or viewing their nakedness are genuine indications of outrage? Why or why not?

5. What do you think of laws permitting the civil commitment of some sex offenders? Are there any potential dangers for widespread abuse in the practice?

6. Why do you think that the United States retains the death penalty when almost all other democracies eliminated it long ago? Should we eliminate it? Why or why not?

Useful Internet Sites

American University Law Review (Some full-text articles available free of charge): www.wcl.american.edu/pub/journals/lawrev/aulrhome.htm

Cardozo Law Review (Provides full-text articles free of charge): www.cardozolawreview.com

Findlaw (One of the best student-friendly legal sites, but tends toward "legalese"): www.findlaw.com

Law Guru (Perhaps the best site because of its links to other search engines): www.lawguru.com

Nolo's Plain English Law Dictionary: www.nolo.com/dictionary/worldindex.cfm

Supreme Court of the United States (Provides up-to-date access to Court decisions and many other things relevant to the Court: well worth a visit): www.supremecourtus.gov

CHAPTER 14

Correctional Programming and Treatment

❖ Introduction: The Rise and Fall (and Rise Again) of Rehabilitation

As we have seen, there are five primary goals of the correctional system: deterrence, incapacitation, retribution, rehabilitation, and reentry. This chapter deals with the fourth of these goals—rehabilitation. The term **rehabilitation** means to restore or return to constructive or healthy activity (habilitation), but because many correctional clients never experienced anything close to habilitation in the first place, there is little to restore. What correctional treatment or programming has to do is begin at the beginning and try to provide some of the things previously missing from the lives of offenders. Such programming obviously cannot supply the warmth and nurturing so critical in the early years of life, nor the deep sense of attachment and commitment to social institutions that comes from such attachments. However, programming and treatment can provide some of the concrete rewards such as an education and job training that most of us have had, largely thanks to the attachments to the family and other social institutions we enjoyed as children, and they can do their best to change the destructive thinking patterns that infect some criminal minds.

We keep on trying to rehabilitate criminals with the realization that whatever helps the offender helps the community. We are also mindful of former United States Supreme Court Chief Justice Warren Burger's famous lines: "To put people behind walls and bars and do little or nothing to change them is to win a battle but lose a war. It is wrong. It is expensive. It is stupid" (as quoted in Schmalleger, 2001, p. 439). In this chapter, we look at various ways in which treatment personnel have been fighting the war. When reading this chapter, keep in mind that the vast majority of money assigned to correctional agencies is spent on surveillance and control functions. For instance, only between 10% and 15% of jail and prison inmates who need substance abuse treatment actually receive it (Foster, 2006, p. 301), and that surely is wrong, expensive, and stupid.

❖ History of Rehabilitation

Influenced by British pioneers Alexander Maconochie and Walter Crofton, rehabilitation was the goal of the early American prison reformers such as Zebulon Brockway. The ideal of rehabilitation reached the pinnacle of its popularity, however, from about 1950

Photo 14.1

In addition to basic educational courses and vocational training, college classes are also offered as part of the rehabilitative role of correctional programs.

through the 1970s when the medical model of criminal behavior prevailed in corrections. The medical model viewed crime as a moral sickness that required treatment. Under the medical model, prisoners were to remain in custody under indeterminate sentences until "cured" of their criminal ways. The indeterminate sentence, so favored by early prison reformers and condemned by conservatives as coddling criminals, was condemned by the Left in the 1970s and 1980s because it was seen as imposing too great a range of sentences on people who had committed the same kinds of crime. Nevertheless, it was during this period that the American Prison Association changed its name to the American Correctional Association to emphasize its changing role. Consistent with the switch from a punishment role to a more rehabilitative corrections role implied by the name change, classification systems, individual and group counseling, therapeutic milieus, and college classes were added to the usual rehabilitative fare of labor, basic education, and vocational training (Cullen & Gendreau, 2001).

The rehabilitative goal was questioned and then fell apart with the publication in 1974 of Robert Martinson's article "What Works? Questions and Answers About Prison Reform," in which Martinson concluded that "with few and isolated exceptions the rehabilitation efforts that have been reported so far have no appreciable effects on recidivism" (p. 25). Unfortunately, the rhetorical question "what works?" got translated into a definitive "nothing works" and became a taken-for-granted part of corrections lore. Before we can decide if something does or does not work, we have to define thresholds for what we mean. If we demand 100% success, then we can be sure that "nothing works." A program designed to change people is not like a machine that either works or does not. Human nature being what it is, nothing works for everybody; some things work for some people some of the time, and nothing will work for anybody all of the time. High failure rates existed in many fields at their inception, but as practitioners in those fields learned from their mistakes and their successes, failure rates inevitably improved.

❖ The Shift From "Nothing Works" to "What Works?"

Many correctional programs Martinson surveyed did not work for a variety of reasons, such as the fact that some relied on methods such as psychoanalysis that were inappropriate for offenders, they sought to change behaviors unrelated to crime, and they were not intensive enough and inadequately skilled staff ran them. Few of the programs were based on the proper assessment of offender risks and needs, and they were often faddish "let's see what happens" programs, including everything from acupuncture to Zen meditation, both of which are beneficial in their own right, but hardly useful for changing criminal lifestyles. One probation department actually insisted that male offenders should "get in touch" with their feminine side by requiring them to dress in female clothes, and another required "poetry therapy" (Latessa, Cullen, & Gendreau, 2002). Correctional resources are scarce, and some argue that they should only be expended on programs that have proven themselves useful in reducing recidivism.

How have Martinson's conclusions stood up over the last 30 years? Gendreau and Ross (1987) reviewed a number of studies of treatment programs and concluded, "it is downright ridiculous to say that 'Nothing works.' This review attests that much is going on to indicate that offender rehabilitation has been, can be, and will be achieved" (p. 395). Others have stated that properly run community-based programs could result in a 30% to 50% reduction in recidivism (Van Voorhis, Braswell, & Lester, 2000), although on the basis of major literature reviews, reductions in the 10% to 20% range are more realistic expectations (Cullen & Gendreau, 2001).

A review of studies from prison, jail, probation, and parole settings conducted by Pearson and his colleagues (Pearson, Lipton, Cleland, & Yee, 2002) found a 55.7% success rate (measured in terms of recidivism rates) for offenders receiving a type of counseling called cognitive-behavioral therapy versus a 42.7% success rate for control group offenders. This translates into an average of 20.5% greater success for the treatment groups (55.7 − 43.3 = 11.6/55.7 = 20.5). Although there are still plenty of failures, if all treatment programs managed only half that success rate, the financial and emotional savings to society would be truly enormous.

Mark Lipsey and Francis Cullen (2007) reviewed numerous studies of a variety of correctional intervention programs conducted from 1990 to 2006 and concluded that treatment works moderately well in reducing recidivism. Lipsey and Cullen assert that the biggest problem in offender treatment is not that "nothing works," but that correctional systems do not use the available research to determine what works and then to implement it. Rather, they tend to rely on convenience (who is available and the methods they use), custom ("We've always done it this way and see no reason to change"), and ideology ("Criminals are scumbags. Why waste time and money on them?").

❖ Evidence-Based Practices

Moving from the medical to the just deserts/risk management model in corrections did not mean the death of the rehabilitation goal, but psychological terms such as "assessment" and "programming" have la]rgely replaced medical terms such as "diagnosis" and "treatment." The main concern of corrections is to reduce the risk that offenders pose to society, not to improve offenders' lives. Of course, the two goals are not incompatible; if more offenders can be taught to walk the straight and narrow, the risk of community members being victimized by them is reduced proportionately. Even though programs are typically run on a financial shoestring, prison officials like programming because it keeps inmates busy and out of trouble. Inmates also like it because it gives them something to do outside of their cells and looks good on their parole board records.

The movement to a "what works" frame of mind has resulted in the most progressive agencies moving to **evidence-based practices (EBP)**. EBP simply means that in order to reduce offender recidivism, corrections must implement practices that have consistently been shown to be effective. Extensive research has identified the following eight principles of evidence-based programming:

1. **Assess Actuarial Risk/Needs**—Assessing offenders' risk and needs (focusing on dynamic and static risk factors and criminogenic needs) at the individual and aggregate levels is essential for implementing the principles of best practice.

2. **Enhance Intrinsic Motivation**—Research strongly suggests that "motivational interviewing" techniques, rather than persuasion tactics, effectively enhance motivation for initiating and maintaining behavior changes.

3. **Target Interventions**

   a. **Risk Principle**—Prioritize supervision and treatment resources for higher-risk offenders.
   b. **Need Principle**—Target interventions to criminogenic needs.
   c. **Responsivity Principle**—Be responsive to temperament, learning style, motivation, gender, and culture when assigning to programs.
   d. **Dosage**—Structure 40% to 70% of high-risk offenders' time for 3 to 9 months.
   e. **Treatment Principle**—Integrate treatment into full sentence/sanctions requirements.

4. **Skill Train With Directed Practice**—Provide evidence-based programming that emphasizes cognitive-behavior strategies and is delivered by well-trained staff.

5. **Increase Positive Reinforcement**—Apply four positive reinforcements for every one negative reinforcement for optimal behavior change results.

6. **Engage Ongoing Support in Natural Communities**—Realign and actively engage prosocial support for offenders in their communities for positive reinforcement of desired new behaviors.

7. **Measure Relevant Processes/Practices**—An accurate and detailed documentation of case information and staff performance, along with a formal and valid mechanism for measuring outcomes, is the foundation of evidence-based practice.

8. **Provide Measurement Feedback**—Providing feedback builds accountability and maintains integrity, ultimately improving outcomes.

Taking a closer look at these principles; the psychosocial assessment of offenders typically begins with the **risk, needs, and responsivity (RNR) model**. The RNR model is the premier treatment model in corrections today, in the United States and many other countries (T. Ward, Melser, & Yates, 2007). The **risk principle** refers to an offender's probability of reoffending, and the idea that those with the highest risk are targeted for the most intense treatment ("Dosage" under principle 3). The **needs principle** refers to meeting the offenders' needs, the lack of which puts them at risk for reoffending, and suggests that these needs should receive high priority. The **responsivity principle** maintains that if offenders are to respond to treatment in meaningful and lasting ways, counselors must be aware of their different development stages, motivation, and learning styles, as well as their need to be treated with respect and dignity (Andrews, Bonta, & Wormith, 2006). The crux of these three principles is that we can no longer rely on "one size fits all" models, and that treatment must be tailored to individual offenders' risks and needs.

Offenders' risks and needs are assessed by two separate scales, one for risk and one for needs. These scales are used to make predictions about offenders' success/failure based on *actuarial data* (principle #1), that is, what has actually occurred and been recorded over many thousands of cases. It has been found time and time again across many professions that decisions made on the basis of actuarial statistical norms trump decisions based on the insight of individuals the great majority of the time (Andrews et al., 2006). *Offender risk* refers to the probability that a given offender will reoffend, and thus the threat he or she poses to the community. This is assessed by assigning numerical scores to the scale according to the extent that the offender evidences factors known to correlate with recidivism. Risk factors are either static or dynamic. *Static risk* factors are those that cannot change (gender, age, ethnicity, and other background variables). *Dynamic risk* factors (substance abuse, attitudes, values, behavior patterns) are those that are targeted for change.

Offender needs refers to deficiencies in offenders' lives that hinder them making a commitment to a prosocial pattern of behavior. Scores on the risks and on the needs sections of the scale tend to be highly correlated—offenders with high risk tend to have high needs. Table 14.1 identifies and describes risks and dynamic needs that must be addressed; note that identifying needs mirrors the identification of risk. The other principles of EBP are either self-explanatory or addressed elsewhere in this book.

❖ Cognitive-Behavioral Therapy

The therapeutic concepts and methods that proponents of the RNR model find most useful in addressing offender risks and needs are cognitive-behavioral (Ward et al., 2007). Most of today's programming consists of **cognitive-behavioral therapy (CBT)**. CBT is an approach that tries to solve dysfunctional cognitions, emotions, and behaviors in a relatively short time through goal-oriented, systematic procedures, and has been called "the most overtly 'scientific' of all major therapy orientations" (McLeod, 2003, p. 123). CBT is not a specific therapeutic theory such as psychoanalysis, but is a general term for a group of therapies that combine the principles of operant psychology, cognitive theory, and social learning theory. Operant psychology is a theory that asserts that behavior is determined by its consequences (rewards and punishments). Cognitive theory also puts forth that our behavior is shaped by the rewards and punishments we have experienced, but asserts that at a more proximal level, self-defeating behaviors are the result of unproductive thought patterns relating to these past experiences (D. Wilson, Bouffard, & Mackenzie, 2005). We can do nothing about past experiences, but we can do something to put the way we think about those things into proper perspective. Finally, social learning theory is a view of socialization that suggests behavior is learned by modeling and imitation as well as by our history of rewards and punishments.

Albert Ellis (1989) claims that the great religious leaders of the past were cognitive-behavioral therapists that were trying to get people to change their behavior from self-indulgence to temperance, from hatred to love, and from cruelty to kindness, by appealing to their rational long-term self-interest. The common message imparted by religion is the need for personal change and the rewards that such change brings with it: "Do these things and

Table 14.1

Major Risk and/or Need Factors and Promising Intermediate Targets for Reduced Recidivism

Factor	Risk	Dynamic Need
History of antisocial behavior	Early and continuing involvement in a number and variety of antisocial acts in a variety of settings.	Build noncriminal alternative behavior in risky situations.
Antisocial Personality Pattern	Adventurous pleasure seeking, weak self-control, restlessly aggressive.	Build problem-solving skills, self-management skills, anger management and coping skills.
Antisocial cognition	Attitudes, values, beliefs, and rationalizations supportive of crime; cognitive emotional states of anger, resentment, and defiance; criminal versus reformed identity.	Reduce antisocial cognition, recognize risky thinking and feeling, build up alternative less risky thinking and feeling, adopt a reform and/or anticriminal identity.
Antisocial associates	Close association with criminal others and relative isolation from anticriminal others; immediate social support for crime.	Reduce association with criminal others, enhance association with anticriminal others.
Family and/or marital	Two key elements are nurturing and/or caring and monitoring and/or supervision.	Reduce conflict, build positive relationships, enhance monitoring and supervision.
School and/or work	Low levels of performance and satisfaction in school and/or work.	Enhance involvement, rewards, and satisfactions.
Leisure and/or recreation	Low levels of involvement and satisfaction in anticriminal leisure pursuits.	Enhance involvement, rewards, and satisfactions.
Substance Abuse	Abuse of alcohol and/or other drugs	Reduce substance abuse, reduce the personal and interpersonal supports for substance-oriented behavior, enhance alternatives to drug abuse.

Source: Andrews et al. (2006). Reprinted with permission of Sage Publications.

you will feel good about yourself now, and will go to Heaven/attain Nirvana in the future." This is what CBT tries to do: change offenders' antisocial and self-destructive behavior into prosocial and constructive behavior by changing the way they think and by showing them that it is in their best interests to do so.

The first lesson of CBT is that criminals think differently from the rest of us. Yochelson and Samenow (1976) and Samenow (1999) pioneered treatment theories based on challenging criminal thinking errors when they realized that modalities based on "outside circumstances" theories did not work. The task is to understand how criminals perceive and evaluate themselves and their world so that we can change them. Criminal thinking is destructive; it lands them in trouble with family, friends, employers, and the criminal justice system. Habitual offenders tend to perceive the world in fatalistic fashion, believing that there is little they can do to change the circumstances of their lives. To illustrate this fatalism and other criminal thinking patterns, Boyd Sharp (2006) cites the cartoon *Calvin and Hobbes,* where the character of Calvin says,

I have concluded that nothing bad I do is my fault. . . . I'm a helpless victim of countless bad influences. An unwholesome culture panders to my undeveloped values and it pushes me into misbehavior. I take no responsibility for my behavior. I'm an innocent pawn of society. (p. 3)

Some criminals think like Calvin in the context of a society where many people prefer to claim victimhood rather than personal responsibility (McDonald's made me fat, cigarette companies made me smoke, etc.). Challenging and changing maladaptive thought patterns takes on a central role in treatment as corrections workers strive to impress on offenders that whatever influences external factors may have on behavior, before they can affect behavior they have to be evaluated by individuals. The frustrations we experience do influence our behavior, but the important thing is not their presence, but whether we deal with them constructively or destructively. The task of correctional workers is to teach criminals to stop blaming outside circumstances for their problems, and to teach them how to take responsibility for their lives and to deal constructively with adversity.

CBT methods are used to address issues relating to self-control, victim awareness, relapse prevention, critical reasoning, and emotional control (Vanstone, 2000). CBT therapy literally "exercises the thinking areas of the brain and thereby strengthens the [neuronal] pathways by which the thinking brain influences the emotional brain" (Restak, 2001, p. 144). A number of brain imaging studies show that CBT changes brain processes exactly the way that drugs such as Prozac do (Linden, 2006). Unfortunately, these studies have only been conducted with individuals with problems such as depression in which patients, unlike most criminals, are motivated to overcome their problems.

❖ Substance Abuse Programming

Alcohol is at the same time our most popular and our most deadly way of drugging ourselves. Police officers spend more than half their law enforcement time on alcohol-related offenses. One-third of all arrests (excluding drunk driving) in the United States are for alcohol-related offenses, about 75% of robberies and 80% of homicides involve a drunken offender and/ or victim, and about 40% of other violent offenders in the United States were drinking at the time of the offense (Mustaine & Tewksbury, 2004). Alcohol is a very powerful and addictive drug, and is the biggest curse of the criminal justice system despite the system's current obsession with illegal drugs. Illegal drug usage presents almost as big a problem, with about 67% of state and 56% of federal prisoners having been regular drug users prior to their imprisonment (Seiter, 2005, p. 432). Clearly, mind-altering substances, both legal and illegal, are strongly associated with criminal behavior, and as such the tendency for criminals to over-indulge in them must be addressed by correctional agencies.

Substance abuse problems are extremely difficult to treat because individuals most at risk for becoming addicted share many of the same traits associated with chronic criminal behavior, with many of these traits being strongly genetic (Vaughn, 2009). For instance, alcoholism researchers divide alcoholics into two types: Type I and Type II. Type II alcoholics start drinking and using other drugs earlier; become more rapidly addicted; and exhibit many more character disorders, behavior problems, and criminal involvement, both prior to and subsequent to their alcoholism, than Type I alcoholics (Crabbe, 2002). Genetic researchers maintain that genes are much more heavily involved in Type II than in Type I alcoholism (Crabbe, 2002).

Researchers have also shown that drug addiction and criminality are part of a broader propensity to engage in many forms of deviant and antisocial behavior (Fishbein, 2003; Vaughn, 2009). For instance, the U.S. government's Arrestee Drug Abuse Monitoring (ADAM) program collects urine samples from arrestees across the country to test for the presence of drugs. Figure 14.1 shows the percentage of adult arrestees in 10 large U.S. cities who tested positive for illicit drugs over a 3-year period. The numbers show that illicit drug abuse is clearly strongly *associated* with criminal behavior, but the association is not necessarily a *causal* one. A large body of research indicates that drug abuse does not appear to *initiate* a criminal career, although it does increase the extent and seriousness of one (Menard, Mihalic, & Huizinga, 2001). In other words, research seems to point to the fact that chronic drug abuse and criminality are part of a broader tendency of some individuals to engage in a variety of deviant and antisocial behaviors. Numerous studies have shown that traits characterizing antisocial individuals such as conduct disorder, impulsiveness, and psychopathy also characterize drug addicts (Fishbein, 2003; McDermott et al., 2000). The large amount of research indicating a strong genetic vulnerability to alcoholism/drug addiction helps to explain why some of the many millions who drink or experiment with drugs do not descend into the hell of addiction and why others are "sitting ducks" for it (T. Robinson & Berridge, 2003).

❖ Anger Management

A central component of many treatment programs in corrections is anger management, particularly in violent, drug, and sex offender treatment programs. *Anger management* programs consist of a number of CBT techniques through which someone with problems controlling anger can learn the cause and consequences of that anger, reduce the degree of anger, and avoid anger-inducing triggers. Anger is often central to violent criminal behavior, and the frustrations resulting from being in custody or under correctional supervision in the community often lead to violence. Anger is a normal and often adaptive human feeling that is aroused when we feel we have been offended or wronged in some way. The tendency to undo that wrongdoing by retaliating is motivated by anger, and is adaptive in the sense that it warns those who have offended or wronged the person that he or she is not to be treated that way. The problem, however, is not anger per se, but rather the inability of some to manage it. These individuals often become excessively angry over minor real or imagined slights to the point of rage.

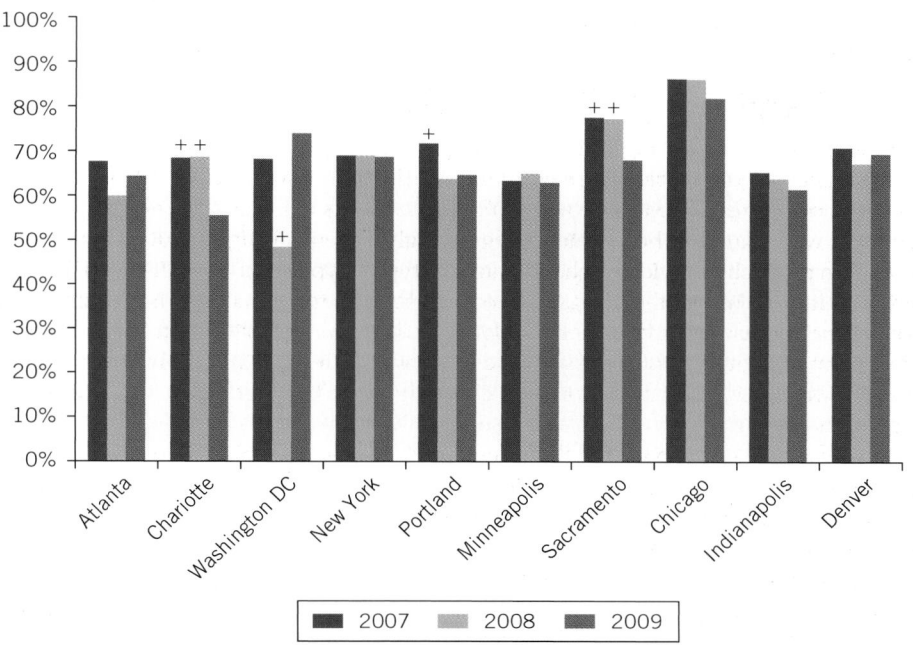

Figure 14.1

Percentage of
Arrestees Testing
Positive for Drugs
in Ten U.S. Cities,
2007–2009.

Source: Office of Drug
Control Policy (2010).

Anger management classes are taught in groups and are designed to increase offenders' responsibility for ownership of their emotions (anger) and their reactions to them. This kind of destructive thinking must be challenged and replaced by individual responsibility. Anger management classes also teach such skills as rational thinking ("Did this person really mean to diss me?"), to increase offenders' ability to react to frustration and conflict in assertive rather than aggressive ways, and to develop effective communication skills (Jolliffe & Farrington, 2009). There appears to be a growing consensus that properly conducted anger management programs reduce inmate violence and reduce violent recidivism for program completers versus control subjects by about 8% to 10% (Jolliffe & Farrington, 2009; Serin, Gobeil, & Preston, 2009). Although this seems like a small return on a corrections investment, even an 8% reduction in violent offenses prevents much needless suffering and millions of dollars in expenses.

❖ Therapeutic Communities

Therapeutic communities (TCs) are residential settings for drug and alcohol treatment that use the community spirit generated by the influence of peers and various group processes to help individuals overcome their addiction and to develop effective social skills. Most such communities offer long-term (typically 6 to 12 months) residence in which opportunities for attitude and behavioral change operate on a hierarchical model whereby treatment stages reflect increased levels of personal insight and social responsibility. The interactions of the residents are both structured and unstructured, but always designed to influence attitudes and behaviors associated with substance abuse (Litt & Mallon, 2003). TCs provide dynamic "mutual self-help" environments in which residents transmit and reinforce one another's acceptance of, and conformity with, the highly structured and stringent expectations of the TC and of the wider community. Life in a TC is extremely hard on people who have never experienced any sort of disciplined expectations from others, and as a consequence, there are many dropouts; some residents withdraw voluntarily, and others are removed by TC staff for noncompliance.

Therapeutic communities also operate within prison walls and are most often known as *residential substance abuse treatment (RSAT)* communities. These RSATs typically last 6 to 12 months and are composed of inmates in need of substance abuse treatment and whose parole dates are set to coincide with the end of the program. RSAT inmates are separated from the negativity and violence of the rest of the prison; are provided with extensive cognitive-behavioral counseling; and attend Alcoholics Anonymous (AA) and Narcotics Anonymous (NA) meetings, as well as many other kinds of rehabilitative classes (Dietz, O'Connell, & Scarpitti, 2003). The majority of participants in these RSATs are positive about many aspects of their experience, with most inmates listing cognitive self-change programs as the strongest positive aspect of their treatment (Stohr, Hemmens, Shapiro, Chambers, & Kelly, 2002). Dietz et al. also found that most inmates in prison-based TCs were positive about the program and that they had significantly fewer rule violations and rates of grievance filing than inmates in the general population.

An interesting program implemented in a prison setting and transitioning into the community is the Delaware Multistage Program (Mathias, 1997). In the beginning stage, offenders spend 12 months in a prison-based TC called Key; in phase 2, they spend 6 months in a pre-release TC called Crest; and finally, in phase 3 (Key-Crest), they receive an additional 6 months counseling while on parole or in work release. Figure 14.2 compares drug use and arrest outcomes for offenders completing all phases (Key, Crest, and Key-Crest) 18 months after release from prison and a comparison group of offenders who did not participate in any of the phases. We see that 76% of Key-Crest members remained drug free and 71% remained arrest free compared with only 19% and 30%, respectively, of the control group. Put another way, 3 times as many Key-Crest participants were drug-free after 18 months as

Eighteen Months After Release From Prison

Figure 14.2

Delaware Multistage Correctional Treatment Program

Source: Mathias (1997).

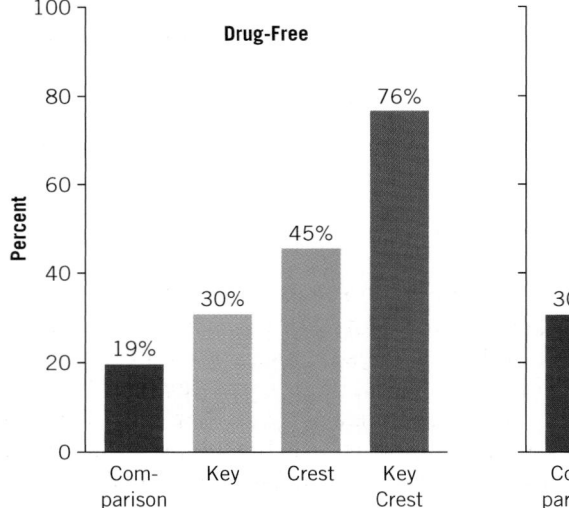
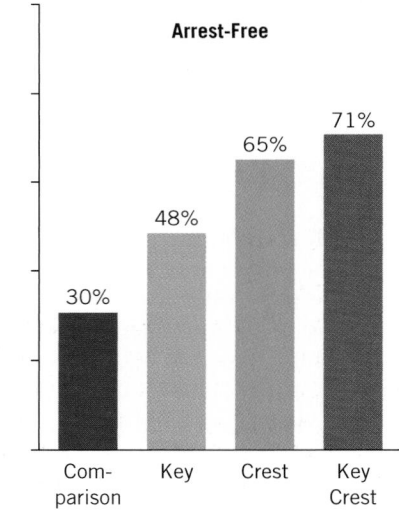

the comparison group, and 2.37 times more Key-Crest participants were arrest-free than the comparison group.

Inciardi, Martin, and Butzin (2004) followed this same group 5 years after release from prison. As expected, the greater the time lapse between treatment and evaluation, the greater was the relapse rate. Over the 5-year period, it was found that 71% of drug abusers who went through a residential treatment program and who received additional treatment upon release (the Key-Crest group) had relapsed, and 52% had been rearrested. However, the contrast with the comparison group still makes the Key-Crest program impressive. Among the comparison subjects, 95% had relapsed and 77% had been rearrested. This study shows how extremely difficult it is to battle addiction even after a long period of forced abstinence and extensive psychosocial treatment.

❖ Pharmacological Treatment

Allan Leshner (1998) informs us that addiction is a brain disease and a "prototypical psychobiological illness, with critical biological, behavioral, and social context elements" (p. 5). As **addiction** is basically a brain chemistry problem, pharmacological treatment with drug antagonists (drugs that work by blocking the effects of other drugs) stabilizes brain chemistry and renders addicts more receptive to psychosocial counseling. Proponents of pharmacological treatment emphasize that it is not a magic bullet and that it augments, not replaces, traditional treatment methods.

There are many drug antagonists, but only one has claimed success in curbing both alcohol and drug addiction—naltrexone. Naltrexone reduces craving among alcohol-/drug-abstinent addicts and reduces the pleasurable effects for those who continue to use (Schmitz, Stotts, Sayre, DeLaune, & Grabowski, 2004). A study of drug addicts on federal probation found that about one-third of probationers who received naltrexone plus counseling relapsed as opposed to two-thirds of those who only received counseling (Kleber, 2003).

Proponents of pharmacological treatment claim that the effects of such treatment are more effective and immediate, and they wonder why the correctional system is relatively uninterested in pharmacological treatment (Kleber, 2003). It could be that corrections professionals received their training primarily in the social sciences, and there are some who have genuine ethical problems regarding chemical treatments for behavioral problems. However, according to the National Institute on Drug Abuse (2006), medication is important for treating many addicts, but it must be combined with counseling, because medication helps them to stabilize their physiology, and counseling helps them to stabilize their lives.

❖ Sex Offenders and Their Treatment

The American public harbors all sorts of very negative images of sex offenders. We lock them up under civil commitment orders after they have completed their prison terms, and all 50 states have sex offender registration laws (Talbot, Gilligan, Carter, & Matson, 2002). However, the term *sex offender* defines a very broad category of offenders ranging from "flashers" to true sexual predators, just as property offenders include everyone from petty shoplifters to career burglars. At least 98% of all sex offenders are either in the community on probation or parole or will be someday (M. Carter & Morris, 2002), making the issue of sex offender treatment of the utmost importance.

Although it is part of popular lore that sex offenders are untreatable and will never stop their offending, as a category of offenders they are actually less likely to reoffend than any other category. Looking at many years of British crime statistics, it was found that burglars are the most likely of all criminals to be reconvicted (76%) within 2 years of being released from prison, with sex offenders being the least likely (19%) (Mawby, 2001, p. 182). A review of 61 studies of sex offender recidivism found an average rate of reconviction for sexual crimes of 13.4% over a 4- to 5-year follow-up (Hanson & Bussiere, 1998). Perhaps the most instructive study of recidivism conducted to date was a study by the Bureau of Justice Statistics whose researchers tracked 9,691 sex offenders released from prisons in 15 U.S. states in 1994 (Langan, Schmitt, & Durose, 2003). Over the 3-year period of the follow-up, sex offenders had a lower rearrest rate (43%) than 272,111 non–sex offenders released at the same time in the same states (68%). Rearrest rates included all types of crimes and technical violations such as failing to register as a sex offender or missing appointments with their parole officers. Only 3.5% of the sex offenders were reconvicted of a new sex crime during the follow-up period. Because recidivism rates include only those offenders who have been caught, in common with rates for other types of offenders, the above figures should be considered bare minimums.

State-of-the-art treatment of sex offenders must include a thorough assessment of psychosocial problem areas, deviant arousal patterns, and polygraph ("lie detector") assessment (Marsh & Walsh, 1995). Deviant arousal patterns are assessed by a device called a penile plethysmograph (PPG), which measures blood flow in the penis (the level of the swelling of the penis) when exposed to deviant sexual images. These measures are then compared with measures in response to consensual adult sexual images.

Counselors are in agreement that effective treatment is impossible until the full extent of the offender's sex-offending history is acknowledged by him or her and made known to treatment personnel (Walsh & Stohr, 2010). But sex offenders are notorious for hiding their sexual histories, so polygraph assessment is needed to access them. Comparing self-reports and pre- and post-polygraph testing across two decades of research, it has been found that child molesters underreport the number of sex crimes they have committed by about 500% and overreport their own childhood sexual victimization (the "I'm a victim too" excuse) by about 250% (Hindman & Peters, 2001). The polygraph may therefore be seen as a very useful tool if the first goal of treatment is to honestly acknowledge one's sexual history.

Unlike treatment for other problems in corrections, there has been a great deal of interest in the pharmacological treatment of sex offenders. Numerous researchers have concluded that optimal treatment (following a thorough psychosocial and physiological assessment) combines the biomedical with cognitive-behavioral approaches (Walsh & Stohr, 2010). The biomedical approach involves so-called chemical castration with a synthetic hormone called Depo-Provera, which is also sold as a method of female birth control. Depo-Provera works in males to reduce sexual thoughts, fantasies, and erections by drastically reducing the production of testosterone, the major male sex hormone. Depo-Provera prevents testosterone production, and it is testosterone activating a part of the brain called the hypothalamus that controls the male sex drive. Depriving the brain of testosterone allows offenders to concentrate on their psychosocial problems without the distracting fantasies and urges (Marsh & Walsh, 1995).

Following the state of California in 1997, several states now mandate chemical castration ("castration" is reversible upon withdrawal from the drug) for repeat offenders. Not all sex offenders should be treated with this drug because there are sometimes negative side effects, and treatment can only be provided by a medical doctor. However, a number of reviews of the literature from Europe and America show that antiandrogen drugs (drugs that inhibit the male sex hormone) such as Depo-Provera result in recidivism rates for repeat rapists and child molesters that are remarkably low (in the 2% to 3% range) when compared to offenders treated with only psychosocial methods (Maletzky & Field, 2003).

Some may decry the use of medications on the grounds that it supports the notion that sexual offending is "about sex" rather than about "violence" and/or "power." However, it is likely about both. The treatment literature is full of studies emphasizing cognitive restructuring for deviant sexual fantasies or medications designed to reduce sexual arousal (Dreznick, 2003; Giotakos, Markianos, Vaidakis, & Christodoulou, 2003; R. Howard; 2002; Lindsay, 2002; Maletzky & Field, 2003). Perhaps sexual offenses are best viewed as a fusion of sex and aggression because both sexual and aggressive behavior are mediated by the same brain areas, and both are activated by the same sex hormones (Grubin, 2007). There certainly are men who enjoy hurting and humiliating their victims, but in the vast majority of cases, forceful coercion is a tactic, not a motive (Figueredo, Sales, Russell, Becker, & Kaplan, 2000).

Comparative Perspective: Offender Treatment

The treatment of criminal offenders in modern Western nations is fairly uniformly centered on the risk-needs-responsivity model. Countries that care little about the civil rights and treatment of their citizens in general obviously are not very concerned about the humane treatment of their prisoners. A modern country concerned with rehabilitation, but with some different assumptions about how to achieve it, is Japan. A large cultural difference between Japan and Western nations such as the United States is that whereas the latter emphasizes individualism and a great degree of personal freedom, the Japanese emphasis is on collectivism and individual conformity to community norms. Unlike the general tendency in American criminology to shift the blame for crime away from the individual and onto "society," the Japanese place responsibility for criminal behavior squarely on the shoulders of the individuals who commit it.

Western CBT tends to play down introspection (gaining insight into one's self) as too time-consuming, and asserts that criminals need the participation and direction of the counselor in order to benefit. Japanese correctional counselors, on the other hand, see their role as a kindly guide on the edge, rather than as a trainer at the center providing definite directions (Bindzus, 2001). One type of counseling favored in Japanese corrections is called *Naikan,* which means "to see oneself," and is designed to get the offender to see him- or herself as others see the person. According to Kanazawa (2007), inmates undergoing *Naikan* therapy will spend many hours alone asking themselves three questions: "What has my mother (and other significant persons in my life) done for me," "what have I done for her (and other significant persons in my life) in return," and "what problems have I caused her (and other significant persons in my life)?" (p. 762). Thinking about these things is supposed to generate feelings of remorse, sadness, empathy, guilt, and consciousness of responsibility. The counselor will enter the offender's cell every hour or so to check on progress. Only if the offender displays tendencies to blame outside forces for his or her criminal behavior or has not adequately explored the questions will the counselor intervene to clarify.

Could such introspective methods work with Western prisoners who exist in cultures that seem to have little respect for individual responsibility? Dieter Bindzus (2001) expresses "doubts about the physical and psychological ability of European prisoners to stick through Naikan for a one-week period, with daily sessions up to sixteen hours" (p. 266). What works in one cultural context may not work in others for a variety of reasons. Of course, inmates require some sort of concrete help as well as introspective self-knowledge, even in Japan. Indeed, Eskridge (1989) lists a large number of vocational training courses available in Japanese penal institutions ranging from auto mechanics, to seamanship, to welding.

❖ Mentally Ill Offenders

Mentally ill offenders under correctional supervision present a particularly difficult treatment problem. Alcoholics and drug addicts ingest substances that alter the functioning of their brains in ways that interfere with their ability to cope with everyday life, although their brains may be normal when not artificially befuddled. Mentally ill persons also have brains that limit their capacity to cope, but that limitation is intrinsic to their brains, not attributable to intoxicating substances. Studies around the world have found that mentally ill persons (mostly schizophrenics and manic depressives) are at least 3 to 4 times more likely to have a conviction for violent offenses than persons in general (Fisher et al., 2006). Most mentally ill persons, however, are more likely to be victims than victimizers, and many of them make their problems worse by abusing alcohol or drugs (Walsh & Ellis, 2007). It is because of their substance abuse and greater propensity for violence, in addition to mental hospital deinstitutionalization, that the mentally ill are overrepresented in the correctional system.

About 1 in 5 inmates in U.S. prisons suffers from some kind of mental illness, which amounts to about 3 times the number of individuals in prison than in mental hospitals in the United States (Cuellar, Snowden, & Ewing, 2007). In addition, the Bazelon Center for Mental Health Law (2008) estimates that about 16% of individuals on probation or parole have some form of mental illness. This state of affairs has resulted from the deinstitutionalization of all but the most seriously ill patients from mental hospitals that occurred in the 1960s. For instance, there were 559,000 persons in U.S. mental hospitals in 1955; in 2000 (with a U.S. population about 80% greater), there were only 70,000 (Gainsborough, 2002). Deinstitutionalization of the mentally ill from mental hospitals has shifted to their institutionalization in jails and prisons, which in essence has resulted in the criminalization of mental illness (2000). Figure 14.3 presents the highlights of a Bureau of Justice Statistics report on the mental health problems of prison and jail inmates (James & Glaze, 2006).

Figure 14.3

Prevalence of Mental Health Problems Among Prison and Jail Inmates

	Percentage of Inmates in—			
	State Prison		Local Jail	
Selected characteristics	**With Mental Problem**	**Without**	**With Mental Problem**	**Without**
Criminal record				
Current or past violent offense	61	58	44	36
Eight or more prior incarcerations	25	19	26	20
Substance dependence or abuse	74	58	76	53
Drug use in month before arrest	63	49	62	42
Family background				
Homelessness in year before arrest	13	6	17	9
Past physical or sexual abuse	27	10	24	8
Parents abused alcohol or drugs	39	25	37	19
Charged with violating facility rules*	58	43	19	9
Physical or verbal assault	24	14	8	2
Injured in a fight since admission	20	10	9	3

*Includes items not shown.

Source: James & Glaze (2006).

Mentally ill offenders in jails and prisons are often victimized by other inmates, who call them "bugs" and exploit them sexually and materially (stealing from them), although most inmates seek to avoid them. Mentally ill offenders are also punished by corrections officers for behavior that, while not pleasant, is symptomatic of their illness. These behaviors include such things as excessive noise, refusing orders or medication, self-mutilation, and poor hygiene. Obviously, correctional facilities are not the ideal place for providing mental health treatment, even assuming that the staff are aware of who the mentally ill are among their charges. Few correctional or probation/parole officers have any training regarding mental health issues, and one nationwide survey of probation departments found that only 15% of them operated special treatment programs for the mentally ill (Lurigio, 2000). No one expects correctional workers to become treatment providers because that is a job for psychologists and psychiatrists. However, they should be expected to recognize signs and symptoms of mental illness, should know how to effectively deal with situations involving mentally ill persons, and possess a basic understanding of the causes of and treatment for the major mental illnesses.

Summary

- Although the vast majority of the correctional budget is spent on security, rehabilitation efforts have not completely ceased. The success rates of many rehabilitation programs are low, but outcomes are significantly better for treated offenders than for similarly situated offenders who did not receive treatment.

- Successful treatment programs implement evidence-based practices that proceed by conducting a thorough assessment of offenders' risks and needs, and then address these issues using cognitive-behavioral techniques along with the principles of responsivity. Treatment is best accomplished for severe substance abusers in therapeutic communities, although even then, there is a significant percentage of failure. Much of this failure has to do with the intense psychological craving for the substance of abuse, which is something that may be significantly alleviated by certain alcohol/drug antagonists such as naltrexone.

- Similar observations were made about sex offenders who have difficulty refraining from acting out their sexual fantasies with inappropriate targets. Repeat sex offenders treated with Depo-Provera combined with cognitive-behavioral counseling have very low recidivism rates compared with offenders treated only psychologically.

- Mentally ill individuals are represented in the correctional system by a factor of at least 3 or 4 times their prevalence in the general population. The correctional system is not equipped to deal with mentally ill people, who are often victimized by other jail/prison inmates or disciplined by corrections officers for exhibiting behavior that is basically part of their mental disease syndromes.

Key Terms

Addiction

Cognitive-behavioral therapy (CBT)

Evidence-based practices (EBP)

Needs principle

Rehabilitation

Responsivity principle

Risk principle

Risk, needs, and responsivity (RNR) model

Therapeutic communities (TCs)

Discussion Questions

1. In your estimation, are the time, effort, and finances spent on rehabilitative efforts worth it, given the low success rates? Would longer periods of incarceration better protect the public? Why or why not?

2. Cognitive-behavioral approaches stress thinking and rationality. How about emotions? Do you think that human behavior is motivated more by emotions than by rationality? Give your reasons.

3. Given the greater involvement of genes in Type II alcoholism, in what ways would you treat Type II alcoholics differently from Type I's if you were a treatment provider? How about if you were a probation/parole officer?

4. Should all sex offenders undergo Depo-Provera treatment? What are the ethical issues involved with such invasive treatment?

5. Discuss the various component parts of the responsivity principle.

Useful Internet Site

U.S. Bureau of Labor (If you are contemplating a career as a correctional treatment specialist, the following website provides a lot of information about job prospects and pay.): www.bls.gov/oco/ocos265.htm

CHAPTER 15

Corrections in the 21st Century

❖ Introduction: Learning From the Past So That We Have Hope for the Future

Americans have a tendency to revisit old themes, efforts, and programming every generation or so, even when such endeavors were clear failures and rejected by generations past. Perhaps it is because we are a relatively new nation and are seemingly remade as new generations of immigrants flood our shores, bringing their histories and cultures that do not include memories of past reform efforts made in this country. Perhaps we keep retrying old endeavors because of the media influence that reduces very complex problems to a brief and simplistic message, and, as a consequence, we do not understand that despite the new packaging and marketing, we have been there and done that before. Or maybe it is a hand-and-glove collusion by the media and politicians in this reductionism of complex topics and collective memory loss.

Whatever the reason, we do not seem to learn much from the experience of those who have come before us, at least as that is related to correctional practice. Or more accurately, we certainly could learn more from our past than we have! It is an oft-cited truism, courtesy of the 20th-century philosopher Santayana, and as already mentioned in this book, that those who do not know their history are likely to repeat it. This adage bears repeating, as it clearly applies here regarding correctional programs, operation, and practice.

❖ Punitive Policies Yield Overuse of Corrections

To illustrate this point, all we need to do is consider the current, but declining in popularity, efforts of the last two decades, namely, the drug war, mandatory sentencing, supermax prisons, and abandonment of treatment programming. Spurred by punitive sentiments that swept the political, social, and economic systems, the statutes, declarations, and practices that derived from these efforts profoundly changed corrections as Americans experienced it (Jackson & Jacobs, 2010; Whitman, 2003).

First, they vastly increased the use of all forms of corrections in this country. Our imprisonment rate (just for prisons, not including jails) was stabilized at about 125 persons per 100,000 residents for 50 years (1920 to 1970) until the drug war, mandatory sentences, and other punitive policies increased it (Ruddell, 2004). At the end of 2009, this number had risen to 502 persons per 100,000 residents, or over 4 times the imprisonment rate of that 50-year period. In raw numbers, the offenders sentenced to prison more than quadrupled for an increase of over 500 percent, from 319,598 in 1980 to 1,613,740 in 2009 (West, Sabol, & Greenman, 2010, p. 1).

A similar steady and swift increase in the use of jails has occurred since punitive policies have been put in place: In 1986, the incarceration rate for jails in the United States was 108 persons per 100,000 residents, and by 2009 it was 250, or almost a 250% increase in the use of jails in more recent years (Minton, 2010, p. 4). Put another way, the number of persons incarcerated in America's jails more than tripled, or increased by 417%, from 183,988 in 1980 to 767,620 in 2009 (Minton, 2010, p. 1).

Similarly astounding increases can be found in the use of probation and parole because of these efforts. From 1980 to 2009, the number of people on probation more than tripled, with an increase of 376% (1,118,097 in 1980 to 4,203,967 in 2009) (Glaze, Bonczar, Zhang, 2010, p. 2). Likewise, the numbers of persons on parole also more than tripled during this time period (from 220,438 to 819,308), even though 16 states and the federal government eliminated all or most of their parole programs (Glaze et al., 2010, p. 4).

Yet, as best we can tell (and as was mentioned in Chapter 1), we do not have proportionately more crime these days than at any other period in our history (see Figures 15.1 and 15.2). In fact, based on victimization and police reports, it appears that by the end of the 2000s, we had the lowest rate of victimization since the National Victimization Survey began in 1973. As indicated in Figures 15.1 and 15.2, violent crime by adults and juveniles did increase in the 1970s and through the early 1990s, but it has

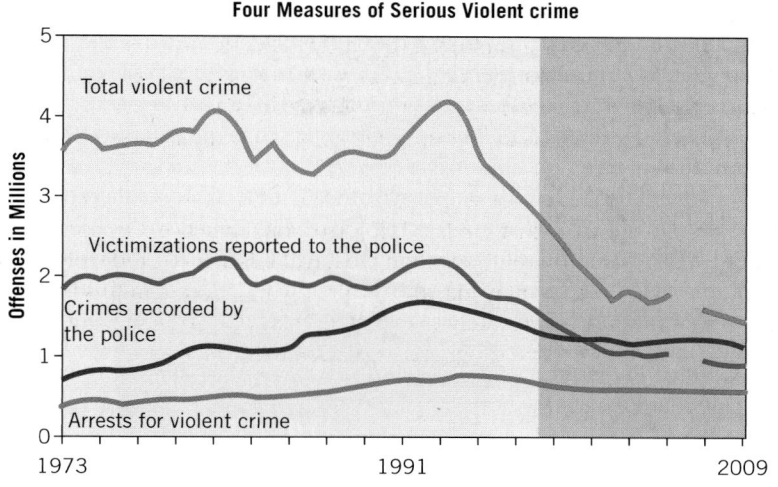

Four Measures of Serious Violent crime

Total violent crime

Victimizations reported to the police

Crimes recorded by the police

Arrests for violent crime

Offenses in Millions

1973   1991   2009

Figure 15.1

Since 1994, violent crime rates have declined, reaching the lowest level ever recorded in 2009.

Source: Bureau of Justice Statistics (2010, p. l).

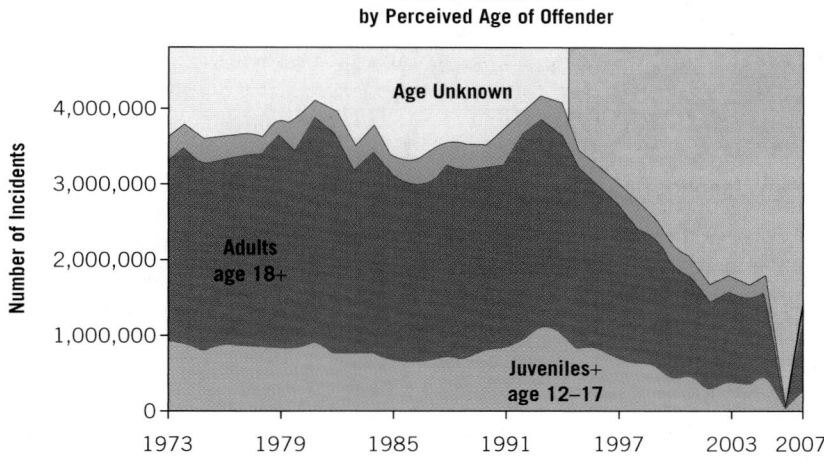

Serious Violent Crime by Perceived Age of Offender

Age Unknown

Adults age 18+

Juveniles+ age 12–17

Number of Incidents

1973   1979   1985   1991   1997   2003   2007

Figure 15.2

The proportion of serious violent crimes committed by juveniles has generally declined since 1993.

* Victimization rate trends exclude NCVS estimates for 2006 because of methodological inconsistencies between the data for that year and the data for sother years.

Source: Bureau of Justice Statistics (2010, p. l).

- Victims perceived that between 1/5 and 1/4 of violent crimes were committed by juveniles.
- According to the victim's perception of the age of the offender, the number of serious violent offenses committed by persons ages 12 to 17 declined 61% from 1993 to 2005, while those committed by persons older than 17 fell 58%.

since dropped precipitously, though clearly the use of incarceration has not mirrored this decrease. Moreover, property crimes have dropped just as consistently and sharply (Bureau of Justice Statistics, 2010).

In addition, our use of corrections is not in sync with what other countries are doing. We have similar crime rates (Farrington, Langan, & Tonry, 2004), yet our incarceration rate is more than 14 times that of Japan; close to 7 times that of Canada, France, Germany, and Italy; and over 5 times that of the United Kingdom (Ruddell, 2004). Russia is the only other developed nation that gets close in its incarceration rate, and the United States still outpaces that nation by almost 100 more people per 100,000 residents. As of 2001, almost 2.7% of Americans served time in prison, a figure that is twice that of 1974 (1.3%) (Bureau of Justice Statistics, 2006a).

Second, interestingly enough, these punitive policies have not had the effect of increasing sentencing length. The average sentence to prison in state courts in 1992 was 6.5 years, as compared to 4 years and 11 months in 2006 (Durose, Farole, & Rosenmerkel, 2009; Langan & Cohen, 1996). It is not clear why sentencing length has decreased at the same time that more punitive policies are in effect. It is possible that decreased sentence length might be one of those unintended consequences of the overuse of incarceration. The capacity of prisons and jails, along with probation and parole caseloads, has been vastly increased over the last 20 years, but it may not have increased enough to accommodate the numbers of processed felons in the courts. What this means is that courts are forced to adjust their sentences to the lower relative capacity of prisons, and parole boards are pressured to release inmates as prisons and jails fill up.

Third, punitive policies, as was discussed in Chapters 10 and 11, have led to an explosion in the number of women and minority group members who are incarcerated or under some form of correctional supervision (Bureau of Justice Statistics, 2006a; Irwin, 2005; Irwin & Austin, 2001; Pollock, 2004; Zimring, Hawkins, & Kamin, 2001). Until the current version of the drug war was resurrected—yes, there were others in American history (Abadinsky, 1993)—the proportion of women to men and racial and ethnic minorities to whites in prisons, in jails, and on parole and probation was somewhat stable (Bureau of Justice Statistics, 2006a).

Photo 15.1

The number of persons incarcerated in America's jails more than tripled, or increased by 417%, from 183,988 in 1980 to 767,620 in 2009.

Fourth, such policies have favored the use of more isolation, "punishment," and warehousing to deal with both bulging correctional populations and recalcitrant inmates. The number of supermax facilities has exploded, as has the number of supermax inmates, as the management of the number of inmates turned to favoring punishment and warehousing over treatment (Irwin, 2005).

As a result of such policies, a fifth outcome has been the abandonment of a core principle of some correctional institutions and practices (e.g., probation, parole, minimum security institutions, work releases)—namely, treatment—based on insufficient evidence. Some correctional programs and institutions were formulated on the premise that treatment is a major goal of corrections. Though the public has continued to believe this (Applegate, Cullen, & Fisher, 2001), for all intents and purposes, real efforts at treatment, beyond basic Alcoholics Anonymous/Narcotics Anonymous, religious, and GED programs, in prisons, jails, and in community corrections received little funding and virtually disappeared for 20 years in many places (mid-1970s to the mid-1990s).

Though some of these endeavors and efforts such as the drug war and mandatory sentencing continue, and even grow in some communities, for the most part scholars and some policy makers have deemed the former a failure and the latter a spectacular waste of money. Furthermore, though the number of supermaxes has grown in the recent past, there is not nearly the hype about their promise for eliciting inmate reform. Finally, all indications are that the belief in, and embrace of, treatment programming, albeit programs that can demonstrate their worth, is on the upswing both in correctional institutions and in the communities (see Chapter 14). Although all of these changes in attitudes and perceptions are positive, it is frustrating to realize that we knew, or should have known from our own past, that drug wars, mandatory sentencing, isolation, and pure punishment in the form of warehousing are not likely to reduce crime in this country, let alone reform those under correctional control. The long and the short of it is, we should have known better because it had all been done before.

❖ Decarceration

Decreased Use of Incarceration

As of 2009 (the latest date for which we have data) the imprisonment rate for adults nationally had declined a few points, from a high of 506 in 2007 to 502 in 2009 (West et al., 2010, p. 2). These declines were at the state level, not the federal, and were more likely to affect women, rather than men. As indicated from Bureau of Justice Statistics data, 24 states experienced a decrease in their imprisonment rates from 2008 to 2009, and these states represented every region of the country, from California to Michigan to Rhode Island to Texas. Only in three northeastern states—New York, New Jersey, and Maryland—were these declines a discernable trend over the last decade, though in Illinois, Michigan, Delaware, and Texas, the 2009 declines had been preceded by anemic growth rates earlier in the decade (West et al., 2010). Notably, not all states experienced such declines—26 states, from all across the United States, experienced increases in their use of imprisonment (West et al., 2010, p. 2). The difference here is that, for the first time in a long time, not all, or even an overwhelming majority, of states experienced such increases; which makes the declines interesting.

Explanations for the Decline in the Use of Incarceration

What might these declines in the use of incarceration be attributed to? Is it possible that declining crime, particularly violent crime, has actually begun to affect the use of prisons in some states? Is it possible that enforcement of the drug war has waned as states have legalized medical marijuana, and the drug hype as tied to criminality (e.g., the crack mothers producing crack babies hysteria) has receded in the minds of the public, politicians, and criminal justice practitioners? Is it possible, in the case of the northeastern states, that declines are tied to reduced state populations, particularly of young people who are most engaged in street crimes? Could it be that the reduced use of incarceration for some of these states is a consequence of the recession that hit the United States in 2008 and the consequent declining tax revenues, and increased debts, faced by states and localities who could no longer afford the bill for spiraling imprisonment?

We think it is possible that all of these scenarios have some value in explaining the decreased use of incarceration in some states in the last few years, and for a few states for the last several years. We know that overall, the number of admissions and parole violators returned to state prisons declined in 2010, signaling a change in the number of people before the courts, a change in sentencing by court actors, a change in parole agents' behavior, or some combination of these (West et al., 2010).

We also know that about 60% of the increase in the use of prison by the states was due to the increased imprisonment of violent offenders, perhaps signaling greater attention to them, even as their proportional numbers decrease, rather than to drug offenders (West et al., 2010, p. 7). Finally, we know that despite their greater contribution to the composition of prisons, these violent offenders were actually "doing" less time. In 2000, the mean length of stay for violent offenders was 46 months, but by 2008 this mean had declined to 44 months—a seemingly small effect, but when multiplied by hundreds of thousands of people, this lower mean can have a huge impact on some prison systems (West et al., 2010, p. 8). Reduced length of stay may be related to the effect of declining state revenues on state prisons.

The Recession and Decreased Use of Incarceration

In 2009, the Pew Center on the States published a report subtitled *The Long Reach of American Corrections* in which the authors made the case that "more prison spending brings lower public safety returns" (Saylor & Iwaszko, 2009, p. 17). They argue that over the last 20 years, we have incarcerated too many first-time and nonviolent offenders who never needed to be incarcerated. Doing so has cost us billions of dollars with no collateral decrease in crime, as most of these people would either not reoffend or could have been handled in a much less expensive and intrusive way in the community. Moreover, the more we incarcerate, the less we achieve the ideal of incarcerating the serious repeat offenders (as there are far fewer of these), and the more the lower-level offenders are "replaced," leading to greater involvement of more people in crime, and the less likely we are to deter current criminals, as the research does not show that longer sentences deter more effectively. Though Saylor and Iwaszko acknowledge that the huge incarceration increase was likely responsible for somewhere between 10% and 35% of the decrease in violent crime since the mid-1990s (depending on which researcher you listen to and which model and assumptions he or she adopts), the authors note that most of the drop in crime was likely attributable to factors outside of incarceration. It makes no sense costwise then, they argue, to continue to incarcerate low-level, aging, and less criminogenic offenders.

There are some recent examples from the news that indicate that some states and localities came to the same conclusion. Washington State, for instance, in April 2011, closed the 135-year-old McNeil Island Prison because of the need to cut that state's budget (The cut is projected to save the state 12.7 million.) (Mulick, 2010, p. 1). Likewise, for the first time in state history, Oregon in October 2010 closed a minimum security facility for the same reason (The cut was projected to save the state 33.8 million.) (Zaitz, 2010, p. 1). In both the Washington and Oregon instances, Department of Corrections officials claimed that none of the inmates would be released early because of the closures, though it remains to be seen whether the decreased capacity of these state prison systems will lead to decreased incarceration in those states in the next few years.

In another example from Washington State, this time involving a county jail and its finances (Thurston County), the county manager declined to open the newly built and $45 million-dollar facility, at least for a year, because the county could not afford to staff it or pay for its operation (Hulings, 2010, p. A4). In addition, the county manager noted that the new jail is not needed as much because of the declining jail population in the county, which happened after the jail construction had begun.

Moreover, in a recent report by The Sentencing Project (Porter, 2011), the author revealed that in 2010, "state legislatures in at least 23 states and the District of Columbia adopted 35 criminal justice policies that may contribute to reductions in prison populations and eliminate barriers to reentry while promoting effective approaches to public safety" (p. 3). Such policies included medical marijuana laws (Arizona, District of Columbia), authorized medical parole for seriously ill or dying inmates (California, Kansas), reduced penalties for drug and property offenses (Colorado, Illinois, Tennessee), increased parole eligibility and reduced revocations (Colorado, Indiana, Massachusetts, Michigan, New Hampshire, New Jersey, Pennsylvania), established "ban the box" law that delay questions regarding the criminal history of an applicant until they are interviewed (Connecticut, Massachusetts, New Mexico), and modified policies regarding juveniles to reduce their incarceration and the severity of it (Colorado, Virginia, Wyoming) among other changes (p. 3). Porter notes that developing alternatives to prison contributed to a 20% drop in incarceration, from 1999 to 2009, in both New York and New Jersey. She argues that though several of these changes in state policies were done to reduce current and future budgets, they were also adopted because policy makers were no longer convinced of the efficacy of mass incarceration as a public safety measure.

Implications of Decarceration and the Need for a Plan of Action

Should these early indications of decarceration turn into a flood of releases—dare we say *mass decarceration*—there will be many positive outcomes, such as less incarceration of low-level offenders resulting in a greater sense of "justice" for community members; less incarceration of minority men and women, also resulting in a greater sense of "justice" for all; fewer tax dollars being devoted to incarcerating people; a reduction in the growth of the corrections industrial complex, at least at the institution level (see a discussion of the complex later in this chapter); and more opportunities for people to age out of crime and contribute in a meaningful way to their communities. However, there are likely some negative outcomes that will result from decarceration, such as greater unemployment, more low-level crime, and increased use of drugs and alcohol by ex-mates, which could occur if policy makers and correctional officials do not plan appropriately. Clearly, money will be saved through decarceration, but some of these monies will likely be needed to fund reentry programs in corrections (see Chapter 8) and work and training programs for the decarcerated in communities, along with the expansion of drug and alcohol treatment in communities so that decarcerated people have the opportunity to rebuild their lives.

❖ Professionalization

As we look to the future, there are a number of concerns, in addition to the amount of incarceration, that should preoccupy those of us concerned about correctional practice. One issue that affects almost all areas of practice is that of professionalism. As indicated in Chapter 9, the effort to professionalize corrections has not yet yielded consistent fruit around the country. Some correctional institutions and programs have moved to enforce professional standards for their new hires, such as required college-level educational background, sufficient training, and pay that is commensurate with job requirements. However, most correctional organizations, perhaps primarily because of a lack of resources, have failed to move in a similar direction.

Yet hiring and keeping a professional staff are key to moving correctional institutions into the 21st century. When the correctional practitioner does not have the kind of education that acquaints him or her with the history, background, concepts, and research regarding corrections, then the correctional organization is simply ill prepared to meet the challenges it faces. Moreover, when turnover is high, because training and pay are insufficient, the organization becomes less stable and less equipped to problem solve regarding pressing concerns. Therefore, if we ever hope to move beyond the past and failed correctional endeavors and perspectives, the ranks of correctional practitioners need to be professionalized.

❖ Corrections Is a Relationship Business

The correctional experience for clients/offenders/inmates and staff, and the success of treatment and probation/parole programming, all hinge on the relationships among the people in these organizations. It is often said that the greatest expense for any public service organization is its staff. A collateral expense for correctional institutions and programs is the care of their inmates and/or clients. Notably, these expenses wax and wane to some degree based on the relationships among the actors. If those relationships are characterized by respect and concern among staff and respect and care (coupled with a healthy degree of control) between staff and clients/offenders, then there are less likely to be costly lawsuits, staff turnover, riots, and just general stress that produces discord in the workplace.

In his groundbreaking work on less explored and identified types of intelligence—emotional and social—Goleman (1995, 2006, p. 4) argues that scientific research on the brain indicates that we are "wired to connect" to others, which means that every time we engage with other human beings, we affect, and are affected by, their thoughts and consequently their behavior. Those relationships that are the most prolonged and intense in our lifetime are most likely to affect us, not just socially or emotionally, but biologically.

To a surprising extent, then, our relationships mold not just our experience but also our biology. The brain-to-brain link allows our strongest relationships to shape us on matters as benign as whether we laugh at the same jokes or as profound as which genes are (or are not) activated in T-cells, the immune system's foot soldiers in the constant battle against invading bacteria and viruses (Goleman, 2006, p. 5).

Goleman (2006) identifies a "double-edged sword" in relationships, in that those that are positive are healthful, but those that are negative can lead to stress, fear, frustration, anger, and despair, all emotions that can manifest themselves in physical ailments (p. 5). Of course, correctional environments are chock-full of stressed, fearful, frustrated, angry, and despairing people, and we are not just referring to the inmates here! So this means that unless correctional environments can foster some positive relationships between and among staff and clients/inmates, both will suffer psychologically and physically.

Recognition of the need to provide opportunities for inmates to "maturely cope" while under correctional supervision (see Chapter 7) would appear to be an acknowledgment that something positive can come out of the decent incapacitation of offenders. Moves to democratize workplaces and give people a voice and choice in their work (as discussed in Chapter 9) may serve to reduce some of the negative emotions associated with working in corrections. More recent attempts to "treat" rather than just "warehouse" inmates in institutions and offenders on probation and parole also represent a move to more positive relationships and so a better future for corrections. As indicated by the findings from the research presented in Chapters 5, 6, 8, and 14, there is reason to believe that some treatment and supervision tactics can work to help offenders as they endeavor to deal with their substance and other abuse issues.

❖ Privatization

The Profit Motive in Corrections

Privatization in corrections is not a new phenomenon. As was discussed in the first chapters of this book, transportation and the convict lease system were both based on privatization. The privatization of parts of prison operations (e.g., health care, food service, or work programs) continues in many public prisons. Since the 1980s, however, the number of completely private prisons has grown at both the state and federal level. Several of those prisons have experienced problems with escapes, violence, staff turnover, inexperienced staff and deficient staff training, brutality and abuse by staff, along with inadequate physical facilities (Camp & Gaes, 2002).

In her 1973 book, *Kind and Usual Punishment: The Prison Business,* Jessica Mitford details the misuse of public monies for prisons; instead of improving the diets and opportunities of inmates after the Attica riot (causes of the riot), for instance, millions of dollars were used by prison officials in New York to purchase riot gear and technology, as well as to hire more staff. Her argument is that closed off institutions, like prisons and jails, with their relatively powerless inmates, are particularly susceptible to graft and corruption of both the legal and the illegal sort. She notes that state legislatures are particularly susceptible to contributions by private entities who want to do business with corrections.

Forms of corruption and abuse of monies, or the illegal sort, are easy to spot, though not always stopped, Mitford (1973) complains, "Convicts will tell you about profitable deals made with local merchants for supplies in which the warden pockets a handsome rakeoff, unexplained shortages in the canteens, the disappearance of large quantities of food from the kitchen" (p. 172). It is the legally sanctioned graft in corrections, however, whereby money from state legislatures is intended for one type of purchase but is diverted to another— such as the hiring of more administrators rather than the provision of adequate food—that fascinated her and caused her to characterize the operation of public sector corrections in this country in the 1970s as a "business." She charges that corrections in many states were out to make a profit for administrators and their supporters (among vendors and state legislatures) and not to attend to their core mission of holding people decently and securely, while assisting them in their reform.

Almost 40 years later, if anything, the operation of corrections in the United States has become even more businesslike, in the worst way, with some commentators characterizing the system as the **corrections industrial complex** (Welch, 2005). In his last speech, President Eisenhower warned the nation about the development of the *military industrial complex,* or the collusion among politicians, defense contractors, and leaders in the military regarding the value of war and military spending as a money-making and power-generating enterprise for all three (Mills, 1956). Similarly, Welch argues that corrections has become such an enterprise for state legislatures, governors, and city and town leaders (politicians) who receive contributions from prison/jail contractors, vendors and private prison corporations (businesses); and directors and secretaries of corrections, wardens, sheriffs, and probation and parole managers (criminal justice officials) who participate in contracting with business. The criminal justice officials contract with the private sector businesses, not necessarily because they provide better services or even less-expensive services than the public sector, but because they are pressured to do so by the politicians who appoint/select and fund them, or they receive kickbacks themselves in terms of remuneration or jobs when they leave public service for the private sector.

The Walnut Grove Correctional Facility

The perfect example of how this kind of collusion among politicians, business and criminal justice officials in this corrections industrial complex can lead to gross injustices can be found in a small Mississippi town. One hour's drive east of Jackson, Mississippi, is the Walnut Grove Correctional Facility for boys and young men (they incarcerate up to age 22 to increase the number, and thus the profit the private corporation can make) in the town of Walnut Grove. It is the largest juvenile prison in the country, housing 1,200 juveniles and young men. The state of Mississippi pays the private corporation GEO Group to operate the prison, which they do for a profit. Before August of 2010, the prison was operated by another private prison corporation, Cornell Companies (August was when GEO Group bought Cornell Companies and so acquired the contract to operate the Walnut Grove facility).

In a two-part investigation of the Walnut Grove facility by National Public Radio (NPR), Burnett (2011) found that violence is set up and encouraged by the correctional officers. They also found that sexual abuse of the young male inmates, by female officers, is rife. Currently, the prison is under investigation by the U.S. Department of Justice and is being sued on behalf of 13 inmates by the Southern Poverty Law Center (SPLC) and the American Civil Liberties Union (ACLU).

> "When we began investigating conditions inside this facility and seeing how these kids were living with the beat downs and the sexual abuse and violence and corruption, it became a no-brainer. It became something we had to do," said Sheila Bedi, the lead attorney on the case and deputy legal director for the SPLC. (Burnett, 2011, p. 1)

According to Burnett (2011), the crux of the problem at Walnut Grove is the "guards." There are too few of them, there is little supervision of them, some are gang members themselves, and others are inclined to abuse the inmates either physically or sexually. The national average of officers to inmates for juveniles is 1 to 10, but the 2009 audit at Walnut Grove determined that the ratio was only 1 to 60 there. As staff are the most expensive item for any correctional entity to fund, cutting staff and their salaries is one way to assure profits for a money-making enterprise like this prison (the GEO Group is traded on the New York Stock Exchange and made $1 billion in 2010) (Burnett, 2011, p. 1). According to the audit done in 2009, "there were three inmate injuries a day. In the first six months of 2010 there was more than one fight a day, an assault on staff at least every other day and nine attempted suicides" (Burnett, 2011, p. 1).

NPR found after a review of public records that the warden and deputy wardens at Walnut Grove were receiving supplemental checks from the federal government for administering educational grants for the juveniles, in the amount of $2,500 to $5,000, when they were already paid by GEO (Burnett, 2011, p. 1). The town of Walnut Grove also "makes money" out of the existence and growth of the prison: The mayor of Walnut Grove claims that it funds the local police department. "'It's been a sweet deal for Walnut Grove,' Sims [the mayor] said. Indeed, every month, the prison pays the town $15,000 in lieu of taxes—which comprises nearly 15 percent of its annual budget" (Burnett, 2011, p. 1). In addition, a vending company owned by the mayor has 18 machines inside the prison. Moreover, the correctional authority that sends the Walnut Grove prison its grant money is given $4,500 per month by GEO. Left unexplored in the investigation is the question of why the Mississippi state legislature and governor's office authorized the private operation of Walnut Grove in the first place.

The Extent of Privatization and Its Problems

In a related report by an NPR reporter, it was noted that 8% of all inmates in prisons are held in private prisons, and in federal prisons that number rises to 16% (Shahani, 2011, p. 1). As the recession deepened and state budgets tightened in 2009 through 2011, the projection was for the use of private prisons to grow and for private corporations to diversify into the "alternatives to corrections" market. Though private prisons do not cost less to operate—in fact, they might be more expensive when all costs are accounted for—they are easier and faster to build, as they do not require the authorization of bonds from state legislatures (Camp & Gaes, 2002; General Accounting Office, 1996).

In a study of private prisons at the state and federal level by Camp and Gaes (2002), the researchers, who were employees of the Federal Bureau of Prisons at the time, found that "the private sector experienced significant problems with staff turnover, escapes and drug use" (p. 427). For instance, in 1999, secure facility private prisons experienced 18 escapes, while out of all of the Bureau of Prison facilities, a system that was larger than all of the private secure prisons combined, there was only 1 escape that year (p. 433). The researchers concluded that "[t]he failures that produce escapes or illegal drug use can result from problems in policy and procedures, in technology, and in staff capabilities" (p. 445).

❖ Concluding Thoughts

It is a stunning realization that much of the future looks like the past, but it is true, in a way. Current trends in corrections mimic those themes we laid out in the early chapters of this book. However, as has been demonstrated by the research presented throughout this book, there is also great progress in refining how we handle correctional practice and programming

There is little doubt that most correctional experiences for clients/inmates are not tinged with violence or brutality. The vast majority of correctional staff, whether in communities or institutions, act professionally, whether the attributes of their work fit that designation or not. Basic health care, clean housing, and nutritious food are provided to most incarcerated persons in the United States. Probation and parole officers do provide referrals to their clients when time permits and programs are available. Despite crowded caseloads, these officers usually make every effort to carefully watch the most dangerous of their charges. Jails, though often overcrowded, or at least overused, are generally helpful at ensuring the safety of suicide-prone or mentally disturbed inmates or people detoxing from drug or alcohol-induced highs. Jails may not represent the best places for such people, but they are usually safer than the streets and do provide a minimum of much-needed services for them. There is much more programming for those incarcerated in prisons, and even jails, and available to probationers and parolees than there was even 10 years ago. In short, though we tend to repeat our past mistakes, there has been some learning from them as well, as that is manifested in improved correctional practice.

Summary

- The current decarceration trend in a few states, which is apparently catching on in others, is also hopeful in that prisons can then be reserved for the truly violent and serious offenders, with the effect of increasing public safety, reducing minority community disruption, making the system "fairer" and more "just" as the punishments fit the crime, increasing the likelihood of successful reentry by offenders, and reducing the monetary costs of corrections for the public.
- Another currently popular movement, that of privatization, should give us pause, however. As a number of

studies and infamous examples indicate, private prisons tend to have more problems in the decent incarceration of inmates and the level of professionalism of their staff. For these reasons, and given the susceptibility of these institutions (and state legislatures and all political entities) to corruption and profiteering, privatization of whole institutions should be reconsidered.
- As the political winds shift away from purely punishment-oriented corrections, it will be interesting to see how correctional organizations, programs, and their actors will adjust in terms of privatization, professionalism, and decarceration.

Key Term

Corrections industrial complex

Discussion Questions

1. Discuss the evidence that indicates our correctional practices do not fit the amount of crime in the United States. Note how we compare with other countries, in terms of the use of incarceration.

2. Review the attributes of a professional and why and how the presence of those characteristics would serve to "improve" correctional operation.

3. What problems do "get tough" policies create for correctional operation? What benefits, if any, do they provide?

4. What is the connection between biology and environment in correctional operation? How do positive and negative environments affect the "biology" of those

who work in corrections and those who are clients and inmates within and outside of them?

5. Why do you think that private prisons have more problems with the operation of their facilities in terms of both inmate and staff management? Would you argue that we as a country should continue to authorize the construction and operation of private prisons and privatization in corrections? Why or why not?

6. In your opinion, what current initiatives in corrections offer the most promise for the future? Support this opinion with research and readings provided in this chapter or the rest of the text.

Useful Internet Sites

American Correctional Association: www.aca.org

American Jail Association: www.corrections.com/aja/

American Probation and Parole Association: www.appa-net.org

Bureau of Justice Statistics (information available on all manner of criminal justice topics): http://bjs.ojp.usdoj.gov/

National Criminal Justice Reference Service: www.ncjrs.gov

National Public Radio: www.npr.org

Vera Institute (information available on a number of corrections and other justice-related topics): www.vera.org

Glossary

Addiction: A psychobiological illness characterized by intense craving for a particular substance.

Amicus curiae briefs: "Friend of the court" briefs presented to the court, arguing in support of one side or the other, by interested parties not directly involved with the case.

Attica Prison Riot (1971): The riot began with a spontaneous act of violence by one inmate against an officer who had tried to break up a fight. The violence quickly spread because inmates were frustrated and angry about the overcrowded conditions, lack of programming, and other conditions of confinement. There were charges of racism by the mostly African American inmates regarding their treatment by the mostly white staff. As negotiations broke down, the prison was stormed by the state police and by correctional staff. As a consequence, 10 hostages and 29 inmates were dead or dying when the prison was secured and another 80 inmates had gunshot wounds. It was the bloodiest riot in American history.

Balanced approach: A three-pronged goal of the juvenile justice system: (1) to protect the community, (2) to hold delinquent youths accountable, and (3) to provide treatment and positive role models.

Big House prisons: According to Irwin (2005), these are fortress stone or concrete prisons, usually maximum security, whose attributes include "isolation, routine, and monotony" (p. 32). Strict security and rule enforcement, at least formally, and a regimented schedule are other hallmarks of such facilities.

Bill of Rights: The first 10 amendments to the United States Constitution.

Bridewells: Workhouses established in every English county in the 16th century. They were constructed to hold and whip or otherwise punish "beggars, prostitutes, and nightwalkers" and later as places of detention; their use began in London in 1553 (Orland, 1975, p. 16).

Bureaucracy: A type of organizational structure that includes these three elements: hierarchy, specialization, and rule of law.

Civil death statutes: Statutes in former times mandating that convicted felons lose all citizenship rights.

Civil rights claim: A "section 1983" claim that a person has been deprived of some legally granted right.

Classical School: The Classical School of penology/criminology was a non-empirical mode of inquiry similar to the philosophy practiced by the classical Greek philosophers—that is, "armchair philosophy."

Client management classification system (CMC): An assessment instrument used in probation mainly to assign probationers to a supervision level based on their risk level and needs.

Co-equal staffing: This practice involves the use of personnel processes that provide comparable pay and benefits for those who work in the jail to that for people who work on the streets as law enforcement in sheriff's departments.

Cognitive-behavioral therapy (CBT): A counseling approach that tries to address dysfunctional cognitions, emotions, and behaviors in a relatively short time through goal-oriented, systematic procedures using a mixture of operant psychology, cognitive theory, and social modeling theory.

Community corrections: A branch of corrections defined as any activity performed by agents of the state to assist offenders to reestablish functional law-abiding roles in the community while at the same time monitoring their behavior for criminal activity.

Community service order: Part of a disposition requiring probationers to work a certain number of hours doing tasks to help their communities.

Community jails: These are jails organized so those inmates engaged in educational, drug, or alcohol counseling or mental health programming in the community will seamlessly receive such services while incarcerated (primarily from community experts in those areas) and again as they transition out of the facility.

Concurrent sentence: Two-separate sentences are served at the same time.

Consecutive sentence: Two or more sentences that must be served sequentially.

Contract and lease systems: Systems devised to use inmates' labor. Inmate labor under Southern states' lease systems was leased by the prison to farmers or other contractors. Inmates under a contract system in northern and midwestern prisons worked in larger groups under private or public employers.

Contrast effect: The effect of punishment on future behavior depends on how much the punishment and the usual life experience of the person being punished differ or contrast.

Convict code: These are the informal rules that inmates live by vis-à-vis the institution and staff and include "1. Do not inform; 2. Do not openly interact or cooperate with the guards or the administration; 3. Do your own time" (Irwin, 2005, p. 33).

Correctional boot camps: Facilities modeled after military boot camps where young and nonviolent offenders are subjected to military-style discipline and physical and educational programs.

Corrections industrial complex: The collusion among politicians, business, and criminal justice officials to make money for themselves and their organizations off of correctional services, construction, and operations.

Correctional institutions: (This term originally applied only to prisons, but now can refer to jails as well.) Correctional institutions (prisons) carefully classify inmates into treatment programs that address their needs and perceived deficiencies. They are also intended to be places where inmates can earn "good time" and eventual parole. Correctional institutions use the medical model to treat inmates who are believed to be "sick" and in need of a treatment regimen, provided by the prison, which will address that sickness and hopefully "cure" the inmates so that they might become productive members of society.

Corrections: A generic term covering a wide variety of functions carried out by government (and, increasingly, private) agencies having to do with the punishment, treatment, supervision, and management of individuals who have been accused (in the case of some jail inmates) or convicted of criminal offenses.

Deferred adjudication: A decision made by certain criminal justice personnel to delay or defer formal court proceedings if a youth follows probation conditions.

Deinstitutionalization of the mentally ill: This happened in the United States as a result of the civil rights movement and the related effort to increase the rights of people involuntarily committed to mental hospitals.

Delinquents: Juveniles who commit acts that are criminal when committed by adults.

Determinate sentence: A prison sentence of a fixed number of years that must be served rather than a range.

Deterrence: A philosophy of punishment aimed at the prevention of crime by the threat of punishment.

Discretion: The ability to make choices and to act or not act on them.

Discretionary parole: Parole granted at the discretion of a parole board for selected inmates who have earned it.

Discrimination: Occurs when a person or group is treated differently because of who they are (e.g., race, ethnicity, gender, age, disability, religion, nationality, sexual orientation or identity, or income) rather than their abilities or something they did.

Disparity: Occurs when one group is treated differently, and unfairly, by governmental actors, as compared to other groups.

Driving While Black or Brown (DWB): Refers to the practice of police focusing law enforcement attention on black- or brown-skinned drivers.

Drug court: A special sentence for drug-related, nonviolent offenders who must then complete an extensive drug treatment program.

Eighth Amendment: Constitutional amendment that forbids cruel and unusual punishment.

Electronic monitoring (EM): A system by which offenders under house arrest can be monitored for compliance using computerized technology such as an electronic device worn around the offender's ankle.

Elmira Reformatory: Founded in 1876 in New York as a model prison in response to calls for the reform of prisons from an earlier era. The inmates were to be younger men. It would encompass all of the rehabilitation focus and graduated reward system (termed the marks system) that reformers were agitating for. The lash was not to be used. Elmira was supposed to hire an educated and trained staff and to maintain uncrowded facilities (Orland, 1975). The operation of Elmira led to the creation of good time, the indeterminate sentence (defined in Chapter 4), a focus on programming to address inmate deficiencies, and the promotion of probation and parole. Ultimately, in practice, it was not the model reformers had hoped it would be.

Enlightenment: Period in history in which a major shift in the way people began to view the world and their place in it occurred, moving from a supernaturalistic to a naturalistic and rational worldview.

Ethics: Refers to right or wrong behavior on the job.

Ethnicity: Refers to the differences between groups of people based on culture. An ethnic group will often have a distinct language, as well as particular values, religion, history, and traditions. Ethnic groups may be made up of several races and have a diverse national heritage.

Evidence-based practices (EBP): EBP means that in order to reduce recidivism, corrections must implement practices that have consistently been shown to be effective.

Fair Sentencing Act of 2010: The act mandated that the amount of crack cocaine subject to the 5-year minimum sentence be increased from 5 grams to 28 grams, thus reducing the 100-to-1 ratio to an 18-to-1 ratio (28 grams of crack gets as much time as 500 grams of powder cocaine).

First Amendment: Guarantees freedom of religion, speech, press, and assembly.

Fourteenth Amendment: Contains the due process clause, which declares that no state shall deprive any person of life, liberty, or property without due process of law.

Fourth Amendment: Guarantees the right to be free from unreasonable searches and seizures.

Galley slavery: This was used as a sentence for crimes or as a means of removing the poor from streets. Galley slavery also served the twin purpose of providing the requisite labor—rowing—needed to propel ships for seafaring nations interested in engagement in trade and warfare. Used by the ancient Greeks and Romans and in the Middle Ages, roughly ending sometime in the 1700s.

Gangs (prison): Groups of people with similar interests who socialize together and support each other, but who also engage in deviant or criminal activities. Prison gangs have a hierarchical organizational structure and a set, and often strict, code of conduct for members.

General deterrence: The presumed preventive effect of the threat of punishment on the general population; that is, deterrence aimed at *potential* offenders.

Global Positioning System (GPS): A system of probation/parole supervision whereby probationers/parolees are required to wear a tracking unit that can be monitored by satellites.

Great Law: William Penn's idea, based on Quaker principles, de-emphasized the use of corporal and capital punishment for all crimes but the most serious.

Habeas corpus: Latin term meaning "you have the body." It is a court order requiring that an arrested person be brought before it to determine the legality of detention.

Habitual offender statutes: Statutes mandating that offenders with a third felony conviction be sentenced to life imprisonment regardless of the nature of the third felony.

Hack: A correctional officer in a prison who is a violent, cynical, and alienated keeper of inmates (Johnson, 2002).

Halfway houses: Transitional places of residence for correctional clients who are "half way" between the constant supervision of prison and the much looser supervision in the community.

Hands-off doctrine: An early American court-articulated belief that the judiciary should not interfere with the management and administration of prisons.

Hedonism: A doctrine maintaining that all goals in life are means to the end of achieving pleasure and/or avoiding pain.

Hedonistic calculus: A method by which individuals are assumed to logically weigh the anticipated benefits of a given course of action against its possible costs.

Hostile environment: Occurs when the workplace is sexualized with jokes, pictures, or in other ways that are offensive to one gender.

House arrest: Programs that require offenders to remain in their homes except for approved periods to travel to work, school, or other approved destinations.

Houses of refuge: Part of the Jacksonian movement (named after President Andrew Jackson) of the early 1800s to use institutions as the solution for social problems. Their stated purpose was to remove impressionable youth, mainly boys, but also girls, from the contamination that association with more hardened adult prisoners might bring.

Hulks: These were and are derelict naval vessels transformed into prisons and jails.

Human service: Describes a correctional officer who provides "goods and services," serves as an "advocate" for inmates when appropriate, and assists them with their "adjustment" and through "helping networks" (Johnson, 2002, p. 242–259).

Importation: This is what occurs when inmates bring aspects of the larger culture into the prison.

Incapacitation: A philosophy of punishment that refers to the inability of criminals to victimize people outside prison walls while they are locked up.

Indeterminate sentence: A prison sentence consisting of a range of years to be determined by the convict's behavior, rather than one of a fixed number of years.

Intensive supervision probation (ISP): Probation that involves more frequent surveillance of probationers and that is typically limited to more serious offenders in the belief that there is a fighting chance that they may be rehabilitated (or to save the costs of incarceration).

Intermediate sanctions: Refers to a number of innovative alternative sentences that may be imposed in place of the traditional prison/probation dichotomy.

Irish system: A prison system used in the 19th century. This system involved four stages, beginning with a 9-month period of solitary confinement, the first 3 months with reduced rations and no work. This period of enforced idleness was presumed to make even the laziest of men yearn for some kind of activity.

Jails: These are local community institutions that hold people who are presumed innocent before trial; convicted people before they are sentenced; convicted minor offenders who are sentenced for terms that are usually less than a year; juveniles (usually in their own jails or separated from adults or before transport to juvenile facilities); women (usually separated from men and sometimes in their own jails); people for the state or federal authorities; and, depending on the particular jail population being served, and the capacity of any given facility, they serve to incapacitate, deter, rehabilitate, punish, and reintegrate.

Jim Crow laws: Laws devised by Southern states following the Civil War, starting in the 1870s and lasting until 1965 and the civil rights movement, to prevent African Americans from fully participating in social/economic and civic life. These laws restricted the rights and liberties of black citizens in employment, housing, education, travel, and voting. Voter disenfranchisement, or preventing African Americans from voting, was a key part of the Jim Crow laws.

Judicial reprieve: British and early American practice of delaying sentencing following a conviction that could become permanent, depending on the offender's behavior.

Justice: A moral concept about just or fair treatment consisting of treating equals equally and unequals unequally according to relevant differences.

Liberal feminist: One who believes that the problem for girls and women involved in crime lies more with the social structure around them (e.g., poverty and lack of sufficient schooling or training, along with patriarchal beliefs), and that the solution lies in preparing them for an alternate existence so that they do not turn to crime.

Liberty interest: Refers to an interest in freedom from governmental deprivation of liberty without due process.

Life without parole (LWOP): A life sentence with the additional condition that the person never be allowed parole—"life" *means* life for those receiving an LWOP sentence.

Mandatory parole: Automatic parole after a set period of time for almost all inmates.

Mandatory sentence: A prison sentence imposed for crimes for which probation is not an option, where the minimum time to be served is set by law.

Marks system: A graduated reward system for prisons, developed by Maconochie in which, if one behaves, it is possible to earn "marks," which in turn entitle one to privileges.

Mature coping: This occurs in prisons when the inmate deals "with life's problems like a responsive and responsible human being, one who seeks autonomy without violating the rights of others, security without resort to deception or violence, and relatedness to others as the finest and fullest expression of human identity" (Johnson, 2002, p. 83).

Maximum security prisons: Prisons where both external and internal security are high. Inmates are often locked up for all or a large part of the day, save for showering or recreation outside of their cell, and they are ideally in single cells, deprived of much contact with others. Programming is limited or nonexistent. Visits and contact with the outside are restricted. The exterior security consists of some combination of layers of razor wire, walls, lights, cameras, armed guards, and attack dogs on patrol. Inmates in maximum security tend to be those who have very serious offenses or those who have problems with adhering to the rules in prisons. Death rows, if a state has capital punishment, are usually located in maximum security prisons.

Medical model: Rehabilitation model that assumes criminals are sick and need treatment.

Medium security prisons: Prisons that hold a mix of people in terms of crime categories, but who program well. Medium security prisons have high external security, but inmates are able to move around more freely within the "walls." Some are built like a college campus, with several buildings devoted to distinct purposes.

Mental disorders: Clinically significant conditions characterized by alterations in thinking, mood, or behavior associated with personal distress or impaired functioning.

Minimum security prisons: Prisons created for lower-level felony offenders and those who are "short timers" or people who are relatively close to a release date. Those sent here are not expected to be an escape or behavioral problem. The ability and willingness to work is often a prerequisite for classification to this type of facility.

Mortification: Process that occurs as inmates enter the prison and they suffer from the loss of the many roles they occupied in the wider world (Goffman, 1961; Sykes, 1958). Instead, only the role of "inmate" is available, a role that is formally powerless and dependent.

Mount Pleasant Prison: The first prison constructed for women in the United States. Built in 1839 close to the Sing Sing (New York) Prison for men, it was in part administered by it, but had its own buildings, staff, and administrator.

Needs principle: A principle that refers to an offender's prosocial needs, the lack of which puts him or her at risk for reoffending, and that suggests these needs should receive attention in program targeting.

Net widening: Occurs when criminal justice system practices or programming brings people into greater contact with the system, when the intent was to decrease such contact.

New generation or podular direct supervision jails: Jails that have two key components: a rounded or "podular" architecture for living units and the "direct," as opposed to indirect or intermittent, supervision of inmates by staff; in other words, staff are to be in the living units full time. Other important facets of these jails are the provision of more goods and services in the living unit (e.g., access to telephones, visiting booths, recreation, library books) and more enriched leadership and communication roles for staff.

New Mexico Prison Riot (1980): This prison exploded in a riot over the conditions of confinement and crowding, which were at epidemic levels. Despite repeated warnings that a riot was going to occur, the administration and staff failed to adequately prepare. The state eventually retook the prison; however, over 3 days, 33 inmates were killed by other inmates. Numerous other inmates, along with staff hostages, were beaten or raped and millions of dollars in damage was done (Useem, 1985; Useem & Kimball, 1989).

New York prison system (Auburn and Sing Sing): These prisons, and those modeled after them, included congregate work and eating arrangements, but silent and separate housing. This mode of operation allowed prison managers to offset the cost of incarceration by allowing inmates to work together and hence produce more. When inmates were allowed out of their cells, they could also perform maintenance jobs and other tasks in the prison, such as cleaning and cooking, which also reduced the cost of incarcerating. The ball and chain, lockstep, and striped uniforms also originated in these prisons.

Newgate Prison in New York City: Built in 1797, the prison was operated based on Quaker ideals and so focused on rehabilitation, religious redemption, work programs to support prison upkeep, and no corporal punishment. The builders of Newgate even constructed a prison hospital and school for the inmates. Because of crowding, single celling was not possible for most, and a number of outbreaks of violence erupted (such as a riot in 1802).

Newgate Prison in Simsbury, Connecticut: According to Phelps (1860/1996), this early colonial "prison" started as a copper mine, and during its 54 years of operation (from 1773 to 1827), some 800 inmates passed through its doors. The mine was originally worked in 1705, and one-third of the taxes it paid to the town of Simsbury at that time were used to support Yale College (p. 15).

Norfolk Island: An English penal colony, operated 1,000 miles off the Australian coast. It was established in 1788 as a place designated for prisoners from England and Australia and was regarded as a brutal and violent island prison where inmates were poorly fed, clothed, and housed and were mistreated by staff and their fellow inmates. Alexander Maconochie, an ex-naval captain, asked to be transferred to Norfolk, usually an undesirable placement, so that he could put into practice some ideas he had about prison reform, which he successfully did, vastly improving conditions at the prison.

Overcrowding: A phenomenon that occurs when the number of inmates exceeds the physical capacity (the beds and space) available.

Pains of imprisonment: Gresham Sykes (1958) described these perils as the "deprivation of liberty, the deprivation of goods and services, the deprivation of heterosexual relationships, the deprivation of autonomy, and the deprivation of security" (pp. 63–83).

Panopticon: An idea of Jeremy Bentham's that ingeniously melded the ideas of improved supervision with architecture (because of its rounded and open and unobstructed views), which would greatly enhance supervision of inmates.

Parens patriae: A legal principle giving the state the right to intercede and act in the best interest of the child or any other legally incapacitated persons, such as the mentally ill.

Parole: The release of prisoners from prison before completing their full sentence.

Parole board: A panel of people presumably qualified to make judgments about the suitability of a prisoner to be released from prison after having served some specified time of his or her sentence.

Patriarchy: Involves the attitudes, beliefs, and behaviors that value men and boys over women and girls (Daly & Chesney-Lind, 1988). Members of patriarchal societies tend to believe that men and boys are worth more than women and girls. They also believe that women and girls, as well as men and boys, should have certain restricted roles to play and that those of the former are less important than those of the latter. Therefore, education, work training that helps one make a living, and better pay are more important to secure for men and boys than for women and girls, who are best suited for more feminine, and by definition in a patriarchal society, less worthy, professions.

Pennsylvania prison system (Walnut Street Jail, Western and Eastern Pennsylvania Prisons): These prisons and those that were modeled after them emphasized silent and separate eating, and working and living arrangements that isolated inmates in their cells, restricted their contact with others, and reinforced the need for penitence. When labor was allowed at all, it was a solitary affair in one's cell.

Penology: Refers to the study of the processes adopted and the institutions involved in the punishment and prevention of crime.

Positivists: Those who believe that human actions have causes, and that these causes are to be found in the thoughts and experiences that typically precede those actions.

Power: The ability to "get people to do what they otherwise wouldn't" (Dahl, 1961).

Predisposition report: A report done in juvenile courts that is analogous to a presentence investigation report in adult courts (see below).

Presentence investigation report (PSI): Report written by the probation officer informing the judge of various aspects of the offense for which the defendant is being sentenced as well as providing information about the defendant's background (educational, family, and employment history), character, and criminal history.

Principle of utility: A principle that posits that human action should be judged moral or immoral by its effects on the happiness of the community and that the proper function of the legislature is to make laws aimed at maximizing the pleasure and minimizing the pains of the population: "the greatest happiness for the greatest number."

Prison Rape Elimination Act of 2003: Act that mandated that the Bureau of Justice Statistics collect data on sexual assaults in adult and juvenile jails and prisons and that they identify facilities with high levels of victimization.

Prison subculture: This is the norms, values, beliefs and even language of the prison.

Prisoner's Litigation Reform Act (1996): The primary intention of the PLRA was to free state prisons and jails from federal court supervision and to limit prisoners' access to the federal courts.

Prisonization: The adopting of the inmate subculture by inmates.

Prisons: Correctional facilities that have a *philosophy of penitence* (hence "penitentiary") and that were created as a grand reform, as they represented, in theory at least, a major improvement over the brutality of punishment that characterized early Western, English, and American law and practice.

Probation: A sentence imposed on convicted offenders that allows them to remain in the community under the supervision of a probation officer instead of being sent to prison.

Profession: Regarding the positions of corrections officers and staff, a profession is distinguished by prior educational attainment involving college, formal training on the job or just prior to the start of the job, pay and benefits that are commensurate with the work, the ability to exercise discretion, and work that is guided by a code of ethics.

Punishment: The act of imposing some unwanted burden such as fines, probation, imprisonment, or death on convicted persons in recompense for their crimes.

Quid pro quo sexual harassment: Involves something for something, as in you give your boss sexual favors and he allows you to keep your job.

Race: Refers to the skin color and features of a group of people. It is based on biology.

Recidivism: Occurs when an ex-offender commits further crimes.

Reentry: The process of reintegrating offenders back into the community after release from jail or prison. Part of that process is preparing offenders through the use of various programs targeting their risks and needs so that they will have a fighting chance of remaining in the community.

Rehabilitation: A philosophy of punishment aimed at "curing" criminals of their antisocial behavior.

Reintegration: A philosophy of punishment that aims to use the time criminals are under correctional supervision to prepare them to reenter the free community as well-equipped to do so as possible.

Responsivity principle: A principle maintaining that if offenders are to respond to treatment in meaningful and lasting ways, counselors must be aware of their different developmental stages, learning styles, and need to be treated with respect and dignity.

Restitutive justice: A philosophy of punishment driven by simple deterrence and a need to repair the wrongs done.

Restorative justice: A system of justice that gives approximately equal weight to community protection, offender accountability, and the offender. It is oriented to repairing the harm caused by the crime and involves face-to-face confrontation between victim and perpetrator in the hope of arriving at a mutually agreeable solution.

Retribution: A philosophy of punishment that demands that criminals' punishments match the degree of harm they have inflicted on their victims, i.e., what they justly deserve.

Retributive justice: A philosophy of punishment driven by a passion for revenge.

Risk, needs, and responsivity (RNR) model: A treatment correctional model that maintains that offenders and the community are better served if offenders' risks for reoffending and their needs (their deficiencies, such as lack of job skills) are addressed in a way that matches their developmental stage.

Risk principle: A principle that refers to an offender's probability of reoffending and maintains that those with the highest risk should be targeted for the most intense treatment.

Role: What a person does on the job every day.

Section 1983 suits: A mechanism for state prison inmates to sue state officials in federal court regarding their confinement and their conditions of confinement. Section 1983 is part of the Civil Rights Act of 1871, which was initially enacted to protect Southern blacks from state officials and the Ku Klux Klan.

Selective incapacitation: Refers to a punishment strategy that largely reserves prison for a distinct group of offenders composed primarily of violent repeat offenders.

Sentence: A punitive penalty ordered by the court after a defendant has been convicted of a crime either by a jury or in a plea bargain.

Sentencing disparity: Wide variation in sentences received by different offenders that may be legitimate or discriminatory.

Sentencing guidelines: Scales for numerically computing sentences that offenders should receive based on the crime they committed and on their criminal records.

Sex offender: A broad category of offenders ranging from nuisance "flashers" to true sexual predators.

Shock probation: A type of sentence aimed at shocking offenders into going straight by exposing them to the reality of prison life for a short period followed by probation.

Specific deterrence: The supposed effect of punishment on the future behavior of persons who experience the punishment.

Split sentences: Sentences that require convicted persons to serve brief periods of confinement in a county jail prior to probation placement.

Stanford Prison Experiment: A 1971 experiment conducted at Stanford University, which utilized volunteer students, divided into officers and inmates in a makeshift "prison" (Haney, Banks, & Zimbardo, 1981). In the end, about a third of the "officers" engaged in the abuse of "inmates," and other officers stood by while it was going on. The experiment was stopped after a few days and is often referenced as an example of how correctional work, and the subcultures that develop as part of the job, can foster corrupt behavior by officers.

Stateville Prison: Built in Illinois as a panopticon in 1925 in reaction to the deplorable conditions of the old Joliet, Illinois, prison built in 1860 (J. B. Jacobs, 1977).

Status offenders: Juveniles who commit certain actions that are legal for adults, but not for children, such as smoking and not obeying parents.

Status offenses: Offenses that apply only to juveniles such as disobeying parents or smoking.

Supermax prisons: High-security prisons, both internally and externally, that hold those who are violent or disruptive in other prisons in the state or federal system. Inmates are confined to their windowless cells 24 hours a day, except for showers 3 times a week (where they are restrained) and solitary exercise time a couple times a week; they eat in their cell. Visiting and programming are very limited.

Therapeutic communities (TCs): Residential communities providing dynamic "mutual self-help" environments and offering long-term opportunities for attitude and behavioral change and the learning of constructive prosocial ways of coping with life.

Total institution: "A place of residence and work where a large number of like-situated individuals, cut off from the wider society for an appreciable period of time, together lead an enclosed, formally administered round of life" (Goffman, 1961, p. 6).

Transportation: A sentence whereby the convicted person would be sent to another country, via a ship whose captain owned the right to sell the convict's labor for a period of time. The captain would, in turn, sell the convicted person's labor to others in the new land. It was in use for roughly 350 years by English or European countries, from 1607 to 1953.

Truth-in-sentencing laws: Laws that require there be a truthful, realistic connection between the sentences imposed on offenders and the time they actually serve.

Unconditional Release: A type of release from prison for inmates who have completed their entire sentences. They are therefore released unconditionally—with no parole.

United States Sentencing Commission: A commission charged with creating mandatory sentencing guideline to control judicial discretion.

Victim–offender reconciliation programs (VORPs): Programs designed to bring offenders and their victims together in an attempt to reconcile ("make right") the wrongs offenders have caused; an integral component of the restorative justice philosophy.

Waiver: Refers to a process by which a juvenile offender is "waived" (transferred) to an adult court because he or she has committed a particularly serious crime or is habitually delinquent..

Warehouse prison: Large prisons, of any security level, but more likely a super maximum or maximum security prison, where inmates' lives and movement are severely restricted and rule-bound. There is no pretense of rehabilitation in warehouse prisons; punishment, incapacitation, and deterrence are the only justifications for such places. The more hardened and dangerous prisoners are supposed to be sent there, and their severe punishment is to serve as a deterrent to others in lower security prisons.

Work release programs: Programs designed to control offenders in a secure environment while at the same time allowing them to maintain employment.

References

Cases Cited

Atkins v. Virginia, 536 U.S. 304 (2002).
Bell v. Wolfish, 441 U.S. 520 (1979).
Breed v. Jones, 421 U.S. 517 (1975).
Coker v. Georgia, 433 U.S. 584 (1977).
Cooper v. Pate, 378 U.S. 546 (1964).
Eddings v. Oklahoma, 445 U.S. 104 (1982).
Estelle v. Gamble, 429 U.S. 97 (1976).
Ex parte Crouse, 4 Whart. 9 (Pa. 1838).
Ex parte Hull, 312 U.S. 546 (1941).
Ex parte United States, 242 U.S. 27 (1916).
Furman v. Georgia, 408 U.S. 238 (1972).
Graham v. Florida, 560 U.S. ___ (2010).
Gregg v. Georgia, 428 U.S. 153 (1976).
Holt v. Sarver, 309 F. Supp. 362 (1969).
Hudson v. McMillian, 503 U.S. 1 (1992).
Hudson v. Palmer, 468 US 517 (1984).
In re Gault, 387 U.S. 1 (1967).
In re Winship, 397 U.S. 358 (1970).
Jones v. Cunningham, 371 U.S. 236 (1963).
Jordan v. Gardner, 986 F.2d 1521 (9th Cir. 1993).
Kansas v. Hendricks, 521 U.S. 346 (1997).
Kent v. United States. 383 U.S. 541 (1966).
McCleskey v. Kemp, 481 U.S. 279 (1987).
McKeiver v. Pennsylvania. 402 U.S. 528 (1971).
Mempa v. Rhay, 389 U.S. 128 (1967).
Morrissey v. Brewer, 408 U.S. 471 (1972).
Panetti v. Quarterman, 551 U.S. 930 (2007).
Pennsylvania Board of Probation and Parole v. Scott. 524 U.S. 357 (1998).
Penry v. Lynaugh. 492 U.S. 302 (1989).
Pervear v. Massachusetts, 72 U.S. 475 (1866).
Powell v. Alabama, 287 U.S. 45 (1932).
Pulido v. State of California, et al. (Marin Co. Sup. Ct. 1994).
Roper v. Simmons, 112 S.W. 3rd 397 (2005).
Ruffin v. Commonwealth, 62 Va. 790, 796 (1871).
Rummel v. Estelle, 445 U.S. 263, (1980).
Salazar et al. v. City of Espanola et al. (2004).
Sandin v. Conner, 515 U.S. 472 (1995).
Schall v. Martin, 104 U.S. 2403 (1984).
Smith v. Mosley, 532 F.3d 1270 (11th Cir. 2008),
Thompson v. Oklahoma, 487 U.S. (1988).
Turner v. Safley, 482 U.S. 78 (1987).
United States v. Booker, 543 U.S. 220 (2005).
United States v. Knights, 534 U.S. 112 (2001).
Wilson v. Seiter, 501 U.S 294 (1991).
Wolff v. McDonnell, 418 U.S. 539 (1974).
Woodson v. North Carolina, 428 U.S. 280 (1976).

Works Cited

Abadinsky, H. (1993). *Drug abuse: An introduction.* Chicago: Nelson-Hall.
Abadinsky, H. (2009). *Probation and parole: Theory and practice.* Upper Saddle River, NJ: Prentice Hall.
Albanese, J., & Pursley, R. (1993). *Crime in America: Some existing and emerging issues.* Englewood Cliffs, NJ: Regents/Prentice Hall.
Alcock, J. (1998). *Animal behavior: An evolutionary approach* (6th ed.). Sunderland, MA: Sinauer Associates.
Alexander, M. (2010). *The new Jim Crow: Mass incarceration in the age of colorblindness.* New York: The New Press.
Alexander, R. (2004). The United States Supreme Court and the civil commitment of sex offenders. *The Prison Journal, 84,* 361–378.
Alosi, T. (Writer & Director). (2008). *Eastern State: Living behind the walls* [Documentary film]. Folsom, CA: Dark Hollow Films.
Alvarado, J. (2009). Keeping jailers from keeping the keys to the courthouse: The Prison Litigation Reform Act's exhaustion requirement and Section 5 of the Fourteenth Amendment. *Seattle Journal for Social Justice, 8,* 323–365.
American Correctional Association. (1983). *The American prison: From the beginning . . . A pictorial history.* Lanham, MD: Author.
Amnesty International. (2004). *Abuse of women in custody: Sexual misconduct and shackling of pregnant women. A state-by-state survey of policies and practices in the United States.* New York: Author. Available at http://www.amnestyusa.org
Amnesty International. (2005). *Death penalty: 3,977 executed in 2004.* London: Author.

Amnesty International. (2006). *Death sentences and executions in 2005.* Available at http://web.amnesty.org/web/web/nsf/print/

Andrews, D., Bonta, J., & Wormith, J. (2006). The recent past and near future of risk and/or needs assessment. *Crime & Delinquency, 52,* 7–27.

Antonio, M. E., Young, J. L., & Wingeard, L. M. (2009). When actions and attitude count most: Assessing perceived level of responsibility and support for inmate treatment and rehabilitation programs among correctional employees. *The Prison Journal, 89*(4), 363–382.

Applegate, B. K., Cullen, F. T., & Fisher, B. S. (2001). Public support for correctional treatment: The continuing appeal of the rehabilitative ideal. In E. J. Latessa, A. Holsinger, J. W. Marquart, & J. R. Sorensen (Eds.), *Correctional contexts: Contemporary and classical readings* (2nd ed., pp. 506–519). Los Angeles: Roxbury.

Applegate, B. K., & Paoline, E. A., III. (2007). Jail officers' perceptions of the work environment in traditional versus new generation facilities. *American Journal of Criminal Justice, 31,* 64–80.

Applegate, B. K., & Sitren, A. H. (2008). The jail and the community: Comparing jails in rural and urban contexts. *The Prison Journal, 88*(2), 252–269.

Archibold, R. C. (2010, April 23). Arizona enacts stringent law on immigration. *New York Times.* Available at http://www.nytimes.com/2010/04/24/us/politics/24immig.html

Ashcroft, J., Daniels, D., & Herraiz, D. (1997). *Defining drug courts: The key components.* Washington, DC: U.S. Department of Justice.

Austin, J., & Irwin, J. (2001). *It's about time: America's imprisonment binge.* Belmont, CA: Wadsworth.

Balazs, G. (2006). The workaday world of a juvenile probation officer. In A. Walsh, *Correctional assessment, casework & counseling* (4th ed., pp. 414–416). Alexandria, VA: American Correctional Association.

Bard, S. (1997, July 31). Idaho inmates tell of assault in Texas: Nurses, who prisoners say fondled them, have been arrested and fired. *Idaho Statesman,* p. B1.

Barker, V. (2006). The politics of punishing: Building a state governance theory of American imprisonment variation. *Punishment & Society, 8,* 5–32.

Barlow, L. W., Hight, S., & Hight, M. (2006). Jails and their communities: Piedmont regional jail as a community model. *American Jails, 20,* 38–45.

Barton, A. (1999). Breaking the crime/drugs cycle: The birth of a new approach? *The Howard Journal of Criminal Justice, 38,* 144–157.

Baunach, P. J. (1992). Critical problems of women in prison. In I. L. Moyer (Ed.), *The changing role of women in the criminal justice system* (pp. 99–112). Prospect Heights, IL: Waveland Press.

Bazelon Center for Mental Health Law. (2008). *Individuals with mental illness in jail and prison.* Available at http://www.bazelon.org/issues/criminalization/factssheets/criminal3.html

Bazemore, G. (2000). What's "new" about the balanced approach? In P. Kratkoski (Ed.), *Correctional counseling and treatment* (pp. 1–22). Prospect Heights, IL: Waveland Press.

Beaumont, G., & Tocqueville, A. de. (1964). *On the penitentiary system in the United States and its application in France.* Carbondale, IL: Southern Illinois University Press. (Original work published 1833)

Beccaria, C. (1963). *On crimes and punishment* (H. Paulucci, Trans.). Indianapolis, IN: BobbsMerrill. (Original work published 1764)

Beccaria, C. (2003). *On crimes and punishments and other writings.* Cambridge, UK: Cambridge University Press. (Original work published 1764)

Beck, A. J., & Harrison, P. M. (2007). *Sexual violence reported by correctional authorities, 2006.* Bureau of Justice Statistics, Special Report. Washington, DC: U.S. Department of Justice.

Beck, A. J., & Harrison, P. M. (2010). *Sexual victimization in prisons and jails reported by inmates, 2008–2009.* Washington, DC: U.S. Department of Justice, Office of Justice Programs. Available at http:bjs.ojp.usdoj.gov

Beck, A. J., Harrison, P. M., & Adams, D. B. (2007). *Sexual violence reported by correctional authorities, 2006.* Washington, DC: U.S. Department of Justice, Office of Justice Programs. Available at http://www.ojp.usdoj.gov/bjs/pub/pdf/svrca06.pdf

Becker, G. (1997). The economics of crime. In M. Fisch (Ed.), *Criminology 97/98* (pp. 15–20). Guilford, CT: Duskin.

Beckett, K., Nyrop, K., Pfingst, L., & Bowen, M. (2005). Drug use, drug possession arrests, and the question of race: Lessons from Seattle. *Social Problems, 52,* 419–441.

Belbot, B., & Hemmens, C. (2010). *The legal rights of the convicted.* El Paso, TX: LFB Scholarly Publishing.

Belknap, J. (2001). *The invisible woman: Gender, crime, and justice* (2nd ed.). Belmont, CA: Wadsworth.

Bennett, K. (1995). Constitutional issues in cross-gender searches and visual observation of nude inmates by opposite-sex officers: A battle between and within the sexes. *The Prison Journal, 75,* 90–112.

Bentham, J. (1948). *A fragment on government and an introduction to the principles of morals and legislation* (W. Harrison, Ed.). Oxford, UK: Basil Blackwell. (Original work published 1789)

Bentham, J. (1969). *Panopticon papers. A Bentham reader* (M. P. Mack, Ed.). New York: Pegasus. (Original work published 1789)

Bentham, J. (2003). *The rationale of punishment.* London: Robert Heward. (Original work published 1811)

Bergner, D. (1998). *God of the rodeo: The quest for redemption in Louisiana's Angola Prison.* New York: Ballantine Books.

Binder, A., Geis, G., & Bruce, D. (2001). *Juvenile delinquency: Historical, cultural, and legal perspectives.* Cincinnati, OH: Anderson.

Bindzus, D. (2001). The practice of Naikan in Japanese prisons. In E. Fairchild & H. Dammer, *Comparative criminal justice systems* (pp. 264–266). New York: Wadsworth.

Bissonnette, G. (2006). "Consulting" the federal sentencing guidelines after Booker. *UCLA Law Review, 54,* 1497–1547.

Black, M., & Smith, R. (2003). *Electronic monitoring in the criminal justice system.* (Trends and Issues in Crime and Criminal Justice No. 254). Canberra: Australian Institute of Criminology.

Blackburn, A. G., Mullings, J. L., & Marquart, J. W. (2008). Sexual assault in prison and beyond: Toward an understanding of lifetime sexual assault among incarcerated women. *The Prison Journal, 88*(3) 351–377.

Blalock, H. M. (1967). *Toward a theory of minority group relations.* New York: Wiley.

Bleckman, E., & Vryan, K. (2000). Prosocial family therapy: A manualized preventive intervention for juveniles. *Aggression and Violent Behavior, 5,* 343–378.

Blonigen, D. (2010). Explaining the relationship between age and crime: Contributions from the developmental literature on personality. *Clinical Psychology Review, 30,* 89–100.

Bohm R., & Haley, K. (2007). *Introduction to criminal justice.* New York: McGraw-Hill.

Bonczar, T. P. (2008). *Characteristics of state parole supervising agencies, 2006.* Washington, DC: U.S. Department of Justice, Office of Justice Programs, Bureau of Justice Statistics. Available at http://bjs.ojp.usdoj.gov

Bookman, C. R., Lightfoot, C. A., & Scott, D. L. (2005). Pathways for change: An offender reintegration collaboration. *American Jails, 19,* 9–14.

Broussard, D., Leichliter, J. S., Evans, A., Kee, R., Vallury, V., & McFarlane, M. M. (2002). Screening adolescents in a juvenile detention center for gonorrhea and chlamydia: Prevalence and reinfection rates. *The Prison Journal, 82,* 8–18.

Brown, M. (2006). Gender, ethnicity, and offending over the life course: Women's pathways to prison in the aloha state. *Critical Criminology, 14,* 137–158.

Bureau of Indian Affairs. (2011). *Indian affairs.* Washington, DC: U.S. Department of the Interior. Available at http://www.bia.gov

Bureau of Justice Statistics. (1998). *Correctional populations in the United States, 1997.* Washington, DC: U.S. Department of Justice, Office of Justice Programs. Available at http://www.ojp.usdoj.gov/bjs/jails

Bureau of Justice Statistics (2005a). *Since 1994, violent crime rates have declined, reaching the lowest level ever in 2005.* Washington, DC: U.S. Department of Justice, Office of Justice Programs. Available at http://www.ojp.usdoj.gov/bjs

Bureau of Justice Statistics (2005b). *Sourcebook of criminal justice statistics online.* Washington, DC: U.S. Department of Justice, Office of Justice Programs.

Bureau of Justice Statistics. (2006a). *Prevalence of imprisonment in the United States.* Washington, DC: U.S. Department of Justice, Office of Justice Programs. Available at http://www.ojp.usdoj.gov/bjs

Bureau of Justice Statistics. (2007). *Prison and Jail Inmates at Midyear series, 1998–2006.* Washington, DC: U.S. Department of Justice, Office of Justice Programs. Available at http://bjs.ojp.usdoj.gov/index.cfm?ty=pbse&sid=38

Bureau of Justice Statistics. (2008). *Jail statistics: Summary of findings.* Washington, DC: U.S. Department of Justice, Office of Justice Programs. Available at http://bjs.ojp.usdoj.gov/index.cfm?ty=pbse&sid=38

Bureau of Justice Statistics. (2010). *Key facts at a glance: Correctional populations.* Washington, DC: U.S. Department of Justice, Office of Justice Programs. Available at http://www.ojp.usdoj.gov/bjs

Bureau of Labor Statistics. (2009). *Occupational Employment Statistics Survey Program.* Washington, DC: U.S. Department of Labor. Available at http://www.bls.gov/oes/

Burnett, J. (2011). *Town relies on troubled youth prison for profits.* National Public Radio. Available at http://www.npr.org/2011/03/25/134850972/town-relies-on-troubled-youth-prison-for-profits

Burns, H. (1975). *Corrections: Organization and administration.* St. Paul, MN: West Publishing.

Burrell, W. (2006). *APPA caseload standards for probation and parole.* Lexington, KY: American Probation and Parole Association. Available at http://cdpsweb.state.co.us/cccjj/PDF/Research%20Documents/APPA%20Caseload_Standards_PP_0906.pdf

Butterworth, B. (1995, May 25). *News release.* Available at http://myfloridalegal.com/newsrel.nsf/pv/D08D904689EEEFF0852561E7006D0E77

Butts, J., & Mitchell, O. (2000). Brick by brick: Dismantling the border between juvenile and adult justice. *National Institute of Justice 2000. Vol. 2: The nature of crime: Continuity and change.* Washington, DC: National Institute of Justice.

Byrne, J. M., & Hummer, D. (2008). The nature and extent of prison violence. In J. M. Byrne, D. Hummer, & F. S. Taxman (Eds.), *The culture of prison violence.* Boston: Pearson.

Byrne, J. M., Hummer, D., & Stowell, J. (2008). Prison violence, prison culture, and offender change: New directions for research, theory, policy, and practice. In M. Byrne, D. Hummer, & F. S. Taxman (Eds.), *The culture of prison violence.* Boston: Pearson.

Cahalan, M. W. (1986). *Historical corrections statistics in the United States, 1850–1984.* U.S. Department of Justice, Bureau of Justice Statistics. Rockville, MD: U.S. Government Printing Office.

Camp, S. D., & Gaes, G. G. (2002). Growth and quality of U.S. private prisons: Evidence from a national survey. *Criminology & Public Policy, 1*(3), 427–450.

Camp, S. D., Steiger, T. L., Wright, K. N., Saylor, W. G., & Gilman, E. (1997). Affirmative action and the "level playing field": Comparing perceptions of own and minority job advancement opportunities. *The Prison Journal, 77*(3), 313–334.

Carroll, L. (1974). *Hacks, blacks and cons: Race relations in a maximum security prison.* Lexington, MA: Lexington Books.

Carroll, L. (1982). Race, ethnicity and the social order of the prison. In R. Johnson & H. Toch (Eds.), *The pains of imprisonment* (pp. 181–203). Beverly Hills, CA: Sage.

Carter, K. (2006). Restorative justice and the balanced approach. In A. Walsh, *Correctional assessment, casework, and counseling* (4th ed., pp. 8–11). Lanham, MD: American Correctional Association.

Carter, M., & Morris, L. (2002). *Managing sex offenders in the community.* Washington, DC: Center for Sex Offender Management.

Caspi, A., & Moffitt, T. (1995). The continuity of maladaptive behavior: From description to understanding in the study of antisocial behavior. In D. Ciccheti & D. Cohen (Eds.), *Manual of developmental psychology* (pp. 472–511). New York: Wiley.

Champion, D. (2005). *Probation, parole, and community corrections* (5th ed.). Upper Saddle River, NJ: Prentice Hall.

Chen, X. (2010). The Chinese sentencing guideline: A primary analysis. *Federal Sentencing Reporter, 22,* 213–216.

Chesney-Lind, M., & Irwin, K. (2006). Still "the best place to conquer girls": Girls and the juvenile justice system. In A. V. Merlo & J. M. Pollock (Eds.), *Women, law, and social control* (pp. 271–291). Boston: Pearson.

Chesney-Lind, M., & Shelden, R. G. (1998). *Girls, delinquency, and juvenile justice* (2nd ed.). Belmont, CA: West/Wadsworth Company.

Clark, J. (1991). Correctional health care issues in the nineties: Forecast and recommendations. *American Jails Magazine, 5,* 22–23.

Clear, T., Cole, G., & Reisig, M. D. (2011). *American corrections* (9th ed.). Belmont, CA: Wadsworth.

Clear, T., Rose, D., & Ryder, J. (2001). Incarceration and the community: The problem of removing and returning offenders. *Crime & Delinquency, 47,* 335–351.

Clemmer, D. (2001). The prison community. In E. J. Latessa, A. Holsinger, J. W. Marquart, & J. R. Sorensen (Eds.), *Correctional contexts: Contemporary and classical readings* (2nd ed., pp. 83–87). Los Angeles: Roxbury.

Coates, R. (1990).Victim-offender reconciliation programs in North America. In B. Galaway & J. Hudson (Eds.), *Criminal justice, restitution, and reconciliation* (pp. 177–182). Monsey, NY: Criminal Justice Press.

Cohen, F. (2008). Correctional health care: A retrospective. *Correctional Law Reporter, 20,* 5–12.

Colbert, D. (Ed.). (1997). *Eyewitness to America: 500 years of America in the words of those who saw it happen.* New York: Pantheon Books.

Collins, R. (2004). Onset and desistence in criminal careers: Neurobiology and the age–crime relationship. *Journal of Offender Rehabilitation, 39,* 1–19.

Comack, E. (2006). Coping, resisting, and surviving: Connecting women's law violations to their history of abuse. In L. F. Alarid & P. Cromwell (Eds.), *In her own words* (pp. 33–44). Los Angeles: Roxbury.

Conly, C. (1998). *The Women's Prison Association: Supporting women offenders and their families.* Washington, DC: National Institute of Justice.

Conover, T. (2001). *Newjack: Guarding Sing Sing.* New York: Vintage Books.

Cooper, A. (2011, February). Throwing away the key: Michelle Alexander on how prisons have become the new Jim Crow. *The Sun, 422,* 4–12.

Cornelius, G. F. (2007). *The American jail: The cornerstone of modern corrections.* Upper Saddle River, NJ: Pearson/Prentice Hall.

Correctional Service of Canada. (2003). Women correctional officers in male institutions, 1978. Available at http//www.csc-scc.gc.ca

Corrections Compendium 2003. (2003). Correctional officer education and training. *Compendium, 28*(2), 11–12.

Cowburn, M. (1998). A man's world: Gender issues in working with male sex offenders in prison. *The Howard Journal of Criminal Justice, 37,* 234–251.

Cox, N. R., & Osterhoff, W. E. (1991). Managing the crisis in local corrections: A public–private partnership approach. In J. Thompson & G. L. Mays (Eds.), *American jails: Public policy issues* (pp. 227–239). Chicago: Nelson-Hall.

Crabbe, J. (2002). Genetic contributions to addiction. *Annual Review of Psychology, 53,* 435–462.

Craig, S.C. (2009). A historical review of mother and child programs for incarcerated women. *The Prison Journal, 89*(1), 35S–53S.

Crouch, B. (1993). Is incarceration really worse? Analysis of offenders' preferences for prison over probation. *Justice Quarterly,* 10, 67–88.

Cuellar, A., Snowden, L., & Ewing, T. (2007). Criminal records of persons served in the public mental health system. *Psychiatric Services, 58,* 114–120.

Cullen, F. T. (2006). Assessing the penal harm movement. In E. J. Latessa & A. M. Holsinger (Eds.), *Correctional contexts: Contemporary and classical readings* (3rd ed., pp. 61–74). Los Angeles: Roxbury.

Cullen, F., & Gendreau, P. (2001). Assessing correctional rehabilitation: Policy, practice, and prospects. In *Criminal Justice 2000* (Vol. 3, pp. 109–175). Washington, DC: National Institute of Justice.

Currie, E. (1999). Reflections on crime and criminology at the millennium. *Western Criminology Review, 2* [Online]. Available at http://wcr.sonoma.edu/v2n1/currie.html

Dahl, R. (1961). *Who governs? Democracy and power in an American city.* New Haven, CT: Yale University Press.

Daly, K., & Chesney-Lind, M. (1988). Feminism and criminology. *Justice Quarterly, 5,* 497–535.

Dammer, H., & Fairchild, E. (2006). *Comparative criminal justice systems* (3rd ed.). Belmont, CA: Wadsworth.

Damrosch, L. (2010). *Discovery of America.* New York: Farrar, Straus and Giroux.

Davis, K. C. (2008). *America's hidden history: Untold tales of the first pilgrims, fighting women, and forgotten founders who shaped a nation.* New York: HarperCollins.

Death Penalty Information Center. (2009). *The death penalty: An international perspective.* Available at http://www.deathpenaltyinfo.org/death-penalty-international-perspective

Death Penalty Information Center (2010). *Facts about the death penalty.* http://www.deathpenaltyinfo.org

DeFina, R., & Arvanites, T. (2002). The weak effect of imprisonment on crime: 1971–1998. *Social Science Quarterly, 83,* 635–653.

Del Carmen, R., Parker, V., & Reddington, F. (1998). *Briefs of leading cases in juvenile justice.* Cincinnati, OH: Anderson.

Delany, P., Fletcher, B., & Shields, J. (2003). Reorganizing care for the substance using offender—the case for collaboration. *Federal Probation, 67,* 64–69.

DeLisi, M. (2005). *Career criminals in society.* Thousand Oaks, CA: Sage.

Diamond, J. (1997). *Guns, germs, and steel: The fates of human societies.* New York: Norton & Company.

Diaz-Duran, C. (2010, December 31). The Scott sisters' life sentence for $11. *The Daily Beast.* Available at http://www.thedailybeast.com

Dickens, C. (1842). *American notes, chapter 7.* The Literature Network. Available at http://www.online-literature.com

Dietz, E., O'Connell, D., & Scarpitti, F. (2003). Therapeutic communities and prison management: An examination of the effects of operating an in-prison therapeutic community on levels of institutional disorder. *International Journal of Offender Therapy and Comparative Criminology, 47,* 210–223.

Dilworth, D. (1995). Prisoners' lawsuits burden federal civil courts. *Trial, 31,* 98–100.

Director addresses changes within BOP. (2006, March). *The Third Branch, 38.* Available at http://www.uscourts.gov/trb/03-06/interview/index.html

Ditton, P., & Wilson, D. (1999). *Truth in sentencing in state prisons.* Washington, DC: Bureau of Justice Statistics.

Dix, D. (1967). *Remarks on prisons and prison discipline in the United States.* Montclair, NJ: Patterson Smith. (Original work published 1843; Rev. ed. published 1845)

Dodgson, K., Goodwin, P., Howard, P., Llewellyn-Thomas, S., Mortimer, E., Russell, N., et al. (2001). *Electronic monitoring of released prisoners: An evaluation of the home detention curfew scheme.* London: Home Office Research.

Dowden, C., & Tellier, C. (2004). Predicting work-related stress in correctional officers: A meta-analysis. *Journal of Criminal Justice, 32*(1), 31–47.

Dreznick, M. (2003). Heterosexual competence of rapists and child molesters: A meta-analysis. *Journal of Sex Research, 40,* 170–178.

Duran, J. (1996). Anne Viscountess Conway: A seventeenth-century rationalist. In L. L. McAlister (Ed.), *Hypatia's daughters: Fifteen hundred years of women philosophers* (pp. 92–108). Bloomington: Indiana University Press.

Durant, W., & Durant, A. (1967). *Rousseau and revolution*. New York: Simon & Schuster.

Durkheim, E. (1964). *The division of labor in society*. New York: Free Press. (Original work published 1893)

Durnescu, I. (2008). An exploration of the purposes and outcomes of probation in European jurisdictions. *Probation Journal, 55,* 273–281.

Durose, M. R., Farole, D. J., & Rosenmerkel, S. P. (2009). *Felony sentences in state courts, 2006—statistical tables*. Washington, DC: U.S. Department of Justice, Office of Justice Programs, Bureau of Justice Statistics. Available at http://bjs .ojp.usdoj.gov

Durose, M. R., Farole, D. J., & Rosenmerkel, S. P. (2010). *Felony sentences in state courts, 2006*. Washington, DC: Bureau of Justice Statistics

Dworkin, A. (1993). Against the male flood: Censorship, pornography, and equality. In P. Smith (Ed.), *Feminist jurisprudence* (pp. 449–465). Oxford, UK: Oxford University Press.

Ellis, A. (1989). The history of cognition in psychotherapy. In A. Freeman, K. Simon, L. Beutler, & H. Arkowitz (Eds.), *Comprehensive handbook of cognitive therapy* (pp. 5–19). New York: Plenum.

Ellis, L. (2003). Genes, criminality, and the evolutionary neuroandrogenic theory. In A. Walsh & L. Ellis (Eds.), *Biosocial criminology: Challenging environmentalism's supremacy* (pp. 13–34). Hauppauge, NY: Nova Science.

Eskridge, C. (1989). Correctional practices in Japan. *Journal of Offender Counseling Services & Rehabilitation, 14,* 5–23.

Fairchild, E., & Dammer, D. (2001). *Comparative criminal justice systems*. Belmont, CA: Wadsworth.

Farabee, D., Pendergast, M., & Anglin, M. (1998). The effectiveness of coerced treatment for drug abusing offenders. *Federal Probation, 109,* 3–10.

Farbstein, J., & Associates, with Wener, R. (1989). *A comparison of "direct" and "indirect" supervision of correctional facilities: Final report*. Washington, DC: National Institute of Corrections.

Farkas, M. A. (1999). Inmate supervisory style: Does gender make a difference? *Women & Criminal Justice, 10,* 25–45.

Farkas, M. A. (2000). A typology of correctional officers. *International Journal of Offender Therapy and Comparative Criminology, 44,* 431–449.

Farkas, M. A. (2001). Correctional officers: What factors influence work attitudes. *Correctional Management Quarterly, 5*(2), 20–26.

Farkas, M. A., & Rand, K. R. L. (1999). Sex matters: A gender-specific standard for cross-gender searches of inmates. *Women & Criminal Justice, 10*(3), 31–56.

Farkas, M., & Stichman, S. (2002). Can treatment, punishment, incapacitation, and public safety be reconciled? *Criminal Justice Review, 27,* 256–283.

Farrington, D. P., Langan, P. A., & Tonry, M. (2004). *Cross-national studies in crime and justice*. U.S. Department of Justice, Office of Justice Programs, Bureau of Justice Statistics. Washington, DC: U.S. Government Printing Office.

Federal Bureau of Investigation. (2007). *2006, Crime in the United States*. Available at http://www.fbi.gov/ucr/cius2006/arrests/index.html/

Federal Bureau of Investigation. (2009). *Crime in the United States, 2008: Uniform Crime Reports*. Washington, DC: U.S. Government Printing Office.

Federal Bureau of Investigation. (2010). *Crime in the United States, 2009*. Washington, DC: U.S. Government Printing Office.

Federerman, C. (2004). Who has the body? The paths to habeas corpus reform. *The Prison Journal, 84,* 317–339.

Feeley, M. M. (1991). The privatization of prisons in historical perspective. *Criminal Justice Research Bulletin, 6*(2), 1–10.

Fehr, E., & Gachter, S. (2002). Altruistic punishment in humans. *Nature, 415,* 137–140.

Ferri, E. (1917). *Criminal sociology*. Boston: Little, Brown. (Original work published 1897)

Fielding, H. (1967). *Inquiry into the causes of the late increase of robbers*. Oxford, UK: Oxford University Press. (Original work published 1751)

Figueredo, A., Sales, B., Russell, K., Becker, J., & Kaplan, M. (2000). A Brunswickian evolutionary-developmental theory of adolescent sex offending. *Behavioral Sciences and the Law, 18,* 309–329.

Finn, P., & Kuck, S. (2005). *Stress among probation and parole officers and what can be done about it*. Washington, DC: Department of Justice.

Firestone, J. M., & Harris, R. J. (1994). Sexual harassment in the military: Environmental and individual contexts. *Armed Forces and Society, 21,* 25–43.

Firestone, J. M., & Harris, R. J. (1999). Changes in patterns of sexual harassment in the U.S. military: A comparison of the 1988 and 1995 DOD surveys. *Armed Forces and Society, 25,* 613–632.

Fishbein, D. (2003). Neuropsychological and emotional regulatory processes in antisocial behavior. In A. Walsh & L. Ellis (Eds.), *Biosocial criminology: Challenging environmentalism's supremacy* (pp. 185–208). Hauppauge, NY: Nova Science.

Fisher, W., Roy-Bujnowski, K., Grudzinskas, A., Clayfield, J., Banks, S., & Wolff, N. (2006). Patterns and prevalence of arrest in a statewide cohort of mental care consumers. *Psychiatric Services, 57,* 1623–1628.

Flavin, J., & Desautels, A. (2006). Feminism and crime. In C. M. Renzetti, L. Goodstein, & S. L. Miller (Eds.), *Rethinking gender, crime, and justice* (pp. 11–28). Los Angeles: Roxbury.

Fleisher, M. S., & Krienert, J. L. (2009). *The myth of prison rape: Sexual culture in American prisons*. Lanham, MD: Rowman & Littlefield.

Florida Department of Corrections. (2010). *Gang and security threat group awareness*. Available at http://www.dc.state.fl.us

Foster, B. (2006). *Corrections: The fundamentals*. Upper Saddle River, NJ: Prentice Hall.

Foucault, M. (1979*). Discipline and punish: The birth of the prison*. New York: Vintage.

Frankel, L. (1996). Damaris Cudworth Masham: A seventeenth-century feminist philosopher. In L. L. McAlister (Ed.), *Hypatia's daughters: Fifteen hundred years of women philosophers* (pp. 128–138). Bloomington: Indiana University Press.

Freudenberg, N. (2006). Coming home from jail: A review of health and social problems facing U.S. jail populations and of opportunities for reentry interventions. *American Jails, 20,* 9–24.

Gainsborough, J. (2002). *Mentally ill offenders in the criminal justice system: An analysis and prescription*. Washington, DC: The Sentencing Project.

Gard, R. (2007). The first probation officers in England and Wales. *British Journal of Criminology, 47,* 1–17.

Gardner, L. (1996, March 1). Lessons from abroad: Japan's parole models. *Policy Review, 75* [Online]. Available at http://www .hoover.org/publications/policy-review/article/7679

Garland, B., & Frankel, M. (2006). Considering convergence: A policy dialogue about behavioral genetics, neuroscience, and law. *Law and Contemporary Problems, 69,* 101–113.

Garland, D. (1990). *Punishment and modern society: A study in social theory.* Chicago: University of Chicago Press.

Garofalo, R. (1968). *Criminology.* Montclair: NJ: Patterson Smith. (Original work published 1885)

Gendreau, P., & Ross, R. R. (1987). Revivification of rehabilitation: Evidence from the 1980s. *Justice Quarterly, 4,* 349–407.

General Accounting Office. (1996). *Private and public prisons: Studies comparing operational costs and/or quality of service.* Washington, DC: Author.

Gettinger, S. H. (1984). *New generation jails: An innovative approach to an age-old problem.* Washington, DC: U.S. Department of Justice, National Institute of Corrections.

Giallombardo, R. (1966). *Society of women: A study of a women's prison.* New York: Wiley.

Gibbs, A., & King, D. (2003). The electronic ball and chain? The operation and impact of home detention with electronic monitoring in New Zealand. *Australian and New Zealand Journal of Criminology, 36,* 1–17.

Gilliard, D. K., & Beck, A. J. (1997). *Prison and inmates at midyear 1996.* Washington, DC: U.S. Department of Justice, Office of Justice Programs. Available at http://www.ojp.usdoj.gov/bjs/pub/pdf/mhppji.pdf

Giotakos, O., Markianos, M., Vaidakis, N., & Christodoulou, G. (2003). Aggression, impulsivity, plasma sex hormones, and biogenic amine turnover in a forensic population of rapists. *Journal of Sex and Marital Therapy, 29,* 215–225.

Glaze, L., & Bonczar, T. (2009). Probation and parole in the United States, 2008. *Bureau of Justice Statistics Bulletin.* Washington, DC: U.S. Department of Justice.

Glaze, L. E., Bonczar, T. P., & Zhang, F. (2010). *Probation and parole in the United States, 2009.* Washington, DC: U.S. Department of Justice, Office of Justice Programs, Bureau of Justice Statistics. Available at http://bjs.ojp.usdoj.gov

Glaze, L., & Palla, S. (2005). Probation and parole in the United States, 2004. *Bureau of Justice Statistics Bulletin.* Washington, DC: U.S. Department of Justice.

Glenn, S. (2010, October 7). Board suspends license of former McNeil Island nurse. *The News Tribune* (Tacoma, WA), p. A4.

Goffman, E. (1961). *Asylums: Essays on the social situation of mental patients and other inmates.* Garden City, NY: Anchor Books.

Goldfarb, R. (1975). *Jails: The ultimate ghetto.* Garden City, NY: Anchor Press.

Goleman, D. (1995). *Emotional intelligence:* New York: Bantam Books.

Goleman, D. (2006). *Social intelligence: The new science of human relationships.* New York: Bantam Books.

Gonnerman, J. (2001). Remembering Attica: Thirty years later, the story of America's worst prison riot continues. *The Village Voice.* Available at http://www.villagevoice.com

Goodstein, L. (2006). Introduction: Gender, crime, and criminal justice. In C. M. Renzetti, L. Goodstein, & S. L. Miller (Eds.), *Rethinking gender, crime, and justice* (pp. 1–10). Los Angeles: Roxbury.

Gottfredson, M., & Hirschi, T. (2002). The nature of criminality: Low self-control. In S. Cote (Ed.), *Criminological theories: Bridging the past to the future* (pp. 210–216). Thousand Oaks, CA: Sage.

Gray, T., Mays, G. L., & Stohr, M. K. (1995). Inmate needs and programming in exclusively women's jails. *The Prison Journal, 75*(2), 186–202.

Grinfield, M. (2005). Sexual predator ruling raises ethical moral dilemmas. *Psychiatric Times, 16,* 1–4.

Grubin, D. (2007). Sexual offending and the treatment of sex offenders. *Psychiatry, 6,* 439–443.

Grusec, J., & Hastings, P. (2007). *Handbook of socialization: Theory and research.* New York: Guilford Press.

Haney, C., Banks, C., & Zimbardo, P. (1981). Interpersonal dynamics in a simulated prison. In R. R. Ross (Ed.), *Prison guard/correctional officer* (pp. 137–168). Toronto, ON, Canada: Butterworth.

Hanson, K., & Bussiere, M. (1998). Predicting relapse: A meta-analysis of sexual offenders recidivism studies. *Journal of Consulting and Clinical Psychology, 66,* 348–362.

Harrington, P. E. (2002). Advice to women beginning a career in policing. *Women & Criminal Justice, 9,* 1–21.

Harris, J. (1973). *Crisis in corrections: The prison problem.* New York: McGraw-Hill.

Harrison, P. M., & Beck, A. J. (2005). *Prisoners in 2004.* Bureau of Justice Statistics, Office of Justice Programs, U.S. Department of Justice. Washington, DC: U.S. Government Printing Office.

Hassine, V. (1996). *Life without parole: Living in prison today.* Los Angeles: Roxbury.

Hassine, V. (2009). *Life without parole* (4th ed.). New York: Oxford University Press.

Havercamp, R., Mayer, M., & Levy, R. (2004). Electronic monitoring in Europe. *European Journal of Crime, Criminal Law, and Criminal Justice, 12,* 36–45.

Hawkes, M. Q. (1998). Edna Mahan: Sustaining the reformatory tradition. *Women & Criminal Justice, 9,* 1–21.

Hawkins, H. C. (2005). Race and sentencing outcomes in Michigan. *Journal of Ethnicity in Criminal Justice, 3*(1/2), 91–109.

Hayes, L. M. (2010). *National study of jail suicide—20 years later.* National Center on Institutions and Alternatives. Washington, DC: U.S. Department of Justice National Institute of Corrections. Available at http://www.nicic.gov

Hefferman, E. (1972). *Making it in prison: The square, the cool, and the life.* New York: Wiley.

Hemmens, C., & Atherton, E. (1999). *Use of force: Current practice and policy.* Lanham, MD: American Correctional Association.

Hemmens, C., & Marquart, J. W. (2000). Race, age, and inmate perceptions of inmate–staff relations. *Journal of Criminal Justice, 28,* 297–312.

Hemmens, C., Steiner, B., & Mueller, D. (2003). *Significant cases in juvenile justice.* Los Angeles: Roxbury

Hemmens, C., Stohr, M. K., Schoeler, M., & Miller, B. (2002). One step up, two steps back: The progression of perceptions of women's work in prisons and jails. *Journal of Criminal Justice, 30,* 473–489.

Henriques, Z. W. (1996). Imprisoned mothers and their children: Separation-reunion syndrome dual impact. *Women & Criminal Justice, 8*(1), 77–95.

Henriques, Z. W., & Gilbert, E. (2003). Sexual abuse and sexual assault of women in prison. In R. Muraskin (Ed.), *It's a crime: Women and justice* (pp. 258–272). Upper Saddle River, NJ: Prentice Hall.

Hersh, S. M. (2004, May 10). Torture at Abu Ghraib: American soldiers brutalized Iraqis. How far does the responsibility go? *New Yorker Magazine.* Available at http://www.newyorker.com/archive/2004/05/10/040510fa_fact?currentPage=4#ixzz11iYHNcFX

Hickman, M. J., & Reeves, B. A. (2006). *Local police departments, 2003.* Washington, DC: Department of Justice, Bureau of Justice Statistics.

Hindman, J., & Peters, J. (2001). Polygraph testing leads to better understanding adult and juvenile sex offenders. *Federal Probation, 65,* 1–15.

Hirsch, A. J. (1992). *The rise of the penitentiary.* New Haven, CT: Yale University Press.

Hirschi, T. (1969). *The causes of delinquency.* Berkeley: University of California Press.

History Channel. (2010). *Who really discovered America?* [TV program first aired on the History Channel on June 22, 2010].

Hogan, N. L., Lambert, E. G., Hepburn, J. R., Burton, V. S., & Cullen, F. T. (2004). Is there a difference? Exploring male and female correctional officers' definition of and response to conflict situations. *Women & Criminal Justice, 15*(3/4), 143–165.

Holcomb, I. (2008). *The carrying of firearms by probation officers.* Paper presented at the annual meeting of the American Criminal Justice Society, Cincinnati, OH.

Howard, J. (2000). *The state of prisons in England and Wales, with preliminary observations, and an account of some foreign prisons.* London: Routledge/Thoemmes Press. (Original work published 1775)

Howard, R. (2002). Brain waves, dangerousness, and deviant desires. *Journal of Forensic Psychiatry, 13,* 367–384.

Hughes, K. (2006). *Justice expenditure and employment in the United States, 2003.* U.S. Department of Justice, Office of Justice Programs, Bureau of Justice Statistics. Washington, DC: U.S. Government Printing Office.

Hughes, R. (1987). *The fatal shore.* New York: Vintage Books.

Hughes, T., Wilson, D., & Beck, A. (2001). *Trends in state parole, 1990–2000.* Washington, DC: U.S. Department of Justice, Bureau of Justice Statistics.

Hulings, N. (2010, October 5). Jail ready but still too costly. *The News Tribune* (Tacoma, WA), p. A4.

Ignatieff, M. (1978). *A just measure of pain: The penitentiary in the industrial revolution, 1750–1850.* New York: Columbia University Press.

Inciardi, J. (2007). *Criminal justice.* New York: McGraw-Hill.

Inciardi, J., Martin, S., & Butzin, C. (2004). Five-year outcomes of therapeutic community treatment of drug-involved offenders after release from prison. *Crime & Delinquency, 50,* 88–111.

Inderbitzin, M. (2006). Guardians of the state's problem children: An ethnographic study of staff members in a juvenile correctional facility. *The Prison Journal, 86*(4), 431–451.

Ireland, C., & Berg, B. (2006). Women in parole: Gendered adaptations of female parole agents in California. *Women & Criminal Justice, 18*(1/2), 131–150.

Irwin, J. (1985). *The jail: Managing the underclass in American society.* Berkeley: University of California Press.

Irwin, J. (2005). *The warehouse prison: Disposal of the new dangerous class.* Los Angeles: Roxbury.

Irwin, J., & Austin, J. (2001). *It's about time: America's imprisonment binge* (3rd ed.). Belmont, CA: Wadsworth.

Jacobs, B., & Wright, R. (1999). Stick-up, street culture, and offender motivation. *Criminology, 37,* 149–173.

Jacobs, J. B. (1977). *Stateville: The penitentiary in mass society.* Chicago: University of Chicago Press.

Jacoby, S. (1983). *Wild justice: The evolution of revenge.* New York: Harper & Row.

Jafee, S., Moffitt, T., Caspi, A., & Taylor, A. (2003). Life with (or without) father: The benefits of living with two biological parents depend on the father's antisocial behavior. *Child Development, 74,* 109–126.

James, D. J., & Glaze, L. E. (2006). *Mental health problems of prison and jail inmates.* Washington, DC: U.S. Department of Justice, Office of Justice Programs.

Johnson, R. (2002). *Hard time: Understanding and reforming the prison* (3rd ed.). Belmont, CA: Wadsworth/Thomson Learning.

Johnston, N. (2009). Evolving function: Early use of imprisonment as punishment. *The Prison Journal, 89*(1), 10S–34S.

Johnston, N. (2010). Early Philadelphia prisons: Amour, alcohol, and other forbidden pleasures. *The Prison Journal, 90*(1), 4–11.

Jolliffe, D., & Farrington, D. (2009). *Effectiveness of interventions with adult male violent offenders.* Stockholm: Swedish Council for Crime Prevention.

Joseph, J., & Taylor, D. (Eds.). (2003). *With justice for all: Minorities and women in criminal justice.* Upper Saddle River, NJ: Prentice Hall.

Junger-Tas, J. (1996). Delinquency similar in Western countries. *Overcrowded Times, 7,* 10–13.

Jurik, N. (1985a). Individual and organizational determinants of correctional officer attitudes toward inmates. *Criminology, 23,* 523–539.

Jurik, N. C. (1985b). An officer and a lady: Organizational barriers to women working as correctional officers in men's prisons. *Social Problems, 33,* 375–388.

Jurik, N. C. (1988). Striking a balance: Female correctional officers, gender-role stereotypes, and male prisons. *Sociological Inquiry, 58,* 291–305.

Jurik, N. C., & Halemba, G. J. (1984). Gender, working conditions, and the job satisfaction of women in a non-traditional occupation: Female correctional officers in a men's prison. *Sociological Quarterly, 25,* 551–566.

Kanazawa, Y. (2007). Psychotherapy in Japan: The case of Mrs. A. *Journal of Clinical Psychology, 63,* 755–763.

Kansal, T. (2005). *Racial disparity in sentencing.* The Sentencing Project. Available at http://www.sentencingproject.org

Karberg, J., & James, J. (2005). *Substance dependence, abuse, and treatment of jail inmates, 2002.* Washington, DC: U.S. Department of Justice, Office of Justice Programs, Bureau of Justice Statistics.

Keil, T. J., & Vito, G. F. (2009). Lynching and the death penalty in Kentucky, 1866–1934. *Journal of Ethnicity in Criminal Justice, 7,* 53–68.

Kerle, K. (1991). Introduction. In J. Thompson & G. L. Mays (Eds.), *American jails: Public policy issues* (pp. 1–3). Chicago: Nelson-Hall.

Kerle, K. (2003). *Exploring jail operations.* Hagerstown, MD: American Jail Association.

Kim, A.-S., DeValve, M., DeValve, E. Q., & Jonson, W. W. (2003). Female wardens: Results from a national survey of state correctional executives. *Prison Journal, 83,* 406–425.

King, K., Steiner, B., & Breach, S. R. (2008). Violence in the supermax: A self-fulfilling prophecy. *The Prison Journal, 88*(1), 144–168.

Kirshon, J. (2010). Attica prison riot ends in bloodshed. *Examiner.com.* Available at http://www.examiner.com/history-in-new-york/attica-prison-riot-ends-bloodshed

Kitano, H. H. L. (1997). *Race relations* (5th ed.). Upper Saddle River, NJ: Prentice Hall.

Kleber, H. (2003). Pharmacological treatments for heroin and cocaine dependence. *American Journal on Addictions, 12,* S5–S18.

Klofas, J. M. (1991). Disaggregating jail use: Variety and change in local corrections over a ten-year period. In J. Thompson & G. L. Mays (Eds.), *American jails: Public policy issues* (pp. 40–58). Chicago: Nelson-Hall.

Kluger, J. (2007). The paradox of supermax. *Time, 169*(6), 52–53.

Kramer, J., & Ulmer, J. (2009). *Sentencing guidelines: Lessons from Pennsylvania.* Boulder, CO: Lynne Rienner.

Kuhn, T. (1962). *The structure of scientific revolutions.* Chicago: University of Chicago Press.

Lambert, E. G., Hogan, N. L., & Tucker, K. A. (2009). Problems at work: Exploring the correlates of role stress among correctional staff. *The Prison Journal, 89*(4), 460–481.

Langan, P. A., & Cohen, R. L. (1996). *State court sentencing of convicted felons, 1992.* Washington, DC: U.S. Bureau of Justice Statistics.

Langan, P. A., & Farrington, D. (1998). *Crime and justice in the United States and England and Wales, 1981–1996.* Washington, DC: Bureau of Justice Statistics.

Langan, P. A., Greenfield, L. A., Smith, S. K., Durose, M. R., & Levin, D. J. (2001). *Contacts between police and the public: Findings from the 1999 national survey.* Washington, DC: U.S. Department of Justice.

Langan, P. A., & Levin, D. (2002). *Recidivism of prisoners released in 1994.* Bureau of Justice Statistics Special Report. Washington, DC: U.S. Department of Justice.

Langan, P. A., Schmitt, E., & Durose, M. (2003). *Recidivism of sex offenders released from prison in 1994.* Washington, DC: U.S. Department of Justice, Bureau of Justice Statistics.

Latessa, W., Cullen, F., & Gendreau, P. (2002). Beyond correctional quackery—Professionalism and the possibility of effective treatment. *Federal Probation, 66,* 43–50.

Latimer, J., Dowden, C., & Muise, D. (2005). The effectiveness of restorative justice practices. *The Prison Journal, 85,* 127–144.

Lawes, L. E. (1932). *Twenty thousand years in Sing Sing.* New York: Long & Smith.

Lawrence, R., & Mahan, S. (1998). Women corrections officers in men's prisons: Acceptance and perceived job performance. *Women & Criminal Justice, 9*(3), 63–86.

Leigey, M., & Bachman, R. (2007). The influence of crack cocaine on the likelihood of incarceration for a violent offense: An examination of a prison sample. *Criminal Justice Policy Review, 18,* 335–352.

Leshner, A. (1998). Addiction is a brain disease—and it matters. *National Institute of Justice Journal, 237,* 2–6.

Library of Congress. (2010). *Old jail, State Route 6A and Old Jail Lane, Barnstable, Barnstable County, MA.* Washington, DC: Author. Available at http://www.loc.gov/pictures/search

Lieb, S., Fallon, S. J., Friedman, S. R., Thompson, D. R., Gates, G. J., Liberti, T. M., et al. (2011). Statewide estimation of racial/ethnic populations of men who have sex with men in the U.S. *Public Health Reports, 126,* 60–72.

Lightfoot, C., Zupan, L. L., & Stohr, M. K. (1991). Jails and the community: Modeling the future in local detention facilities. *American Jails, 5,* 50–52.

Linden, D. (2006). How psychotherapy changes the brain—The contribution of functional neuroimaging. *Molecular Psychiatry, 11,* 528–538.

Linder, C. (2007). Thatcher, Augustus, and Hill—The path to statutory probation in the United States and England. *Federal Probation, 71,* 36–41.

Lindsay, W. (2002). Research and literature on sex offenders with intellectual and developmental disabilities. *Journal of Intellectual Disability Research, 46,* 74–85.

Lipsey, M., & Cullen, F. (2007). The effectiveness of correctional rehabilitation: A review of systematic reviews. *Annual Review of Law and Social Science, 3,* 297–320.

Lipsky, M. (1980). *Street-level bureaucracy: Dilemmas of the individual in public services.* New York: Russell Sage.

Litt, M., & Mallon, S. (2003). The design of social support networks for offenders in outpatient drug treatment. *Federal Probation, 67,* 15–22.

Livsey, S. (2010). *Juvenile delinquency probation caseload, 1985–2007.* Washington, DC: U.S. Department of Justice, Office of Juvenile Justice and Delinquency Prevention.

Logan, C., & Gaes, G. (1993). Meta-analysis and the rehabilitation of punishment. *Justice Quarterly, 10,* 245–263.

Lombardo, L. X. (1982). *Guards imprisoned.* Cincinnati, OH: Anderson.

Lombardo, L. X. (2001). Guards imprisoned: Correctional officers at work. In E. J. Latessa, A. Holsinger, J. W. Marquart, & J. R. Sorensen (Eds.), *Correctional contexts: Contemporary and classical readings* (2nd ed., pp. 153–167). Los Angeles: Roxbury.

Loper, A. B., & Tuerk, E. H. (2006). Parenting programs for incarcerated parents: Current research and future directions. *Criminal Justice Policy Review, 14*(4), 407–427.

Lowenkamp, C., & Latessa, E. (2002). *Evaluation of Ohio's community-based corrections facilities and halfway house programs: Final report.* Cincinnati, OH: University of Cincinnati, Center for Criminal Justice Research,.

Lowenkamp, C., & Latessa, E. (2004). Understanding the risk principle: How and why correctional interventions can harm low-risk offenders. In D. Faust (Ed.), *Assessment issues for managers* (pp 3–7). Washington, DC: U.S. Department of Justice, National Institute of Corrections.

Lu, H., & Miethe, T. (2002). Legal representation and legal processing in China. *British Journal of Criminology, 42,* 267–280.

Lubitz, R., & Ross, T. (2001, June). Sentencing guidelines: Reflections on the future. *Sentencing & Corrections: Issues for the 21st Century, No. 10.* Washington, DC: U.S. Department of Justice, National Institute of Justice.

Lurigio, A. (2000). Persons with serious mental illness in the criminal justice system: Background, prevalence, and principles of care. *Criminal Justice Policy Review, 11,* 312–328.

Lurigio, A., & Loose, P. (2008). The disproportionate incarceration of African Americans for drug offenses: The national and Illinois perspective. *Journal of Ethnicity in Criminal Justice, 6*(3), 223–247.

Lutze, F. E., & Murphy, D. W. (1999). Ultra-masculine prison environments and inmates' adjustment: It's time to move beyond the "boys will be boys" paradigm. *Justice Quarterly, 16,* 709–734.

Macher, A. M. (2007). Issues in correctional HIV care: Neurological manifestations of patients with primary HIV infection. *American Jails, 21,* 49–55.

MacKenzie, D., & Brame, R. (2001). Community supervision, prosocial activities, and recidivism. *Justice Quarterly, 18,* 429–448.

Mackin, J., Lucas L., & Lambarth, C. (2010). *Baltimore County Juvenile Drug Court outcome and cost evaluation.* Baltimore: NPC Research.

Maletzky, B., & Field, G. (2003). The biological treatment of dangerous sexual offenders: A review and preliminary report of the Oregon pilot Depo-Provera program. *Aggression and Violent Behavior, 8,* 391–412.

Mann, C. C. (2006). *1491: New revelations of the Americas before Columbus.* New York: Vintage Books.

Marion, N. (2002). Effectiveness of community-based programs: A case study. *The Prison Journal, 82,* 478–497.

Marquart, J. W., Barnhill, M. B., & Balshaw-Biddle, K. (2001). Fatal attraction: An analysis of employee boundary violations in a southern prison system, 1995–1998. *Justice Quarterly, 18,* 877–910.

Marsh, R., & Walsh, A. (1995). Physiological and psychosocial assessment and treatment of sex offenders: A comprehensive victim-oriented program. *Journal of Offender Rehabilitation, 22,* 77-96.

Martin, G. (2005). *Juvenile justice: Process and systems.* Thousand Oaks, CA: Sage.

Martin, S., & Jurik, N. (1996). *Doing justice, doing gender.* Thousand Oaks, CA: Sage.

Martinez, D. J. (2004). Hispanics incarcerated in state correctional facilities: Variations in inmate characteristics across Hispanic subgroups. *Journal of Ethnicity in Criminal Justice, 2*(1/2), 119–131.

Martinson, R. (1974). What works? Questions and answers about prison perform. *The Public Interest, 35,* 22-54.

Maruschak, L. M. (2006). *Medical problems of jail inmates.* Washington, DC: U.S. Department of Justice, Office of Justice Programs.

Maruschak, L. M. (2008). *Medical problems of prisoners.* Washington, DC: U.S. Department of Justice, Office of Justice Programs, Bureau of Justice Statistics.

Maschke, K. J. (1996). Gender in the prison setting: The privacy–equal employment dilemma. *Women & Criminal Justice, 7,* 23–42.

Maslow, A. H. (1998). *Maslow on management.* New York: Wiley.

Maslow, A. H. (2001). A theory of human motivation. In J. M. Shafritz & J. S. Ott (Eds.), *Classics of organization theory* (pp. 152–157). Fort Worth, TX: Harcourt College Publishers. (Original work published 1943)

Mathias, R. (1997, July). Correctional treatment helps offenders stay drug and arrest free. *National Institute on Drug Abuse and Addiction: NIDA Notes.*

Mattick, H. W. (1974). The contemporary jails of the United States: An unknown and neglected area of justice. In D. Glaser (Ed.), *Handbook of criminology.* Chicago: Rand McNally.

Mauer, M. (2005). *Comparative international rates of incarceration: An examination of causes and trends.* Washington, DC: The Sentencing Project.

Mauer, M., King, R., & Young, M. (2004). *The meaning of "life": Long prison sentences in context.* Washington, DC: The Sentencing Project.

Mawby, R. (2001). *Burglary.* Colompton, Devon, UK: Willan Publishing.

May, D., Wood, P., Mooney, J., & Minor, K. (2005). Predicting offender-generated exchange rates: Implications for a theory of sentence severity. *Crime & Delinquency, 51,* 373–399.

McAuliffe, K. (2010). Are we still evolving? In J. Groopman (Ed.), *The best science writing of 2010* (pp. 218–231). New York: HarperCollins.

McDermott, P. A., Alterman, A. I., Cacciola, J. S., Rutherford, M. J., Newman, J. P., & Mulholland, E. M. (2000). Generality of Psychopathy Checklist–Revised factors over prisoners and substance-dependent patients. *Journal of Consulting and Clinical Psychology, 68*(1), 181–186.

McLean, R. L., Robarge, J., & Sherman, S. G. (2006). Release from jail: Moment of crisis or window of opportunity for female detainees? *Journal of Urban Health, 83,* 382–393.

McLeod, J. (2003). *An introduction to counseling.* Buckingham, UK: Open University Press.

McNeese, T. (2010). Inmate rights: Getting the courts to listen. In A. Walsh & M. Stohr, *Correctional assessment, casework and counseling* (5th ed., pp. 321–323). Alexandria, VA: American Correctional Association.

McNiel, D. E., Binder, R. L., & Robinson, J. C. (2005, July). Incarceration associated with homelessness, mental disorder, and co-occurring substance abuse. *Psychiatric Services, 56,* 840–846.

Mears, D. P. (2008). An assessment of supermax prisons using an evaluation research framework. *The Prison Journal, 88*(1), 43–68.

Menard, S., Mihalic, S., & Huizinga, D. (2001). Drugs and crime revisited. *Justice Quarterly, 18,* 269–299.

Menninger, K. (1968). *The crime of punishment.* New York: Penguin Books.

Mercier, M. (2010). *Japanese Americans in the Columbia River Basin.* Columbia River Basin Ethnic History Project. Available at http://archive.vancouver.wsu.edu/crbeha/projteam

Merlo, A. V., & Benekos, P. J. (2000). *What's wrong with the criminal justice system?* Cincinnati, OH: Anderson.

Mills, C. W. (1956). *The power elite.* New York: Oxford University Press.

Minton, T. D. (2007). *Jails in Indian country.* Washington, DC: U.S. Department of Justice, Office of Justice Programs, Bureau of Justice Statistics. Available at http://bjs.ojp.usdoj.gov

Minton, T. D. (2010). *Jail inmates at midyear 2009—statistical tables.* Washington, DC: U.S. Department of Justice, Office of Justice Programs, Bureau of Justice Statistics. Available at http://bjs.ojp.usdoj.gov/index.cfm?ty=pbdetail&iid=2195

Minton, T. D. (2011). *Jails in Indian country, 2009.* Washington, DC: National Institute of Justice, Office of Justice Programs, Bureau of Justice Statistics. Available at http://bjs.ojp.usdoj.gov

Mitford, J. (1973). *Kind and usual punishment: The prison business.* New York: Knopf.

Moe, A. M., & Ferraro, K. J. (2003). Malign neglect or benign respect: Women's health care in a carceral setting. *Women & Criminal Justice, 14*(4), 53–80.

Moffitt, T. (2005). The new look of behavioral genetics in developmental psychopathology: Gene-environment interplay in antisocial behavior. *Psychological Bulletin, 131,* 533–554.

Moffitt, T., & Walsh, A. (2003). The adolescence-limited/life-course persistent theory of antisocial behavior: What have we learned? In A. Walsh & L. Ellis (Eds.), *Biosocial criminology: Challenging environmentalism's supremacy* (pp. 125–144). Hauppauge, NY: Nova Science.

Moore, N., May, D., & Wood, P. (2008). Offenders, judges, and officers rate the relative severity of alternative sanctions compared to prison. *Probation and Parole: Current Issues, 1,* 49–70.

Morash, M. (2006). *Understanding gender, crime, and justice.* Thousand Oaks, CA: Sage.

Morash, M., Haarr, R., & Rucker, L. (1994). A comparison of programming for women and men in U.S. prisons in the 1980s. *Crime & Delinquency, 40,* 197–221.

Morris, N. (2002). *Maconochie's gentlemen: The story of Norfolk Island and the roots of modern prison reform.* Oxford, UK: Oxford University Press.

Moyers, B. (2008). *Becoming American: The Chinese experience.* Public Broadcasting Service. Available at http://www.pbs.org

Moynihan, J. M. (2002, July/August). What to do with historic jails. *American Jails,* 14–16.

Mueller, C. W., De Coster, S., & Estes, S. (2001). Sexual harassment in the workplace. *Work and Occupations, 28,* 411–446.

Mulick, S. (2010, November 20). Budget ax falls on McNeil. *The News Tribune* (Tacoma, WA), p. 1.

Mumola, C. J. (2000). *Incarcerated parents and their children.* Washington, DC: Bureau of Justice Statistics.

Mumola, C. J. (2005). *Suicide and homicide in state prisons and local jails.* Washington, DC: U.S. Department of Justice, Office of Justice Programs. Available at http://www.ojp.usdoj.gov/bjs/pub/pdf/shplj.pdf

Mumola, C. J. (2007). Medical causes of death in state prisons, 2001–2004. *Bureau of Justice Statistics Data Brief.* Washington, DC: U.S. Department of Justice.

Muraskin, R. (2003). Disparate treatment in correctional facilities. In R. Muraskin (Ed.), *It's a crime: Women and justice* (3rd ed., pp. 220–231). Upper Saddle River, NJ: Prentice Hall.

Mustaine, E., & Tewksbury, R. (2004). Alcohol and violence. In S. Holmes & R. Holms (Eds.), *Violence: A contemporary reader* (pp. 9–25). Upper Saddle River, NJ: Prentice Hall.

Naday, A., Freilich, J. D., Mellow, J. (2008). The elusive data on supermax confinement. *The Prison Journal, 88*(1), 69–93.

Nagel, W. G. (1973). *The new red barn: A critical look at the modern American prison.* New York: Walker & Company.

Nagin, D. (1998). Criminal deterrence research at the onset of the twenty-first century. In M. Tonry (Ed.), *Crime and justice: A review of research* (Vol. 23, pp. 1–42). Chicago: University of Chicago Press.

National Center for Policy Analysis. (1998, August 17). Does punishment deter? *NCPA Policy Backgrounder, 148.*

National Institute of Corrections. (2008). *Today's jails: Are you looking for a way to coordinate jail and community health services?* Washington, DC: National Institute of Corrections, Jails Division,. Available at http://nicic.org/JailsDivision

National Institute on Drug Abuse. (2006). *Principles of drug abuse treatment for criminal justice populations: A research-based guide.* Washington, DC: National Institutes of Health, U. S. Department of Health and Human Services. Available at http://www.drugabuse.gov/PODAT_CJ/

Nellis, A. (2010). Throwing away the key: The extension of life without parole sentences in the United States. *Federal Sentencing Reporter, 23,* 27–32.

Nelson, W. R., & Davis, R. M. (1995). Podular direct supervision: The first twenty years. *American Jails, 9,* 11–22.

Neubauer, D. (2008). *American courts and the criminal justice system.* Belmont, CA: Wadsworth.

O'Donohue, W., Downs, K., & Yeater, E. (1998). Sexual harassment: A review of the literature. *Aggression and Behavior, 3,* 111–128.

Office of Drug Control Policy. (2010). *ADAM II 2009 annual report.* Available at http://www.whitehousedrugpolicy.gov/publications/pdf/adam2009.pdf

O'Keefe, M. L. (2008). Administrative segregation from within: A corrections perspective. *The Prison Journal, 88*(1), 123–143.

Olivero, J. M., & Roberts, J. B. (1987). Marion Federal Penitentiary and the 22-month lockdown: The crisis continues. *Crime and Social Justice, 27–28,* 234–253.

Olson, S., & Dzur, A. (2004). Revisiting informal justice: Restorative justice and democratic professionalism. *Law and Society Review, 38,* 139–176.

Orland, L. (1975). *Prisons: Houses of darkness.* New York: Free Press.

Orland, L. (1995). Prisons as punishment: An historical overview. In K. C. Haas & G. P. Alpert (Eds.), *The dilemmas of corrections: Contemporary readings* (3rd ed.). Prospect Heights, IL: Waveland Press.

Osher, F. C. (2007). Short-term strategies to improve reentry of jail populations: Expanding and implementing the APIC model. *American Jails, 20,* 9–18.

Oshinsky, D. M. (1996). *Worse than slavery: Parchman Farm and the ordeal of Jim Crow justice.* New York: Free Press.

O'Sullivan, S. (2006). Representations of prison in nineties Hollywood cinema: From *Con Air* to *The Shawshank Redemption.* In R. Tewksbury (Ed.), *Behind bars: Readings on prison culture* (pp. 483–498). Upper Saddle River, NJ: Pearson/Prentice Hall.

Owen, B. (1998). *In the mix: Struggle and survival in a women's prison.* Albany: State University of New York Press.

Owen, B. (2005). Afterword: The case of the women. In J. Irwin (Ed.), *The warehouse prison: Disposal of the new dangerous class* (pp. 261–289). Los Angeles: Roxbury.

Owen, B. (2006). The contexts of women's imprisonment. In A. V. Merlo & J. M. Pollock (Eds.), *Women, law, and social control* (pp. 251–270). Boston: Pearson.

Owen, B., & Bloom, B. (1995). Profiling women prisoners: Findings from national surveys and a California sample. *The Prison Journal, 75,* 165–185.

Paoline, E. A., III, Lambert, E. G., & Hogan, N. L. (2006). A calm and happy keeper of the keys: The impact of ACA views, relations with coworkers, and policy views on the job stress and job satisfaction of correctional staff. *The Prison Journal, 86*(2), 182–205.

Paparozzi, M., & Gendreau, P. (2005). An intensive supervision program that worked: Service delivery, professional orientation, and organizational supportiveness. *The Prison Journal, 85,* 445–466.

Paparozzi, M., & Guy, R. (2009). The giant that never woke: Parole authorities as the lynchpin to evidence-based practices and prisoner reentry. *Journal of Contemporary Criminal Justice, 25,* 397–411.

Payne, B., DeMichele, M., & Button, D. (2008). Understanding the electronic monitoring of sex offenders. *Corrections Compendium, 33,* 1–5.

Payne, B., & Gainey, R. (2004). The electronic monitoring of offenders released from jail or prison: Safety, control, and comparisons to the incarceration experience. *The Prison Journal, 84,* 413–435.

Pearson, F., Lipton, D., Cleland, C., & Yee, D. (2002). The effects of behavioral/cognitive behavioral programs on recidivism. *Crime & Delinquency, 48,* 438–452.

Penn, W. (1981). *The papers of William Penn, Volume One, 1644–1679* (M. M. Dunn & R. S. Dunn, Eds.). Philadelphia: University of Pennsylvania Press. (Original work written circa 1679)

Penn, W. (1983). *William Penn and the founding of Pennsylvania 1680–1684: A documentary history* (J. R. Soderland, Ed.). Philadelphia: University of Pennsylvania Press. (Original work written circa1 684)

Perroncello, P. (2002). Direct supervision: A 2001 odyssey. *American Jails, 15,* 25–32.

Perry, S. W. (2004). *American Indians and crime.* Washington, DC: U.S. Department of Justice, Office of Justice Programs, Bureau of Justice Statistics. Available at http://bjs.ojp.usdoj.gov

Perry, S. (2008). *Justice expenditure and employment statistical extracts, 2006.* Washington, DC: U.S. Department of Justice, Office of Justice Programs, Bureau of Justice Statistics. Available at http://bjs.ojp.usdoj.gov

Petersilia, J. (1998). Probation in the United States, Part 1. *Perspectives, 22,* 30–41.

Petersilia, J. (2001). Prisoner reentry: Public safety and reintegration challenges. *The Prison Journal, 81,* 360–375.

Petersilia, J. (2004). What works in prison reentry? Reviewing and questioning the evidence. *Federal Probation, 68,* 1–8.

Phelps, R. (1996). Newgate of Connecticut: Its origin and early history. Camden, ME: Picton Press. (Original work published 1860)

Piquero, A., & Blumstein, A. (2007). Does incapacitation reduce crime? *Journal of Quantitative Criminology, 23,* 267–285.

Pitts, L. (2010, November 21). Sisters may be guilty; State of Mississippi assuredly is. *The Tacoma News Tribune,* p. B6.

Pizarro, J. M., & Narag, R. E. (2008). Supermax prisons: What we know, what we do not know, and where we are going. *The Prison Journal, 88*(1), 23–42.

Pogrebin, M. R., & Poole, E. D. (1997). The sexualized work environment: A look at women jail officers. *Prison Journal, 77,* 41–57.

Polcin, D. (2009). Community living settings for adults recovering from substance abuse. *Journal of Groups in Addiction and Recovery, 4,* 7–22.

Pollock, J. (2002a). Parenting programs in women's prisons. *Women & Criminal Justice, 14*(1), 131–154.

Pollock, J. (2002b). *Women, prison & crime* (2nd ed.). Belmont, CA: Wadsworth Thomson Learning.

Pollock, J. M. (2004). *Prisons and prison life: Costs and consequences.* Los Angeles: Roxbury.

Porter, N. D. (2011). *The state of sentencing 2010: Developments in policy and practice.* Washington, DC: The Sentencing Project. Available at http://www.sentencingproject.org

Powell, K. (2006). How does the teenage brain work? *Nature, 442,* 865–867.

Pryor, J., &. Stoller, L. M. (1994). Sexual cognition processes in men high in the likelihood to sexually harass. *Personality and Social Psychology Bulletin, 20,* 163–169.

Public Broadcasting Service. (2000). *American Experience: The Rockefellers* [TV program]. Available at http://www.pbs.org/wgbh/amex/rockefellers/peopleevents/e_attica

Public Broadcasting Service. (2011). *The U.S.–Mexican War.* Available at http://www.pbs.org

Pugh, R. B. (1968). *Imprisonment in medieval England.* Cambridge, UK: Cambridge University Press.

Puzzanchera, C. (2009). *Juvenile arrests 2008.* Washington, DC: Office of Juvenile Justice and Delinquency Prevention.

Puzzanchera, C., & Kang, W. (2007). *Juvenile court statistics databook.* Available at http://www.ojjdp.gov/ojstatbb/ezajcs/asp/process.asp

Radelet, M., & Borg, M. (2000). The changing nature of death penalty debates. *Annual Review of Sociology, 26,* 43–61.

Radzinowicz, L., & King, J. (1979). *The growth of crime: The international experience.* Middlesex, UK: Penguin Books.

Rafter, N. H. (1985). *Partial justice: Women in state prisons, 1800–1935.* Boston: Northeastern University Press.

Rafter, N. H. (2009). "Much and unfortunately neglected": Women in early and mid-nineteenth-century prisons. In M. K. Stohr, A. Walsh, & C. Hemmens (Eds.), *Corrections: A text/reader.* Thousand Oaks, CA: Sage.

Reichel, P. (2005). *Comparative justice systems.* Upper Saddle River, NJ: Prentice Hall.

Reid, S. (2006). *Criminal justice* (7th ed.). Cincinnati, OH: Atomic Dog Publishers.

Reisig, M. (1998). Rates of disorder in higher-custody state prisons: A comparative analysis of managerial practices. *Crime & Delinquency, 44,* 229–244.

Rengert, G., & Wasilchick, J. (2001). *Suburban burglary: A tale of two suburbs.* Springfield, IL: Charles C Thomas.

Restak, R. (2001). *The secret life of the brain.* New York: Dana Press and Joseph Henry Press.

Reuffer, D. M. (2007). Arizona jail sex results in charges for guards, prisoner. *Prison Legal News,* pp. 18, 31.

Reynolds, R. (2009). Equal justice under law: Post-Booker, should federal judges be able to depart from the federal sentencing guidelines to remedy disparity between codefendants' sentencing? *Columbia Law Review, 109,* 538–570.

Rice, S. K., Reitzel, J. D., & Piquero, A. R. (2005). Shades of brown: Perceptions of racial profiling and the intra-ethnic differential. *Journal of Ethnicity in Criminal Justice, 3*(1/2), 47–70.

Richards, S. C. (2008). USP Marion: The first federal supermax. *The Prison Journal, 88*(1), 6–22.

Rideau, W. (2010). *In the place of justice: A story of punishment and deliverance.* New York: Knopf.

Rigby, M. (2007). Dallas, Texas, jail pays $950,000 for neglecting mentally ill prisoners. *Prison Legal News, 18,* 22.

Roberts, J. W. (1997). *Reform and retribution: An illustrated history of American prisons.* Lanham, MD: American Correctional Association.

Robertson, J. (2010). Recent legal developments: Correctional case law, 2009. *Criminal Justice Review, 35,* 260–272.

Robinson, M. (2005). *Justice blind: Ideals and realities of American criminal justice.* Upper Saddle River, NJ: Prentice Hall.

Robinson, T., & Berridge, K. (2003). Addiction. *Annual Review of Psychology, 54,* 25–53.

Rodney, E., & Mupier, R. (1999). Behavioral differences between African American male adolescents with biological fathers and those without biological fathers in the home. *Journal of Black Studies, 30,* 45–61.

Rodriguez, N. (2007). Restorative justice at work: Examining the impact of restorative justice resolutions on juvenile recidivism. *Crime & Delinquency, 53,* 355–379.

Rosenfeld, R. (2000). Patterns in adult homicide. In A. Blumstein & J. Wallman (Eds.), *The crime drop in America* (pp. 130–163). Cambridge, UK: Cambridge University Press.

Ross, J. L., & Richards, S. C. (2002). *Behind bars: Surviving prison*. Indianapolis, IN: Alpha Books.

Roth, M. (2006). *Prisons and prison systems: A global encyclopedia*. Westport, CT: Greenwood.

Rothman, D. J. (1980). *Conscience and convenience: The asylum and its alternatives in progressive America*. Glenview, IL: Scott Foresman.

Rubin, J. (2010). Justice Department warns LAPD to take a stronger stance against racial profiling. *Los Angeles Times*. Available at http://www.latimes.com/nov/14/local/la-me-lapd-bias-20101114

Ruddell, R. (2004). *America behind bars: Trends in imprisonment, 1950 to 2000*. New York: LFB Scholarly Publishing.

Sabol, W. J., & Minton, T. D. (2008). *Jails at midyear 2007*. Washington, DC: U.S. Department of Justice, Office of Justice Programs, Bureau of Justice Statistics. Available at http://bjs.ojp.usdoj.gov

Sabol, W. J., West, H. C., & Cooper, M. (2009). *Prisoners in 2008*. Washington, DC: Bureau of Justice Statistics. Available at http://bjs.ojp.usdoj.gov/content/pub/pdf/p08.pdf

Samenow, S. (1999). *Before it's too late: Why some kids get into trouble and what parents can do about it*. New York: Times Books.

Sampson, R., & Laub, J. (1999). Crime and deviance over the lifecourse: The salience of adult social bonds. In F. Scarpitti & A. Nielsen (Eds.), *Crime and criminals: Contemporary and classical readings in criminology* (pp. 238–246). Los Angeles: Roxbury.

Santayana, G. (1905). *The life of reason: Reason in common sense*. New York: Dover Publications.

Saylor, W., & Iwaszko, A. (2009). *One in 31: The long reach of American corrections*. The PEW Center on the States. Washington, DC: The PEW Charitable Trusts.

Scalia, J. (2002). *Prisoner petitions filed in U.S. district courts, 2000, with trends 1980–2000*. Washington, DC: U.S. Department of Justice, Bureau of Justice Statistics.

Scheidegger, K. (2006). *Supreme Court decisions on habeas corpus*. The Federalist Society. Available at http://www.fed-soc.org

Schmalleger, F. (2001). *Criminal justice today* (6th ed.). Upper Saddle River, NJ: Prentice Hall.

Schmitt, J., Warner, K., & Gupta, S. (2010). *The high budgetary cost of incarceration*. Washington, DC: Center for Economic and Policy Research.

Schmitz, J., Stotts, A., Sayre, S., DeLaune, K., & Grabowski, J. (2004). Treatment of cocaine-alcohol dependence with naltrexone and relapse prevention therapy. *American Journal on Addictions, 13*, 333–341.

Schofield, M. (1997, August 24). Tape puts for-profit prisons back in spotlight: Video shows guards abusing inmates—and jeopardizes system. *Idaho Statesman*, pp. A1, A16.

Seiter, R. (2005). *Corrections: An introduction*. Upper Saddle River, NJ: Prentice Hall.

Seiter, R., & Kadela, K. (2003). Prisoner reentry: What works, what does not, and what is promising. *Crime & Delinquency, 49*, 360–390.

Sentencing Project, The. (2011). *Racial disparity*. Available at http://www.sentencingproject.org

Serin, R., Gobeil, R., & Preston, D. (2009). Evaluation of the persistently violent offender program. *International Journal of Offender Therapy and Comparative Criminology, 53*, 57–73.

Serrano, R. A. (2006, November 20). 9/11 prisoner abuse suit could be landmark: Rounded up, Muslim immigrants were beaten in jail. *Los Angeles Times*. Available at http://www.latimes.com

Severson, M. (2004). Mental health needs and mental health care in jails: The past, the present, and hope for the future. *American Jails, 28*, 9–18.

Sexton, L., Jenness, V., & Sumner, J. M. (2010). Where the margins meet: A demographic assessment of transgender inmates in men's prisons. *Justice Quarterly, 27*(6), 835–866.

Shahani, A. (2011). *What is GEO Group?* National Public Radio. Available at http://www.npr.org/2011

Sharp, B. (2006). *Changing criminal thinking: A treatment program*. Alexandria, VA: American Correctional Association.

Sharp, P., & Hancock, B. (1995). *Juvenile delinquency: Historical, theoretical, and societal reactions to youth*. Englewood Cliffs, NJ: Prentice Hall.

Sharp, S. F., & Marcus-Mendoza, S. T. (2001). It's a family affair: Incarcerated women and their families. *Women & Criminal Justice, 12*(4), 21–49.

Sherman, L. (2005). The use and usefulness of criminology, 1751–2005: Enlightened justice and its failures. *Annals of the American Academy of Political and Social Science, 600*, 115–135.

Shilton, M. (2003). *Increasing public safety through halfway houses*. Center for Community Corrections. Available at http://www.communitycorrectionsworks.org

Sickmund, M., & Wan, Y. (2003). *Census of juveniles in residential placement: 2003 databook*. Washington, DC: Office of Juvenile Justice and Delinquency Prevention.

Siegel, L. (2006). *Criminology*. Belmont, CA: Thomson/Wadsworth.

Skeem, J., & Manchak, S. (2008). Back to the future: From Klockars' model of effective supervision to evidence-based practice in probation. *Journal of Offender Rehabilitation, 47*, 220–247.

Skolnick, J. (1993). Shut the door again on sociopaths [Editorial]. *Los Angeles Times*. Available at http://articles.latimes.com/1993-12-16/local/me-2264_1_richard-allen-davis

Slate, R., & Vogel, R. (1997). Participative management and correctional personnel: A study of perceived atmosphere for participation in correctional decision-making and its impact on employee stress and thoughts about quitting. *Journal of Criminal Justice, 25*(5), 397–408.

Slate, R., Wells, T., & Wesley Johnson, W. (2003). Opening the manager's door: State probation officer stress and perceptions of participation in workplace decision-making. *Crime & Delinquency, 49*, 519–541.

Slate, R. N., & Wesley Johnson, W. (2008). *Criminalization of mental illness*. Durham, NC: Carolina Academic Press.

Smith, P., & Smith, W. A. (2005). Experiencing community through the eyes of young female offenders. *Journal of Contemporary Criminal Justice, 21*, 364–385.

Smykla, J., & Williams, J. (1996). Co-corrections in the United States of America, 1970–1990: Two decades of disadvantages for women prisoners. *Women & Criminal Justice, 8*, 61–76.

Snell, T. (2008). Capital punishment, 2007—statistical tables. *Bureau of Justice Statistics Bulletin*. Available at http://bjs.ojp.usdoj.gov/index.cfm?ty=pbdetail&iid=687

Snyder, H., Espiritu, R., Huizinga, D., Loeber, R., & Petechuck, D. (2003, March). Prevalence and development of child delinquency. *Child Delinquency Bulletin Series*. Washington, DC: U.S. Department of Justice, Office of Juvenile Justice and Delinquency Prevention. Available at http://www.ncjrs.gov/pdffiles1/ojjdp/193411.pdf

Snyder, H., & Sickmund, M. (2006). *Juvenile offenders and victims: 2006 national report*. Washington, DC: National Center for Juvenile Justice.

Solomon, A., Dedel Johnson, K., Travis, J., & McBride, E. (2004). *From prison to work: The employment dimensions of prisoner reentry*. Washington, DC: Urban Institute, Justice Policy Center.

Solomon, A., Kachnowski, V., & Bhati, A. (2005). *Does parole work?* Washington, DC: Urban Institute.

Solomon, E., & Garside, R. (2009). *Ten years of Labour's youth justice reforms: An independent audit*. London: Center for Crime and Justice Studies, The Hadley Trust.

Sourcebook of criminal justice statistics. (2000). Washington, DC: U.S. Department of Justice, Office of Justice Programs, Bureau of Justice Statistics. Available at http://www.albany.edu/sourcebook/index.html

Sourcebook of criminal justice statistics. (2003). Washington, DC: U.S. Department of Justice, Office of Justice Programs, Bureau of Justice Statistics. Available at http://www.albany.edu/sourcebook/index.html

Sourcebook of criminal justice statistics. (2004). Washington, DC: U.S. Department of Justice, Office of Justice Programs, Bureau of Justice Statistics. Available at http://www.albany.edu/sourcebook/index.html

Sourcebook of criminal justice statistics. (2008). Table 6.0001.2006: Estimated number and percent distribution of prisoners under jurisdiction of state correctional authorities. Available at http://www.albany.edu/sourcebook/pdf/t600012006.pdf

Spelman, W. (2000). The limited importance of prison expansion. In A. Blumstein & J. Wallman (Eds.), *The crime drop in America* (pp. 97–129). Cambridge, UK: Cambridge University Press.

Springer, C. (1987). *Justice for juveniles*. Washington, DC: Office of Juvenile Justice and Delinquency Prevention.

Stannard, D. E. (1992). *American holocaust: Columbus and the conquest of the New World*. New York: Oxford University Press.

Staton-Tindall, M., Garner, B. R., Morey, J. T., Leukefeld, C., Krietemeyer, J., Saum, C. A., et al. (2007). Gender differences in treatment engagement among a sample of incarcerated substance abusers. *Criminal Justice and Behavior, 34*(9), 1143–1156.

Steiner, B., & Giacomazzi, A. (2007). Juvenile waiver, boot camp, and recidivism in a northwestern state. *The Prison Journal, 87,* 227–240.

Stephan, J. J. (2001). *Census of state and federal correctional facilities*. Washington, DC: U.S. Department of Justice, Office of Justice Programs, Bureau of Justice Statistics. Available at http://www.bjs.ojp.usdoj.gov

Stephan, J. J. (2008). *Census of state and federal correctional facilities, 2005*. Washington, DC: U.S. Department of Justice, Office of Justice Programs, Bureau of Justice Statistics. Available at http://bjs.ojp.usdoj.gov

Stephan, J., & Karberg, J. (2003). *The census of state and federal correctional facilities*. Washington, DC: U.S. Department of Justice.

Stohr, M. K. (2006). Yes, I've paid the price, but look how much I gained! In C. M. Renzetti, L. Goodstein, & S. L. Miller (Eds.). *Rethinking gender, crime, and justice* (pp. 262–277). Los Angeles: Roxbury.

Stohr, M. K., & Collins, P. A. (2009). *Criminal justice management: Theory and practice in justice-centered organizations*. New York: Oxford University Press.

Stohr, M. K., Hemmens, C., Shapiro, B., Chambers, B., & Kelly, L. (2002). Comparing inmate perceptions of two residential substance abuse treatment programs. *International Journal of Offender Therapy and Comparative Criminology, 46,* 699–714.

Stohr, M. K., Lovrich, N. P., & Mays, G. L. (1997). Service v. security focus in training assessments: Testing gender differences among women's jail correctional officers. *Women & Criminal Justice, 9,* 65–85.

Stohr, M. K., Lovrich, N. P., Menke, B. A., & Zupan, L. L. (1994). Staff management in correctional institutions: Comparing DiIulio's "control model" and "employee investment model" outcomes in five jails. *Justice Quarterly, 11*(3), 471–497.

Stohr, M. K., Lovrich, N. P., & Wilson, G. L. (1994). Staff stress in contemporary jails: Assessing problem severity and the payoff of progressive personnel practices. *Journal of Criminal Justice, 22,* 313–328.

Stohr, M. K., Lovrich, N. P., & Wood, M. (1996). Service v. security concerns in contemporary jails: Testing behavior differences in training topic assessments. *Journal of Criminal Justice, 24,* 437–448.

Stohr, M. K., & Mays, L. G. (1993). *Women's jails: An investigation of offenders, staff, administration and programming* (Final report for grant number 92J04GHP5 awarded by the National Institute of Corrections, Jails Division). Available at http://nicic.gov/Library/008747

Stohr, M. K., Mays, G. L., Beck, A. C., & Kelley, T. (1998). Sexual harassment in women's jails. *Journal of Contemporary Criminal Justice, 14*(24), 135–155.

Streib, V. (2003). *The juvenile death penalty today: Death sentences and executions for juvenile crimes, January 1, 1973–June 30, 2003*. Available at http://www.law.onu.edu/faculty/streib

Sturgess, A., & Macher, A. (2005). Issues in correctional HIV care: Progressive disseminated histoplasmosis. *American Jails, 19,* 54–56.

Sundt, J. L., Castellano, T. C., & Briggs, C. S. (2008). The sociopolitical context of violence and its control: A case study of supermax and its effect in Illinois. *The Prison Journal, 88*(1), 94–122.

Surratt, H. L. (2003). Parenting attitudes of drug-involved women inmates. *The Prison Journal, 83*(2), 206–220.

Swedish Ministry of Justice. (2004). Information about the Swedish prison and probation service. *Fact sheet*. Stockholm: Author. Available at http://www.sweden.gov.se/content/1/c6/01/61/94/0602f648.pdf

Sykes, G. M. (1958). *The society of captives: A study of a maximum security prison*. Princeton, NJ: Princeton University Press.

Sykes, G. M., & Messinger, S. (1960). The inmate social system. In R. Cloward & D. R. Dressey (Eds.), *Theoretical studies in social organization of the prison* (pp. 5–19). New York: Social Science Research Council.

Talbot, T., Gilligan, L., Carter, M., & Matson, S. (2002). *An overview of sex offender management*. Washington, DC: Center for Sex Offender Management.

Tartaro, C. (2002). Examining implementation issues with new generation jails. *Criminal Justice Policy Review, 13,* 219–237.

Tartaro, C. (2006). Watered down: Partial implementation of the new generation jail philosophy. *The Prison Journal, 86,* 284–300.

Tartaro, C., & Ruddell, R. (2006). Trouble in Mayberry: A national analysis of suicides and attempts in small jails. *American Journal of Criminal Justice, 31,* 81–101.

Taylor, R., Fritsch, E., & Caeti, T. (2007). *Juvenile justice: Policies, programs, and practices*. New York: McGraw-Hill.

Terrill, R. (2003). *World justice systems*. Cincinnati, OH: Anderson.

Tewksbury, R., & Collins, S. C. (2006). Aggression levels among correctional officers: Reassessing sex differences. *The Prison Journal, 86*(3), 327–343.

Tewksbury, R., & Potter, R. H. (2005). Transgender prisoners: A forgotten group. In S. Stojkovic (Ed.), *Managing special populations in jails and prisons* (pp. 15-1–15-14). New York: Civic Research Institute.

Thompson, E. (1975). *Whigs and hunters: The origin of the Black Act*. New York: Pantheon.

Thompson, J. A., & Mays, G. L. (1991). Paying the piper but changing the tune: Policy changes and initiatives for the American jail. In J. A. Thompson & G. L. Mays (Eds.), *American jails: Public policy issues.* (pp. 240–246). Chicago: Nelson-Hall.

Tocqueville, A. de. (1956). *Democracy in America* (H. Hefner, Ed.). New York: Norton. (Original work published 1838)

Tocqueville, A. de, & Goldhammer, A. (2004). *Tocqueville: Democracy in America*. New York: Library of America. (Original work published 1835)

Travis, J. (2000, May). But they all come back: Rethinking prisoner reentry. *Sentencing & Corrections: Issues for the 21st century*. Washington, DC: U.S. Department of Justice, National Institute of Justice.

Travis, J. (2005). *But they all come back: Facing the challenges of prison reentry*. Washington, DC: Urban Institute.

Travis, J., & Lawrence, S. (2002). *Beyond the prison gates: The state of parole in America*. Washington, DC: The Urban Institute, Justice Policy Center.

Travis, J., & Petersilia, J. (2001). Reentry reconsidered: A new look at an old question. *Crime & Delinquency, 47,* 291–313.

Tutu, D. (1999). No future without forgiveness. New York: Doubleday.

Umbreit, M. (1994). *Victim meets offender: The impact of restorative justice and mediation*. Monsey, NY: Criminal Justice Press.

United States Sentencing Commission. (1995). *Annual report*. Washington, DC: Author.

U.S. Sentencing Commission. (2011). *January 28, 2011, memorandum*. Available at http://www.ussc.gov

U.S. Census Bureau. (2001). *Population estimates*. Available at http://www.census.gov

U.S. Census Bureau. (2006). *Statistical abstract of the United States*. Available at http://www.census.gov/prod/www/abs/statab2001_2005.html

U.S. Census Bureau. (2010a). *Census of governmental employment and payroll*. Available at http://www.census.gov

U.S. Census Bureau. (2010b). DP-1 general demographic characteristics. *2009 population estimates*. Available at http://www.factfinder.census.gov

U.S. Department of Justice. (2010). *Prison gangs*. Available at http://www.justice.gov/criminal/gangunit/gangs/prison

U.S. Department of Justice. (2010). *Prison gangs*. Available at www.justice.gov/criminal/gangunit/gangs/prison

U.S. Department of Labor. (2010). *The National Compensation Survey: Occupational earnings in the United States, 2009*. Washington, DC: U.S. Bureau of Labor Statistics. Available at http://www.bls.gov/ncs/ocs/sp/nctb1349.pdf

Useem, B. (1985). Disorganization and the New Mexico prison riot. *American Sociological Review, 50*(5), 677–688.

Useem, B., & Kimball, P. A. (1989). *States of siege: U.S. prison riots 1971–1986*. New York: Oxford University Press.

Van Tongeren, D. R., & Klebe, K. J. (2010). Reconceptualizing prison adjustment: A multidimensional approach exploring female offenders' adjustment to prison life. *The Prison Journal, 90*(1), 48–68.

Van Voorhis, P., Braswell, M., & Lester, D. (2000). *Correctional counseling and rehabilitation*. Cincinnati, OH: Anderson.

Vanstone, M. (2000). Cognitive-behavioural work with offenders in the UK: A history of influential endeavour. *The Howard Journal of Criminal Justice, 39,* 171–183.

Vanstone, M. (2004). Mission control: The origins of humanitarian service. *Probation Journal, 51,* 34–47.

Vanstone, M. (2008). The international origins and initial development of probation. *British Journal of Criminology, 48,* 735–755.

Vaughn, M. (2009). Substance abuse and crime: Biosocial foundations. In A. Walsh & K. M. Beaver (Eds.), *Biosocial criminology: New directions in theory and research* (pp. 176–189). New York: Routledge.

Vaughn, M. S., & Carroll, L. (1998). Separate and unequal: Prison versus free-world medical care. *Justice Quarterly, 15,* 3–10.

Vaughn, M., & del Carmen, R. (1995). Civil liability against prison officials for inmate-on-inmate assault: Where are we and where have we been? *The Prison Journal, 75,* 69–89.

Vaughn, M. S., & Smith, L. G. (1999). Practicing penal harm medicine in the United States: Prisoners' voices from jail. *Justice Quarterly, 16,* 175–231.

Victoria Department of Justice. (2010). *Statistical profile of the Victorian prison system: 2005–06, 2009–10*. Melbourne, Australia: Author. Available at http://www.justice.vic.gov.au

Vito, G., Allen, H., & Farmer, G. (1981). Shock probation in Ohio: A comparison of outcomes. *International Journal of Offender Therapy and Comparative Criminology, 25,* 70–76.

Vogel, R. (2004). Silencing the cells: Mass incarceration and legal repression in U.S. prisons. *Monthly Review, 56,* 1–8.

Walker, E. (2002). Adolescent neurodevelopment and psychopathology. *Current Directions in Psychological Science, 11,* 24–28.

Walker, S., Spohn, C., & DeLone, M. (1996). *The color of justice: Race, ethnicity, and crime in America*. Belmont, CA: Wadsworth.

Walsh, A. (2000). Evolutionary psychology and the origins of justice. *Justice Quarterly, 17,* 841–864.

Walsh, A., & Ellis, L. (2007). *Criminology: An interdisciplinary approach*. Thousand Oaks, CA: Sage.

Walsh, A., & Hemmens, C. (2011). *Law, justice, and society: A sociolegal approach* (2nd ed.). New York: Oxford University Press.

Walsh, A., & Stohr, M. (2010). *Correctional assessment, casework and counseling* (5th ed.). Alexandria, VA: American Correctional Association.

Wang, X., Mears, D., Spohn, C., & Dario, L. (2009, December 3). Assessing the differential effects of race and ethnicity on sentence outcomes under different sentencing systems. *Crime & Delinquency*. doi: 10.1177/0011128709352234.

Ward, D., & Kassebaum, G. (2009). *Alcatraz: The gangster years*. Berkeley: University of California Press.

Ward, T., Melser, J., & Yates, P. (2007). Reconstructing the risk-need-responsivity model: A theoretical exploration and evaluation. *Aggression and Violent Behavior, 12,* 208–228.

Weber, M. (1946). Bureaucracy. In H. H. Gerth & C. W. Mills (Eds.), *From Max Weber: Essays in sociology*. Oxford, UK: Oxford University Press.

Wei, W. (1999). The Chinese American experience: 1857–1892. *Harper's Weekly.* Available at http://www.Harpweek.com

Welch, M. (1996). *Corrections: A critical approach.* New York: McGraw-Hill.

Welch, M. (2004). *Corrections: A critical approach* (2nd ed.). Boston: McGraw-Hill.

Welch, M. (2005). *Ironies of imprisonment.* Thousand Oaks, CA: Sage.

Welch, M. (2011). *Corrections: A critical approach* (3rd ed.). London: Routledge.

Welsh, W. (1995). *Counties in court: Jail overcrowding and court-ordered reform.* Philadelphia: Temple University Press.

Wener, R. (2005). The invention of direct supervision. *Corrections Compendium, 30*(4–7), 32–34.

Wener, R. (2006). Effectiveness of the direct supervision system of correctional design and management: A review of the literature. *Criminal Justice and Behavior, 33,* 392–410.

Wennerberg, I., & Pinto, S. (2009). *Sixth European Electronic Monitoring Conference—analysis of questionnaires.* Available at http://www.cepprobation.org/uploaded_files/EM2009%20Questionnaire%20summary.pdf

West, H. C. (2010a). *Prison inmates at midyear 2009—statistical tables.* Washington, DC: U.S. Department of Justice, Office of Justice Programs, Bureau of Justice Statistics. Available at http://bjs.ojp.usdoj.gov

West, H. C., Sabol, W. J., & Greenman, S. J. (2010). *Prisoners in 2009.* Washington, DC: U.S. Department of Justice, Office of Justice Programs, Bureau of Justice Statistics. Available at http://www.bjs.ojp.usdoj.gov

Western, B. (2003). *Incarceration, employment, and public policy.* New Jersey Institute for Social Justice. Available at http://www.njisj.org/reports/western_report.html

White, A. (2004). *Substance use and the adolescent brain: An overview with the focus on alcohol.* Durham, NC: Duke University Medical Center.

White, M. D., Goldkamp, J. S., & Campbell, S. P. (2006). Co-occurring mental illness and substance abuse in the criminal justice system. *The Prison Journal, 86*(3), 301–326.

Whitehead, J., & Lab, S. (1996). *Juvenile justice: An introduction.* Cincinnati, OH: Anderson.

Whitman, J. Q. (2003). *Harsh justice: Criminal punishment and the widening divide between America and Europe.* New York: Oxford University Press.

Whittick, A. (1979). *Woman into citizen.* Santa Barbara, CA: ABC-Clio.

Williams, N. H. (2007). Prison health and the health of the public: Ties that bind. *Journal of Correctional Health Care, 13,* 80–92.

Wilson, D., Bouffard, L., & Mackenzie, D. (2005). A quantitative review of structured, group-oriented, cognitive-behavior programs for offenders. *Criminal Justice and Behavior, 32,* 172–204.

Wilson, J. (1975). *Thinking about crime.* New York: Vintage.

Wilson, S., & Lipsey, M. (2000). Wilderness challenge programs for delinquent youth: A meta-analysis of outcome evaluations. *Evaluation and Program Planning, 23,* 1–12.

Winfree, L. T., & Wooldredge, J. D. (1991). Exploring suicides and deaths by natural causes in America's large jails: A panel study of institutional change, 1978 and 1983. In J. Thompson & G. L. Mays (Eds.), *American jails: Public policy issues* (pp. 63–78). Chicago: Nelson-Hall.

Winter, M. M. (2003). County jail suicides in a midwestern state: Moving beyond the use of profiles. *The Prison Journal, 83,* 130–148.

Wolff, N., Shi, J., & Blitz, C. L. (2008). Racial and ethnic disparities in types and sources of victimization inside prison. *The Prison Journal, 88*(4), 451–472.

Wolfgang, M., Figlio, R., & Selling, T. (1972). *Delinquency in a birth cohort.* Chicago: University of Chicago Press.

Wood, P. B., & May, D. C. (2003). Research notes: Racial differences in perceptions of the severity of sanctions: A comparison of prison with alternatives. *Justice Quarterly, 20*(3), 605–632.

Wright, E. M., Salisbury, E. J., & Van Voorhis, P. (2007). Predicting the prison misconducts of women offenders. *Journal of Contemporary Criminal Justice, 23*(4), 310–340.

Wright, R. (1999). The evidence in favor of prisons. In F. Scarpitti & A. Nielson (Eds.), *Crime and criminals: Contemporary and classic readings in criminology* (pp. 483–493). Los Angeles: Roxbury.

Yates, H. M. (2002). Margaret Moore: African American feminist leader in corrections. *Women & Criminal Justice, 13,* 9–26.

Yochelson, S., & Samenow, S. (1976). *The criminal personality: A profile for change.* Livingston, NJ: Jason Aronson.

Young, V. D. (1994). Race and gender in the establishment of juvenile institutions: The case of the South. *The Prison Journal, 74,* 244–265.

Young, V. D. (2001). All the women in the Maryland State Penitentiary: 1812–1869. *Prison Journal, 81*(1), 113–132.

Zaitz, L. (2010, October 30). Budget crunch reaches prison. *The Oregonian,* p. 1.

Zimmer, L. (1986). *Women guarding men.* Chicago: University of Chicago Press.

Zimmer, L. (1989). Solving women's employment problems in corrections: Shifting the burden to administrators. *Women & Criminal Justice, 1,* 55–79.

Zimring, F. E., Hawkins, G., & Kamin, S. (2001). *Punishment and democracy: Three strikes and you're out in California.* Oxford, UK: Oxford University Press.

Zupan, L. L. (1991). *Jails: Reform and the new generation philosophy.* Cincinnati, OH: Anderson.

Zupan, L. L. (1992). The progress of women correctional officers in all-male prisons. In I. L. Moyer (Ed.), *The changing roles of women in the criminal justice system: Offenders, victims, and professionals* (2nd ed., pp. 323–343). Prospect Heights, IL: Waveland Press.

Index

About the Authors

Mary K. Stohr is a professor in the Department of Criminology and Criminal Justice at Missouri State University. She received a PhD (1990) in political science, with specializations in public administration and criminal justice, from Washington State University. Stohr has published over 75 academic works of one or another in the areas of correctional organizations and operation, correctional personnel, inmate needs and assessment, program evaluation, gender, and victimization. Her coauthored or -edited books include *Correctional Assessment, Casework & Counseling, Corrections: A Text Reader, Criminal Justice Management: Theory and Practice in Justice-Centered Organizations,* and *The Prison Experience.*

Anthony Walsh received his PhD in criminology and statistics from Bowling Green State University, Ohio, and is currently a professor at Boise State University where he teaches criminology, criminal law, statistics, and correctional counseling. He has field experience in both corrections and law enforcement. He is the author or coauthor of 27 books, including *Correctional Assessment, Casework & Counseling* with Mary Stohr, and over 100 articles.

SAGE Research Methods Online

The essential tool for researchers

**Sign up now at
www.sagepub.com/srmo
for more information.**

An expert research tool

- An **expertly designed taxonomy** with more than 1,400 unique terms for social and behavioral science research methods

- **Visual and hierarchical search tools** to help you discover material and link to related methods

- Easy-to-use navigation tools
- Content organized by complexity
- Tools for citing, printing, and downloading content with ease
- Regularly updated content and features

A wealth of essential content

- The most comprehensive picture of quantitative, qualitative, and mixed methods available today

- More than **100,000 pages of SAGE book and reference material** on research methods as well as editorially selected material from SAGE journals

- More than **600 books** available in their entirety online

Launching 2011!

SAGE research methods online